Lecture Notes in Control and Information Sciences 425

Editors: M. Thoma, F. Allgöwer, M. Morari

Maciej Patan

Optimal Sensor Networks Scheduling in Identification of Distributed Parameter Systems

Springer

Author
Maciej Patan
The University of Zielona Góra
Poland

ISSN 0170-8643
e-ISSN 1610-7411
ISBN 978-3-642-28229-4
e-ISBN 978-3-642-28230-0
DOI 10.1007/978-3-642-28230-0
Springer Heidelberg New York Dordrecht London

Library of Congress Control Number: 2012932480

Printed on acid-free paper

Springer is part of Springer Science+Business Media (www.springer.com)

To my wonderful $\mathcal{J}$

Preface

It is well known that the experimental setting, being an integral part of any identification experiment for a distributed physical system, significantly influences the quality of measurement data. For such systems it is usually impossible to observe their states over the entire spatial domain, therefore optimal spatial deployment of discrete sensors and their proper temporal activation scheduling, which allows estimating the unknown system parameters as accurately as possible are of fundamental significance. The precision of parameter estimation is important as parameters are often used in optimal control, environmental monitoring, industrial quality tests, fault diagnosis, or they give an indication of the mechanism of a process.

In the context of rapidly growing developments in the area of sensor networks representing a modern class of monitoring systems, whose immanent feature is the distributed character of measurement data gathering, effective systematic approaches could be of great importance and potential in engineering applications. On the other hand, sensor networks, being, in fact, a new class of computing systems, in the near future promise to be an efficient platform for solving large-scale practical problems which even several years ago would have been out of the question. Special motivation comes from increasing possibilities enabled by robotic networks consisting of mobile nodes, each equipped with sensors and monitoring its environment.

Therefore, the main objective of this monograph is to provide a coherent approach to sensor network scheduling for parameter estimation in dynamic distributed systems modelled with partial differential equations. Nowadays, such sophisticated mathematical modelling tools become of a great significance in the area of the control of industrial processes as the increasing quality requirements are simply out of reach for classical lumped parameter models. In order to properly compare the quality for different solutions, the notion of the so-called Fisher information matrix related to the covariance matrix of parameter estimates is applied. Particular emphasis is put on providing efficient algorithmic solutions dedicated to scanning and moving sensor nodes in various experimental settings. Thus, another objective is to produce useful results which, despite nontrivial mathematical foundations, will lead to the relatively easy computer implementation. This should attract the attention of engineers interested in reliable but not very complex techniques.

A great majority of material covered in this monograph results from my original research that has been conducted for the last eight years within the subject of sensor network configuration for distributed parameter systems identification located at the interface of several fields such as optimum experimental design, partial differential equations, nonlinear programming, optimal control, distributed algorithms and estimation.

The book is divided into ten chapters. The first one introduces the reader into the field of sensor network scheduling for monitoring spatio-temporal systems. Chapter 2 provides a detailed description of the problem of optimal measurement strategy design in the context of system identification and indicates the main difficulties making this problem really hard to solve. Efficient original policies of activating scanning and moving sensors are presented in Chapters 3 and 4. In particular, the concept of optimal design measures is adopted to the setting of network scheduling, and the formulation in terms of a state-constrained optimal control problem is reported as well. The extension of the developed ideas to decentralized network scheduling is reported in Chapter 5. In turn, Chapter 6 presents a combinatorial approach to sensor activation together with generalizations for the parallel and multi-objective optimization with algorithmic analysis. Some methods to overcome the difficulties related to the parametric and sensor location uncertainties are contained in Chapter 7. Then, Chapter 8 establishes a general framework for model-based fault detection in spatio-temporal systems. Chapter 9 deals with extensions toward the problems generally considered to be 'hard' and related to the presence of correlated measurement errors and grouping observations for large scale experiments. Finally, some concluding remarks are made in Chapter 10. Throughout the consecutive chapters, all the concepts described in the book are illustrated with tangible application examples close to real-world engineering problems.

Taking this opportunity, I would like to express my great thanks to Professor Dariusz Uciński, who introduced me into the field of experimental design for distributed-parameter systems, for continuous support and excellent collaboration within our team. For many years of my carrier he has been an inspiring source of many ideas and helpful suggestions.

This monograph comprises a great amount of results of research supported by the Polish Ministry of Science and Higher Education (grants N N514 230537 and N N519 297133). Some results from Chapters 8 and 9 have been also supported by the Air Force Office of Scientific Research, AF Material Command, USAF, (grant number FA8655-08-1-3014) and the UK Engineering and Physical Sciences Research Council under the grant EP/C541715/1.

Zielona Góra, Maciej Patan
August, 2011

Contents

Acronyms

ARS	Adaptive Random Search
BE	Bayesian Estimation
DPS	Distributed Parameter System
DSN	Distributed Sensor Network
FIM	Fisher Information Matrix
LMI	Linear Matrix Inequalities
LPS	Lumped Parameter System
LSE	Least Squares Estimation
MLE	Maximum Likelihood Estimation
MIMO	Multiple-Input Multiple-Output
MISO	Multiple-Input Single-Output
ODE	Ordinary Differential Equation
OED	Optimum Experimental Design
PDE	Partial Differential Equation
SDP	Semi-Definite Programming
SIP	Semi-Infinite Programming
SQP	Sequential Quadratic Programming
UAV	Unmanned Aerial Vehilces
UGV	Unmanned Ground Vehicle

Chapter 1
Introduction

1.1 Identification of Spatio-temporal Systems

Systems with spatio-temporal dynamics, commonly known as Distributed-Parameter Systems (DPSs), constitute one of the most general and important classes of systems which are widely used in modelling for a great variety of real-world engineering problems. As a matter of fact, rapidly growing demands of recent control and quality monitoring systems stimulate engineers to search for more precise mathematical models of the phenomena considered. This leads directly to the description of the system at hand using Partial Differential Equations (PDEs), usually expressing physical laws governing the behavior of the system. The reason is that lumped descriptions of system dynamics basically become unsatisfactory as they cannot provide a sufficient approximation of the distributed nature of the investigated process. Despite the more sophisticated formulation in terms of PDEs, such models achieve high quality and efficiency of simulations and control techniques [19, 47, 87, 145].

Thus, close attention should be paid to this subject, which requires more effective and robust analysis methods driven especially by the ongoing progress in computational and applied mathematics combined with the availability of rapidly increasing computer power. These developments lead to new challenges in the field of modelling and steadily extend the range of applications that can be simulated numerically. In consequence, nowadays DPSs hold a very important position in systems analysis and control theory, and in the more general area of infinite dimensional systems, they established a separate field of research with a plenty of relevant publications [19, 47, 66, 67, 87, 89, 120, 121, 122, 129, 131, 132, 133, 134, 145, 153, 158, 160, 167, 177, 180, 231, 234, 258, 276, 281, 318]. One of essential issues in DPSs is system identification, and from the point of view of engineering applications two major tasks should be addressed:

- the *forward problem*, which consists in determining the system state when the system parameters, spatial region, time observation interval, subsidiary conditions and control variables are known;

M. Patan: Optimal Sensor Networks Scheduling in Identification of DPSs, LNCIS 425, pp. 1–10.
springerlink.com

- the *inverse problem*, which consists in recovering some of the model parameters from the collected observational data (e.g., system states and other available information) provided that the model structure is given; it is also called *parameter identification.*

In the framework of DPSs, inverse problems are especially challenging as they may consist in estimation of physical parameters, forcing inputs or initial and boundary conditions. The main difficulty is that the estimation problem is often ill-conditioned in the Hadamard sense even if the forward problem is well-posed [105, 252, 264]. The paramount importance of solving inverse problems cannot be overestimated as the simulation model should be precisely calibrated based on the measurements, otherwise the forecast of system behavior based on the solution of the forward problem will be unreliable.

The literature related to the subject of parameter estimation in DPSs is considerable. Classical surveys by Kubrusly [137] and Polis [215] systematize the various techniques in this field. The interested reader can be also referred to the book by Banks and Kunish [14], where a broad class of estimation techniques for DPSs is presented, and further to works [12, 38, 86, 140, 287]. However, most contributions are focused on off-line approaches, and only few results concern on-line methods [2, 50].

A fundamental problem in parameter estimation of DPSs is to properly design the process of data acquisition from various sources. This task comprises determination of the spatio-temporal schedule of collecting measurements for a limited number of sensors over a given spatial domain in such a way as to maximize the estimation efficiency of the system parameters. Most often, the dependence between the sensor placement and the system performance is not intuitive and has confounding character. Due to the cost of data acquisition, constraints associated with measurements and requirements of increased efficiencies of identifiers, there is a strong necessity for developing some systematic methods of selecting appropriate sensor configurations and measurement strategies.

1.2 Application Areas

It has already been indicated that motivations for studying the sensor location problem stem directly from real-world engineering problems. One of the most interesting ones is Computer-Assisted Tomography (CAT), which consist in reconstructing material parameters that characterize the inaccessible interior of an examined object based on the measurements taken at the boundary without any damage to the subject of examination [101, 252, 308]. In general, CAT exploits various physical phenomena and data media such as photons, magnetic fields, gamma rays, X-rays, ultrasounds and, finally, the electric currents or electron beams. Depending on the phenomenon used, the data are processed with a computer system in such a way as to obtain the appropriate spatial map (two or three dimensional) of the corresponding material parameters [101, 252]. For example, for X-rays it can be the coefficient of their suppression, for gamma rays the concentration of the radioactive contrast injected into the object, and impedance in electrical tomography. A proper description of underlying physical processes requires an accurate model in the form of PDEs, which are essential for efficient image reconstruction. Since the observations have to be noninvasive, the problem of proper data acquisition becomes extremely difficult and the locations of the measurement electrodes and supply sources are of great significance as they should provide possibly the most informative measurements.

Another inspiring application concerns optimization of air quality monitoring networks since, due to the fast development of industry around the world, air pollution becomes a great societal problem. Protecting and restituting the natural environment require high accuracy of forecasts and diagnoses. To provide them, proper calibration of models which describe the pollutant emission processes is necessary [20, 112, 113, 155, 217, 266, 267]. Usually, changes in pollutant concentrations over a given area are described by PDEs of the advection–diffusion type. Since some coefficients of the equations are not measurable, accurate modelling becomes extremely difficult. In addition, monitoring stations are rather costly and the problem of choosing an appropriate observation strategy is of great practical relevance, indicated in many publications [5, 20, 109, 156, 170, 175, 263, 267].

Similar problems can be found in numerous engineering areas, e.g., in groundwater sources management [130, 264], in gathering measurement data for calibration of models used in meteorology and oceanography [18, 48, 99, 159], in automated inspection of hazardous environments [127, 129], in prediction of radioactive contamination [105], emerging smart material systems [15, 144], structural mechanics in civil engineering or aircraft industry [15, 77], signal transmissions lines in electrical engineering [49, 58], and many others. Some of these are also considered as examples illustrating the algorithms and approaches presented within the scope of the book.

1.3 Sensor Scheduling Problem: State of the Art

Although the importance of the sensor location problem is well understood, till the mid 1990s the results communicated by most authors were limited to selection of stationary sensor positions in the context of state estimation [4, 63, 64, 65, 126, 138]. Extending state estimation methods to the parameter estimation problem is not straightforward and rather difficult, since in the latter case the current state usually depends nonlinearly on unknown parameters [126] (even if the PDE is linear in these parameters), in contrast to the dependence of the current state on the initial one, which is linear if only the system is linear. This essential difference makes the task of parameter estimation much harder and for that reason the number of corresponding results was much smaller [231, 276, 279].

From the perspective of the optimal sensor location problem for parameter identification in DPSs, the existing approaches can be classified into three main groups:

1. methods leading to state estimation,
2. methods making use of random field theory,
3. methods originating in optimum experimental design.

The methods of the first group transform the original problem into a state estimation one in such a way as to raise possibility of applying well-developed methods of optimal sensor location to state estimation. There is rich literature concerning such problems [11, 63, 126, 128, 129, 138, 161, 277]. The main drawback of such an approach is that simultaneous state and parameter estimation usually leads to the high complexity of the resulting nonlinear problem. Some attempts involving sequential linearizations at the consecutive state trajectories [161] were conducted along with suboptimal filtering [128]. Nevertheless, the approach depreciates due to the well-known difficulties encountered in nonlinear state estimation analysis reducing its applicability only to the simplest situations.

The second group of methods makes use of random field theory. However, their usefulness in the context of DPSs is rather limited, since in general transformation between system descriptions in the form of PDEs and suitable random field characteristics is not simple, making the whole approach strongly case-specific. Additionally, such a conversion relies only on statistics up to a given order (e.g., mean, covariance, skewness, kurtosis and higher order functions). Despite those limitations, methods from within this group can be successfully applied in specific cases [116, 169, 170, 264].

The classical theory of optimum experimental design [8, 68, 70, 73, 118, 210, 222, 233, 307] and its extensions to models for dynamic systems [88, 135, 136, 139, 165, 180, 231, 269, 276, 281, 307] constitutes a basis for the third class of methods. Within this group of approaches, the problem is stated as an optimization one, where the performance index is defined in the form of some scalar measure operating on the Fisher Information Matrix (FIM), whose inverse, based on the Cramér–Rao inequality, plays the role of the lower bound of the parameter covariance matrix. Such an idea leads to great simplifications, because even if the exact covariance matrix is impossible to obtain, the inverse of the FIM can be computed with relative ease.

As regards dynamic DPSs, the first treatment of this type for the sensor-location problem was proposed by Uspenskii and Fedorov [296], who maximized the D-optimality criterion, being the determinant of the FIM associated with the estimated parameters characterizing the source term in a simple one-dimensional linear diffusion equation. The authors observed that the linear dependence of the observed outputs on these parameters makes it possible to directly apply the machinery of optimum experimental design theory. A similar technique was used in by Quereshi et al. [227] to find sensor locations and boundary perturbations in dynamic DPSs (a heat-diffusion process and a vibrating string) and Rafajłowicz [228] for optimization of both sensor positions and distributed control for parameter estimation of static linear DPS. The delineated approach was further extended by Rafajłowicz [229] to cover a class of DPSs described by linear hyperbolic equations with known eigenfunctions and unknown eigenvalues. The aim was to find conditions for the optimality of the measurement design and the spectral density of the stochastic input. It was indicated that common numerical procedures from classical experimental design for linear regression models could be adopted to find optimal sensor location. A similar problem was studied [230] in a more general framework of DPSs which can be described in terms of Green's functions.

Over the past two decades, this methodology has been substantially refined to extend its applicability. In the monograph by Sun [265], the optimum experimental design techniques are used to solve inverse problems in groundwater modelling based on the extensive exploitation of the PDEs describing the underlying DPS. A comprehensive treatment of both theoretical and algorithmic aspects of the resulting sensor location strategies were given by Uciński [276] and further extended by numerous authors [139, 180, 234, 281] for various challenging experimental settings. The potential of the approach for generalizations was exploited, e.g., in [194, 200], where a fault detection scheme for DPSs based on the maximization of the power of a parametric hypothesis test regarding the nominal state of a given DPS, or in the works [200, 291] describing a branch-and-bound method applied to the setting in which a large number of possible sites at which to locate a sensor are given, but cost constraints allow only some proper subset of them to be selected.

A natural generalization and an appealing alternative to stationary sensors is to apply spatially movable ones. The complexity of the resulting optimization problem is compensated by a number of benefits. Specifically, sensors are not assigned to fixed positions, which are optimal only on the average, but are capable of tracking points which provide at a given time instant the best information about the parameters to be identified. Consequently, by actively reconfiguring a sensor system, we can expect the minimal value of an adopted design criterion to be lower than the one for the stationary case. The first substantial work in this subject is the seminal article by Rafajłowicz [232] where the D-optimality criterion is applied and an optimal time-dependent measure is sought, rather than the trajectories themselves. A breakthrough contribution was given by Uciński [276] (see also [279, 280, 288] and comprehensive monograph [281]), where apart from generalizations, some computational algorithms are developed based on the FIM. The problem is reduced to a state-constrained optimal-control one for which solutions are obtained via the method of successive linearizations, which is capable of handling various constraints imposed on sensor motions. In turn, Uciński and Chen [284] attempted to properly formulate and solve the time-optimal problem for moving sensors which observe the state of a DPS so as to estimate some of its parameters. This line of research was substantially extended in papers [181, 184, 186, 193, 273] towards different control-oriented and resource-constrained objectives for sensor path planning, and further in the monograph [259], providing some applications involving sensor-actuator networks.

Recently, as an attractive alternative to the mainstream of approaches for direct sensor path planning, the so-called scanning observations, attained a lot of interest as they allow substantially reducing the problem complexity. In this measurement scheme only a subset of sensors selected from among all available sensors, whose positions are fixed, take measurements during a given time interval whilst the other sensors become dormant [50]. Such sensor scheduling can be also interpreted in terms of several sensors which are mobile. One of the major difficulties in this task is its combinatorial nature. An extremely efficient approach based on directly constrained design measures was proposed by Uciński and Patan [289] (see also [180, 281]) and was consecutively refined for different experimental settings in the works [187, 188, 198, 295]. However, the underlying assumptions involve its main limitation which consists in that it can be used only when the number of sensors is relatively high. Recently, to overcome this impediment, as direct generalization of the technique developed in [291], the parallel branch-and-bound approach to sensor scheduling was proposed in [183] and extended further for the case of various constraints on sensor resources [204, 205, 282]. Finally, the problem is additionally complicated, if sensor switchings are allowed to take place in continuous time. As for this issue, an efficient algorithm was demonstrated [180, 182, 281] to solve the scanning sensor scheduling problem using the control parameterization-enhancing technique [150].

At this point, summarizing this concise review of the existing techniques for sensor spatio-temporal configuration, it ought to be emphasized that some fundamental difficulties may be encountered, which make the problem extremely nontrivial. One of them is the dependence of the optimal solutions on the parameters to be identified, being probably the greatest complication from the point of view of practical realization of the identification experiment. Although there exist some approaches developed to overcome those difficulties, cf. [180, 281], they are not free from drawbacks. In addition to this, there is a lack of methods dedicated to a number of tangible practical situations such as the occurrence of correlated measurements, the presence of random fluctuations in system parameters or grouping measurements in series, because the problem itself becomes even more complicated. Therefore, there is still room for new developments and a need for further generalizations toward those issues.

1.4 Observation Scheduling for Sensor Networks

Recent advances in hardware, sensor and networking technologies are enabling large-scale deployment of superior data acquisition systems with adjustable resolutions, called *Distributed Sensor Networks* (DSNs), see [29, 37, 42, 98, 108, 110, 245, 255, 309, 315]. Typically, such a sensor network is a collection of a large number of heterogeneous intelligent sensing devices referred to as sensor nodes distributed spatially over an environment and connected through a communication network. The sensor nodes are able to continuously collect measurements and cooperatively monitor their environment. Then, information that is gathered from all nodes is integrated using some data fusion strategy to synthesize new information such that the whole is greater than the sum of parts from individual nodes. Sensor networks have recently come into prominence because they hold the potential to revolutionize a wide spectrum of both civilian and military applications, including environmental monitoring, scene reconstruction, motion tracking, motion detection, battlefield surveillance, remote sensing, global awareness, etc. The design, implementation and operation for a DSN requires the confluence of many disciplines, including signal processing, networking and protocols, embedded systems, information management, and distributed algorithms [1, 93, 107].

The situation becomes even more interesting when we admit that some of the sensor nodes in the environment may move (in general, in arbitrary ways). In fact, in modern observation systems, sensors can be located on various platforms and these can be highly dynamic in motion. This drastically expands the spectrum of DSNs capabilities and provides considerable flexibility for a dynamic reaction of the network on changes in the monitored environment. What is more, technological advances in communication systems and the ongoing progress in developing small, low power and inexpensive mobile systems lead to various applications involving the swarms of autonomous networked vehicles, see [37, 42, 163, 176, 255]. Obviously, mobility implies an additional layer of complexity [37]. Usually, local computing elements embedded in network nodes suffer from severe limitations with respect to the communication range, energy supplies (batteries) and reliability to unexpected environmental conditions [1, 107, 304].

An ingenious characteristic of DSNs is the distributed nature of data acquisition. Therefore they seem to be ideally prepared for the task of monitoring spatio-temporal processes. A data fusion paradigm makes it clear that careful deployment and activation of sensor nodes are critical for collecting the most valuable information from the observed environment. Consequently, this requires suitable criteria quantifying sensor configurations and systematic approaches for measurement scheduling. Surprisingly, most of the available literature on this subject concentrates on the functionality of the network while a mathematical description of fundamental physical laws governing the observed system is often neglected or very briefly discussed. As a result, the methods focusing on maintaining network connectivity or robustness to nodes failures or providing a dense area coverage dominate this area of research. Although it is clear that those aspects of DSNs are essential for their basic operational reliability, an assumption that this is enough for obtaining qualitative measurements is rather naive from the practical point of view. In fact, much information is lost which could potentially be used to make the operation of the DSN more efficient. At this point, reference should be made to the works [51, 52, 53, 54, 55, 55, 104, 286], related to on-line optimal guidance of actuator/sensor network nodes for control and/or state estimation which demonstrate that inclusion of a DPS model into the optimization setting can substantially improve the quality of the network performance.

This monograph is aimed at providing a systematic approach to the sensor scheduling problem for identification of DPSs, which will be suitable for various experimental settings and will meet the aforementioned needs created by practical applications. Since one of the most important issues distracting engineers from the use of the model-based approach to sensor scheduling is the complexity and large scale of computational tasks, this book is also considered as an attempt to delineate efficient numerical schemes leading to solutions which can be obtained using computational power of standard PCs. Moreover, some potential generalizations toward distributed and parallel computing are indicated as well.

1.5 Conventions and Notation

Notation used throughout the book is more or less standard. Given a set H, $|H|$ and $\bar{H}$ signify its cardinality and closure, respectively. We use $\mathbb{N}, \mathbb{R}$ and $\mathbb{C}$ to denote the sets of integers, real and complex numbers, respectively. Further, $\mathbb{R}_+$ denotes the set of nonnegative real numbers. The n-dimensional Euclidean vector space is denoted by $\mathbb{R}^n$, and the Euclidean matrix space of real matrices with n rows and k columns is denoted by $\mathbb{R}^{n\times k}$. We will write $\mathbb{S}^n$ for the subspace of $\mathbb{R}^{n\times n}$ consisting of all symmetric matrices. The identity matrix of order n is denoted by I_n. In $\mathbb{S}^n$, two sets are of special importance: the cone of nonnegative definite matrices and the cone of positive definite matrices, denoted by $\mathbb{S}^n_+$ and $\mathbb{S}^n_{++}$, respectively. The curly inequality symbol $\succeq$ and its strict form $\succ$ are used to denote the Loewner partial ordering of symmetric matrices: for $A, B \in \mathbb{S}^n$, we have

$$A \succeq B \Longleftrightarrow A - B \in \mathbb{S}^n_+,$$
$$A \succ B \Longleftrightarrow A - B \in \mathbb{S}^n_{++}.$$

Consequently, the notations $A > 0$ and $A \geq 0$ are used to indicate that all elements of A are positive and nonnegative.

Here $\operatorname{rank}(A)$, $\operatorname{trace}(A)$ and $\det(A)$ denote the rank, trace and determinant of a matrix A, respectively. The operator $\operatorname{diag}(A)$ returns a vector of diagonal elements of matrix A and Diag(c) a square matrix with the elements of the vector c put on its main diagonal.

Suppose, that the f is a real-valued function, g is a vector-valued function, α is a scalar, x is a vector and M is a matrix, then:

1. $\partial f/\partial x$ denotes a row vector with i-th element $\partial f/\partial x_i$, x_i being the i-th element of x.
2. $\partial f/\partial M$ is a matrix with the (i,j)-th element $\partial f/\partial m_{ij}$, where m_{ij} is the (i,j)-th element of M.
3. $\partial g/\partial x$ denotes a matrix with the (i,j)-th element $\partial g_i/\partial x_j$, where g_i is the i-th element of g and x_j is the j-th element of x.
4. $\partial M/\partial \alpha$ is a matrix with (i,j)-th element $\partial m_{ij}/\partial alpha$, where m_{ij} is the (i,j)-th element of M.
5. $\nabla_x f$ is a vector $(\partial f/\partial x)^\top$ and we call it the gradient of the function f.

Some useful formulae follow

$$\frac{\partial}{\partial x}(a^\top x) = a^\top, \qquad \forall a \in \mathbb{R}^n, \tag{1.1}$$

$$\frac{\partial}{\partial x}(Ax) = A, \qquad \forall A \in \mathbb{R}^{m\times n}, \tag{1.2}$$

$$\frac{\partial}{\partial x}(x^\top Ax) = 2x^\top A, \qquad \forall A \in \mathbb{S}^n. \tag{1.3}$$

For any $A \in \mathbb{R}^{n\times n}$ which may depend on a parameter α, there holds

$$\frac{\partial}{\partial \alpha}\operatorname{trace}(A) = \operatorname{trace}\left(\frac{\partial A}{\partial \alpha}\right), \tag{1.4}$$

$$\frac{\partial}{\partial \alpha}A^{-1} = -A^{-1}\frac{\partial A}{\partial \alpha}A^{-1}, \tag{1.5}$$

$$\frac{\partial}{\partial \alpha}\ln\det(A) = \operatorname{trace}\left(A^{-1}\frac{\partial A}{\partial \alpha}\right), \tag{1.6}$$

$$\frac{\partial}{\partial \alpha}\det(A) = \det(A)\operatorname{trace}\left(A^{-1}\frac{\partial A}{\partial \alpha}\right). \tag{1.7}$$

The least upper bound (supremum) is denoted by $\sup(\cdot)$ and the greatest lower bound (infimum) by $\inf(\cdot)$. $C(\Omega)$ signifies the class of all continuous real-valued functions on Ω and $C^1(\Omega)$ the class of all continuously differentiable functions on Ω.

We call a point of the form $\alpha_1 u_1 + \cdots + \alpha_\ell u_\ell$, where $\alpha_1 + \cdots + \alpha_\ell = 1$ and $\alpha_i \geq 0$, $i = 1, \ldots, \ell$, a convex combination of the points $u_1, \ldots, u_\ell$ (it can be thought of as a mixture or a weighted average of the points, with α_i as the fraction of u_i in the mixture). Given a set of points U, co(U) stands for its convex hull, i.e., the set of all convex combinations of elements of U,

$$\mathrm{co}(U)=\left\{\sum_{i=1}^{\ell}\alpha_i u_i \,\middle|\, u_i\in U,\ \alpha_i\geq 0,\ i=1,\dots,\ell;\ \sum_{i=1}^{\ell}\alpha_i=1,\ \ell=1,2,3,\dots\right\}.$$

The probability (or canonical) simplex in $\mathbb{R}^n$ is defined as

$$\mathbb{P}_n=\mathrm{co}\big(\{e_1,\dots,e_n\}\big)=\left\{p\in\mathbb{R}^n_+ \,\middle|\, \sum_{i=1}^{n}p_i=1\right\},$$

where e_j is the usual unit vector along the j-th coordinate of $\mathbb{R}^n$.

Finally, we also make some notation associated with probability and statistical concepts and a brief list of the most important notions and abbreviations is given below:

θ, $\hat{\theta}$	unknown parameter vector and its estimate, respectively
$\xi_N(\,\cdot\,)$	exact design of an experiment
$\xi(\,\cdot\,)$	continuous design of an experiment (design measure)
$\mathrm{supp}\,\xi$	support of a measure ξ
P	probability
E	expectation
$\mathrm{E}[\,\cdot\,|\,\cdot\,]$	conditional expectation
var	variance
cov	covariance
$\delta(\,\cdot\,)$	Dirac delta distribution
δ_k	Kronecker delta
pdf	probability density function
cdf	cumulative distribution function

Chapter 2
Experimental Design in Sensor Scheduling

2.1 Parameter Estimation of Lumped Dynamic Systems

Prior to the analysis of more complex situations which are of interest in the context of the present work, it is useful to investigate first the less sophisticated case of a lumped dynamic system. To achieve a possibly high level of generality, we assume that the system under consideration is of the multiple-input and multiple-output (MIMO) type and evolves in a continuous time domain (results for a simpler case of observations in discrete time may be derived in much the same way).

2.1.1 Statistical Model of a Linear System

Let $y(x,t;\theta)$ denote the s-dimensional state of a given DPS at a spatial point $x \in \Omega \subset \mathbb{R}^d$ and time instant $t \in T = [0, t_f]$, t_f being a finite observation horizon. Here θ represents an unknown m-dimensional vector of model parameters. If we assume that the observation conditions in terms of sensor spatial locations are established arbitrarily before the experiment, forming a discretization of spatial domain Ω, then such a system can be treated as a Lumped Parameter System (LPS) with sensor positions playing the role of its additional design parameters which are fixed and are not the subject of estimation. Here it is also assumed that the dependence of the model outputs is linear with respect to the parameters. Thus, at each fixed spatial position for a given output we denote the measurement as

$$z_i^j(t) = y_i(x^j, t; \theta) + \varepsilon_i^j(t) = \sum_{\ell=1}^{m} \theta_\ell f_{i\ell}(x^j, t) + \varepsilon_i^j(t), \quad t \in T = [0, t_f], \quad (2.1)$$

M. Patan: Optimal Sensor Networks Scheduling in Identification of DPSs, LNCIS 425, pp. 11–46.
springerlink.com

where $x^j \in X \subset \mathbb{R}^d$, $j = 1, \dots, N$, stand for fixed design parameters and the set of functions $f_{i\ell}(\cdot, \cdot)$, $i = 1, \dots, s$, $\ell = 1, \dots, m$, is known a priori. The quantities ε_i^j represent stochastic disturbances which are assumed to be realizations of a sequence of random variables.

Introducing a more compact matrix form of the observation equation for the investigated system

$$z(t) = G^{\mathsf{T}}(t)\theta + \varepsilon(t), \quad t \in T = [0, t_f] \tag{2.2}$$

where a continuous time data record $\{z(t)\}_{t=0}^{t_f}$ is the realization of the measurement process. Matrix G is defined here as

$$G(t) = \left[F(x^1, t), \dots, F(x^N, t)\right], \tag{2.3}$$

where

$$F(x, t) = \left[f_1(x, t), \dots, f_s(x, t)\right]$$

and

$$f_i(x, t) = \left[f_{i1}(x, t), \dots, f_{im}(x, t)\right]^{\mathsf{T}}.$$

Moreover, ε is a Gaussian (zero-mean and uncorrelated in time) stochastic process playing the role of the measurement noise. Its covariance is defined by

$$\mathrm{E}\left\{\varepsilon(t)\varepsilon^{\mathsf{T}}(\tau)\right\} = C(t)\delta(t - \tau), \tag{2.4}$$

where δ means Dirac's delta distribution. The matrix $C(\cdot) \in \mathbb{R}^{Ns \times Ns}$ is assumed here to be symmetric and positive-definite for any $t \in T$ and to have the following form:

$$C(t) = \Sigma(t) \otimes \Lambda(t), \tag{2.5}$$

where $\Sigma(t) \in \mathbb{R}^{N \times N}$ and $\Lambda(t) \in \mathbb{R}^{s \times s}$ are positive definite matrices defining correlations between measurements at different settings of x^j and between outputs corresponding to a fixed setting x^j, respectively (the symbol $\otimes$ denotes the Kronecker product of matrices). Such form of $C(t)$ is very convenient as it makes possible to analyze the influence of different sources of correlations independently. On the other hand, from practical point of view, it is not very restrictive as usually such model of correlations can provide a fair approximation of the real situations.

In the remainder of this chapter we shall make the following two standing assumptions:

(A1) X is compact,
(A2) $\forall i,\ f_i \in C(X \times T; \mathbb{R}^m)$.

2.1.2 Methods of Estimation

Weighted least-squares. The Least-Squares Estimation (LSE) is the most popular estimation technique. This is mainly due to fact that it can be directly applied to the deterministic model in an ad hoc manner, without any a priori knowledge about the probability distribution of the observations. Although, in practice, estimates obtained in such a way may be quite unsatisfactory, for many situations nothing better can be provided [88, 250].

Based on the observations $z(\,\cdot\,)$ and the known values of $G(\,\cdot\,)$, the problem of recovering θ reduces to determining the parameter vector $\hat{\theta}$ which minimizes the weighted least-squares criterion,

$$\hat{\theta} = \arg\min_{\theta\in\Theta} \frac{1}{2}\int_0^{t_f} \left[z(t) - G^\mathsf{T}(t)\theta\right]^\mathsf{T} C^{-1}(t)\left[z(t) - G^\mathsf{T}(t)\theta\right] \mathrm{d}t, \tag{2.6}$$

over the set of admissible parameters Θ. Since such a function is a quadratic form with respect to the parameter vector, finding a minimum of (2.6) is equivalent to solving the following system of linear equations often called *normal equations*:

$$\int_0^{t_f} G(t)C^{-1}(t)G^\mathsf{T}(t)\hat{\theta}\,\mathrm{d}t = \int_0^{t_f} G(t)C^{-1}(t)z(t)\,\mathrm{d}t. \tag{2.7}$$

Provided that the matrix

$$M = \int_0^{t_f} G(t)C^{-1}(t)G^\mathsf{T}(t)\,\mathrm{d}t \tag{2.8}$$

is nonsingular, the unique estimate (2.6) exists and is of the form

$$\hat{\theta} = M^{-1}\int_0^{t_f} G(t)C^{-1}(t)z(t)\,\mathrm{d}t, \tag{2.9}$$

where matrix M is called the *Fisher Information Matrix* (FIM).

It is easy to show that the estimator (2.9) is unbiased and its covariance matrix is given by [281],

$$\operatorname{cov}\{\hat{\theta}\} = M^{-1}. \tag{2.10}$$

Furthermore, the Gauss–Markov theorem [16] states that from among all linear unbiased estimators the one described by (2.9) is that with the smallest variance and, if the distribution of $\varepsilon(\,\cdot\,)$ is normal, the estimator is efficient.

Maximum likelihood. In a wider context, where the structure of the joint probability density function of the observations is specified a priori, the Maximum Likelihood Estimation (MSE) method [16, 250] is often readily used by practitioners. This technique, being considered as an extension to the least-squares method, can be applied in the case when the covariance matrix $C(t)$ or its part may be unknown. In fact, if $C(t)$ is known, the least-squares estimator coincides with the maximal likelihood one [88].

Setting the configuration of design parameters as

$$\chi_N = (x^1, \dots, x^N) \tag{2.11}$$

let us denote by $p_T(z(t)|\chi_N, \beta)$ the family of probability measures defining the evolution of the stochastic process $z(t)$ over time interval T when configuration χ_N is applied. Here, β is a vector of the so-called *hyperparameters* which, in general, can contain unknown quantities such as model parameters θ, true values of the observed variables $z^\star(t)$ and/or other parameters describing the structure of the probability distribution.

In the case, when the system state is observed continuously, the general form of the likelihood function for the continuous data record $\{z(t)\}_{t\in T}$ may be obtained directly from the Radon–Nikodym derivative of the relevant probability measures [115, 243]:

$$\mathcal{L}_T(\beta) = \frac{\mathrm{d}p_T(z(t)|\chi_N, \beta)}{\mathrm{d}p_T(z(t)|\chi_N, \beta^\star)}, \tag{2.12}$$

for any T of nonzero length, where $\beta^\star$ is the true value of the vector of hyperparameters. The maximal likelihood estimate is the value of β for which the likelihood function attains its maximum. This makes the problem far more complex as, apart from model parameters, it is necessary to estimate many other factors.

The Radon–Nikodym derivative produces the relevant probability density and can be interpreted as a change of measure among the absolutely continuous probability measures. In general, (2.12) does not suggest how to calculate $\mathcal{L}_T(\beta)$ for a particular model. In order to compute the likelihood Grenander [92] has proposed the following two schemes:

1. Calculate the likelihood ratio (with respect to $p_T(z(t)|\chi_N, \beta^\star)$) for a finite set of time points $0 = t_0 < t_1 < \cdots < t_K = t_f$, and then let the number of time points tend to infinity so that $\max_{1\le k\le K}(t_k - t_{k-1}) \to 0$ as $K \to \infty$;
2. Find a countable set of independent variables (the so-called 'observable coordinates' of the process) and calculate the likelihood based on them.

The first scheme is the one that has been followed by most subsequent authors (e.g., see Billingsley [24] or Karatzas and Shreve [115]) since the second seems to be mainly restricted to Gaussian processes.

Restricting our attention to the model parameters, in one particularly interesting situation when we have observed N independent realizations of the process at different locations x^j, the maximum likelihood estimator has the form

$$\hat{\theta} = \arg\min_{\theta\in\Theta} \prod_{j=1}^{N} \mathcal{L}_T\left(\theta|\{z^j(t)\}_{t\in T}\right), \tag{2.13}$$

where $\mathcal{L}_T\left(\theta|\{z^j(t)\}_{t\in T}\right)$ is the pdf describing the j-th observation.

Although it is possible to apply some procedures such as least-squares without making any probabilistic assumptions, their quality may be poor. By recognizing the role of the hyperparameters it is possible to appropriately estimate them instead of using arbitrary values, which may significantly deteriorate the quality of estimates. Under relatively mild assumptions regarding the form of likelihood function, MLE leads to consistent and asymptotically efficient estimates [88, 92, 115, 250]. This is a strong argument for using it when the sample is large. Although for small samples MLE usually does not have optimal properties, experience shows that it still generates acceptable results.

Bayesian estimation. In the context of the work reported in this monograph, one more estimation method calls for a brief comment. When some prior information is available and should be treated as an integral part of the problem, the Bayesian Estimation (BE) method can be applied. If the prior information about the parameters can be summarized by some probability distribution $p_0(\theta)$, Bayes' theorem provides the so-called posterior distribution of parameters given by

$$p(\theta) = \frac{\mathcal{L}_T(\theta)p_0(\theta)}{\int_\Theta \mathcal{L}_T(\theta)p_0(\theta)\mathrm{d}\theta}. \tag{2.14}$$

To determine an estimate of the model parameters, we usually look for some characteristic value of the posterior distribution, e.g., the mean, median or mode. The last case referred to Maximum of Posterior Distribution (MPD), is probably the most popular variant, being a natural extension of the MLE. Indeed, if a prior distribution is uniform, then the Bayesian estimator coincides with the maximum likelihood one. Moreover, if p_0 is continuous, the MPD asymptotically converges to MLE as the number of observations tends to infinity. In such a case, one who accepts the MLE when no prior information is available will potentially be attracted by BE, because providing a small amount of prior information to the problem should also affect estimates very slightly.

2.1.3 *Measure of Estimation Accuracy*

As is clearly seen from (2.6), (2.13) or (2.14), for each estimation method the employed observation strategy in terms of the sequence of design variables x^j strongly influences the accuracy of the parameter estimates. Therefore, a very natural idea is to improve the quality of the estimates through the appropriate selection of the measurement locations. In order to achieve this, a suitable measure of the information content in the collected observations is necessary. From (2.8) it follows that the information matrix depends on the design parameters χ_N (through matrix G) but not on the observations. The meaning of this significant fact is that it is possible to design experimental conditions by choosing χ_N so as to maximize the information provided by the experiment in the sense of a statistical accuracy of the estimates.

In general, the FIM is defined as [307]

$$M = \mathrm{E}_\theta \left\{ \left[\frac{\partial \log \mathcal{L}_T(\theta)}{\partial \theta} \right]^\mathsf{T} \frac{\partial \log \mathcal{L}_T(\theta)}{\partial \theta} \right\}. \tag{2.15}$$

An essential property of the FIM is that its inverse constitutes an approximation of the covariance matrix for the estimates of the system parameters [88]. More precisely, a classical result known as the Cramér–Rao inequality establishes a lower bound of the covariance matrix [16, 88]

$$\operatorname{cov} \hat{\theta} \succeq M^{-1}. \tag{2.16}$$

Moreover, the inequality becomes an equality provided that the estimator is efficient (in particular this happens in the case of linear models). This leads to great simplification since the inverse of the FIM on the right-hand side of (2.16) can often be easily computed, even when the exact dispersion matrix of the given estimator is very difficult to determine.

Unfortunately, direct use of the covariance matrix or the FIM in searching for the configuration of design parameters maximizing the accuracy of parameter estimates is usually inconvenient. This is because the Loewner ordering of the information matrices used in (2.16) constitutes in general only a partial ordering over the set of all admissible components of χ_N. Thus, for complete comparison of design configurations, a suitable scalar performance index Ψ based on the FIM is necessary. As for a specific form of such a criterion, various options exist [73, 210, 234, 307], including the most popular examples, i.e.,

- the A-optimality criterion,

$$\Psi(M) = \operatorname{trace}(M^{-1}),$$

- the D-optimality criterion,

$$\Psi(M) = \ln \det(M^{-1}),$$

- the sensitivity criterion,

$$\Psi(M) = -\operatorname{trace}(M),$$

- the G-optimality criterion,

$$\Psi(M) = \max_{x \in X} \int_T \operatorname{var}\left(y(x,t;\hat{\theta}) \right) \mathrm{d}t.$$

A more specific discussion on the interpretations of the particular criteria will be provided in Section 2.2.4.

At this point it is convenient to lay the restriction that the measurements corresponding to different values of x^j are independent of each other. It may seem to be quite a strong assumption, but potential benefits justify this fact and it is indeed vindicated in numerous practical situations. Thus, we make the standing assumption that

$$\Sigma(t) = \mathrm{Diag}\left[\sigma^2(x^1,t),\dots,\sigma^2(x^N,t)\right], \tag{2.17}$$

where $\sigma(x^j,t)$, $j=1,\dots,N$ can be interpreted as standard deviations of the measurement errors for different x^js. Although the measurement covariance matrix $\Sigma(t)$ has diagonal form, there is no need to make additional assumptions on the matrix $\Lambda(t)$. In such a way, the FIM can be transformed into a simpler form,

$$M = \sum_{j=1}^{N} \int_0^{t_f} \sigma^{-2}(x^j,t)F(x^j,t)\Lambda^{-1}(t)F^{\mathsf{T}}(x^j,t)\,\mathrm{d}t = \sum_{j=1}^{N} M_j. \tag{2.18}$$

The above equation expresses the additivity of the FIM with respect to the individual settings of x^j, which is the crucial property for the approach presented in what follows.

For convenience, it is often very useful to make the considerations independent of the total number of observations N. Thus, we introduce the so-called *average per observation* (or *normalized*) FIM,

$$\bar{M} = \frac{1}{Nt_f}M = \frac{1}{Nt_f}\sum_{j=1}^{N} \int_0^{t_f} \sigma^{-2}(x^j,t)F(x^j,t)\Lambda^{-1}(t)F^{\mathsf{T}}(x^j,t)\,\mathrm{d}t, \tag{2.19}$$

which equals M up to a constant multiplier. Since most of optimality criteria used in practice satisfy the homogeneity condition

$$\Psi(\alpha M) = \gamma(\alpha)\Psi(M),\ \alpha > 0,$$

where $\gamma(\,\cdot\,)$ is a nondecreasing function, without loss of generality it will be used instead of M (for simplicity, the bar over M will also be omitted).

Owing to the assumption (2.17), we admit of replicated measurements, i.e., some values x^j may appear several times in the optimal solution (this is an inevitable effect of the independence of measurements). Consequently, it is sensible to reformulate the problem so as to operate on the locations $x^1,\dots,x^\ell$ (relabeled distinct sensor locations) in lieu of $x^1,\dots,x^N$. To this end, we introduce $r_1,\dots,r_\ell$ as the numbers of replicated measurements corresponding to the points $x^1,\dots,x^\ell$.

Definition 2.1. The *exact design* of the experiment is the collection of variables

$$\xi_N \overset{\text{def}}{=} \begin{Bmatrix} x^1, & x^2, & \dots, & x^\ell \\ p_1, & p_2, & \dots, & p_\ell \end{Bmatrix}, \tag{2.20}$$

where $p_i = r_i/N$, $N = \sum_{i=1}^{\ell} r_i$. In this formulation, the x^is are said to be the *support* points of the design, and $p_1, \dots, p_\ell$ are called their *weights*.

The proportion p_i of observations performed at x^i can be considered the percentage of experimental effort spent at that point.

On account of the above remarks, we rewrite the normalized FIM in the form

$$M(\xi_N) = \sum_{i=1}^{\ell} p_i \frac{1}{t_f} \int_0^{t_f} \sigma^{-2}(x^i, t) F(x^i, t) \Lambda^{-1}(t) F^{\mathsf{T}}(x^i, t) \, \mathrm{d}t \tag{2.21}$$

Here the p_is are rational numbers, since both r_is and N are integers. This discrete nature of N-observation exact designs causes serious difficulties, as the resultant numerical analysis problem cannot be solved by standard optimization techniques, particularly when N is large. A commonly used device for this problem is to extend the definition of the design. When N is large, the p_is can be considered as real numbers in the interval $[0, 1]$, not necessarily integer multiples of $1/N$. This assumption will also be made in what follows. Obviously, we must have $\sum_{i=1}^{\ell} p_i = 1$, so we may think of the designs as probability distributions on X. This leads to the so-called *relaxed* or *approximated*, designs which constitute the basis of the modern theory of optimal experiments [8, 68, 70, 73, 88, 210, 232, 276, 307].

Definition 2.2. The *continuous design* of the experiment concentrated at the finite number of support points is defined as an array

$$\xi \overset{\text{def}}{=} \begin{Bmatrix} x^1, & x^2, & \dots, & x^\ell \\ p_1, & p_2, & \dots, & p_\ell \end{Bmatrix}, \tag{2.22}$$

where

$$\sum_{i=1}^{\ell} p_i = 1, \quad p_i > 0, \ i = 1, \dots, \ell.$$

A design ξ concentrates approximately Np_1 measurements at x_1, Np_2 at x_2, and so on. It turns out that such an approach drastically simplifies the design problem.

Then, we may define the optimal design as a solution to the optimization problem

$$\xi^\star = \arg \min_{\xi \in \Xi(X)} \Psi[M(\xi)], \tag{2.23}$$

where $\Xi(X)$ denotes the set of all discrete probability measures on X.

At this point it is possible to further generalize the concept of the design to all probability measures ξ over X which are absolutely continuous with respect to the Lebesgue measure and satisfy by definition the condition

$$\int_X \xi(\mathrm{d}x) = 1. \tag{2.24}$$

Such a conceptual extension yields

$$M(\xi) = \int_X \Upsilon(x)\,\xi(\mathrm{d}x), \tag{2.25}$$

where

$$\Upsilon(x) = \frac{1}{t_f}\int_0^{t_f} \sigma^{-2}(x,t)F(x,t)\Lambda^{-1}(t)F^{\mathsf{T}}(x,t)\,\mathrm{d}t, \tag{2.26}$$

and the integration in (2.24) and (2.25) is to be understood in the Lebesgue–Stieltjes sense. The function $\sigma(x,t) \in C(X \times T)$, playing the role of the standard deviation, is assumed to take only positive values.

Introduction of the optimality criterion Ψ makes it possible to formulate the design problem as the optimization one.

Problem 2.1 (Optimal observation strategy problem). Find $\xi^\star \in \Xi(X)$ to minimize

$$\mathcal{P}(\xi) = \Psi[M(\xi)],$$

where $\Xi(X)$ denotes the set of all probability measures on X.

Finally, in some specific situations, it is convenient to study an even more general setting by introducing a fixed measure $\xi(\mathrm{d}t)$ of observation effort in the interval T, i.e., to consider

$$\begin{aligned} M(\xi) &= \int_T \left[\int_X \sigma^{-2}(x,t)F(x,t)\Lambda^{-1}(t)F^{\mathsf{T}}(x,t)\,\xi(\mathrm{d}x|t)\right]\xi(\mathrm{d}t) \\ &= \iint_{X\times T} \sigma^{-2}(x,t)F(x,t)\Lambda^{-1}(t)F^{\mathsf{T}}(x,t)\,\xi(\mathrm{d}x,\mathrm{d}t), \end{aligned} \tag{2.27}$$

where the measure $\xi(\,\cdot\,|t)$ corresponds to a spatial measure at time t,

$$\int_X \xi(\mathrm{d}x|t) = 1 \quad \text{a.e. on } T, \tag{2.28}$$

$$\xi(\mathrm{d}t) = \int_X \xi(\mathrm{d}x,\mathrm{d}t). \tag{2.29}$$

This means that we then focus on designs ξ being non-Cartesian product measures [235] or, in other words, we interpret sensor locations at given time moments as conditional distributions.

2.1.4 Characterization of Optimal Solutions

A number of characterizations of the optimal design $\xi^\star$ can be generalized for the studied case in a rather straightforward manner based on the results reported in [180, 231, 276, 281].

First, let us prove some properties of the FIM given by (2.25)

Lemma 2.1. *For any $\xi \in \Xi(X)$, the matrix $M(\xi)$ is symmetric and nonnegative definite.*

Proof. The symmetry of the matrix is a direct consequence of the definition (2.25) since the matrix $\Lambda^{-1}(t)$ is symmetric. The proof of the nonnegative definiteness starts with the observation that $\Lambda^{-1}(t)$ is positive definite for any t, therefore it can be represented in the form $\Lambda^{-1}(t) = A(t)A^\mathsf{T}(t)$, where $A(t)$ is a lower-triangular square matrix. From this it follows that for any $b \in \mathbb{R}^m$ we have

$$\begin{aligned} b^\mathsf{T} M(\xi) b &= \int_X \left\{ \frac{1}{t_f} \int_0^{t_f} \sigma^{-2}(x,t) b^\mathsf{T} F(x,t) A(t) A^\mathsf{T}(t) F^\mathsf{T}(x,t) b \, \mathrm{d}t \right\} \xi(\mathrm{d}x) \\ &= \int_X \left\{ \frac{1}{t_f} \int_0^{t_f} \sigma^{-2}(x,t) \|\alpha(x,t)\|^2 \, \mathrm{d}t \right\} \xi(\mathrm{d}x), \end{aligned} \tag{2.30}$$

where $\alpha(x,t) = A^\mathsf{T}(t) F^\mathsf{T}(x,t) b$. Since the integrand is nonnegative, the integral itself must be nonnegative, which completes the proof. ■

Let us introduce the notation $\mathbb{M}(X)$ for the set of all admissible information matrices, i.e.,

$$\mathbb{M}(X) = \big\{ M(\xi) : \xi \in \Xi(X) \big\}. \tag{2.31}$$

Lemma 2.2. *$\mathbb{M}(X)$ is compact and convex.*

Proof. According to the assumptions concerning $\sigma(x,t)$, we have $\sigma^{-2}(x,t) \in C(X \times T)$. In combination with (A2) this means that the function $\Upsilon(x)$ in (2.25) is continuous in X. Therefore, the set $\mathcal{U}(X) = \{\Upsilon(x) : x \in X\}$ is compact, being an image of the compact set X, cf. (A1), under the continuous mapping Υ from X into the space of all $m \times m$ matrices. The space $\mathbb{R}^{m\times m}$ is isomorphic with the Euclidean space $\mathbb{R}^{m^2}$. $\mathbb{M}(X)$ is a convex hull of $\mathcal{U}(X)$ and a convex hull of a compact set in the Euclidean space is also compact [242], which is the first claim.

To prove the convexity, let us introduce the design

$$\xi = (1-\alpha)\xi_1 + \alpha\xi_2, \tag{2.32}$$

where $\xi_1, \xi_2 \in \Xi(X)$ and $\alpha \in [0,1]$. Now, constructing the convex combination of the matrices corresponding to ξ_1, ξ_2, we have

$$
\begin{aligned}
(1-\alpha)M(\xi_1)+\alpha M(\xi_2) &= (1-\alpha)\int_X \Upsilon(x)\xi_1(\mathrm{d}x)+\alpha\int_X \Upsilon(x)\xi_2(\mathrm{d}x)\\
&= \int_X \Upsilon(x)[(1-\alpha)\xi_1(\mathrm{d}x)+\alpha\xi_2(\mathrm{d}x)] = M(\xi).
\end{aligned}
\tag{2.33}
$$

According to (2.31), $M(\xi) \in \mathbb{M}(X)$, which proves the second part of the lemma. ∎

Now we can state the following theorem, which complements our knowledge of the information matrices.

Theorem 2.1. *For any matrix $M_0 \in \mathbb{M}(X)$ there exists a design ξ that contains $\ell_0 \le m(m+1)/2+1$ support points and $M(\xi) = M_0$. If M_0 is a boundary point of $\mathbb{M}(X)$, then $\ell_0 \le m(m+1)/2$.*

Proof. The set $\mathbb{M}(X)$ is the convex hull of the set $\mathcal{U}(X) = \{\Upsilon(x) : x \in X\}$. The dimension of this set is $d = m(m+1)/2$ due to the symmetry of the matrix $\Upsilon(x)$ (it is sufficient to use only elements lying over and on the main diagonal). Applying the Carathedéodory theorem[281, p.263] , we may represent M_0 as a convex combination of no more than d_0 points from $\mathcal{U}(X)$,

$$
M_0 = \sum_{i=1}^{d_0} p_i \Upsilon(x^i), \quad \sum_{i=1}^{d_0} p_i = 1, \tag{2.34}
$$

where $d_0 \le d+1$ in the general case and $d_0 \le d$ for boundary points. Choosing

$$
\xi = \begin{Bmatrix} x^1, \dots, x^{d_0} \\ p_1, \dots, p_{d_0} \end{Bmatrix},
$$

we have $M_0 = M(\xi)$, cf. (2.21), and this is precisely the assertion of the theorem. ∎

The result in Theorem 2.1 is of great practical relevance. It allows immense reduction in the dimensionality of the search space for the optimal solution. Our attention can thus be restricted to the designs with a limited (thus not infinite) cardinality of design support. Surprisingly, the abstract concept of continuous designs, despite its complexity, leads to great simplifications of the design problem.

In order to derive necessary and sufficient conditions for the optimality of designs, some additional properties of the optimality criterion $\Psi(\cdot)$ are required [180, 276, 281], i.e.,

(A3) $M_1 \preceq M_2 \Rightarrow \Psi(M_1) \ge \Psi(M_2)$ (monotonicity);

(A4) $\forall \alpha \in [0,1], \quad \Psi[(1-\alpha)M_1+\alpha M_2] \le (1-\alpha)\Psi(M_1)+\alpha\Psi(M_2)$ (convexity);

(A5) $\exists q \in \mathbb{R}, \quad \widetilde{\Xi} = \{\xi : \Psi[M(\xi)] \le q < \infty\} \ne \emptyset$; (boundedness)

(A6) $\forall \xi \in \widetilde{\Xi}$, $\forall \bar{\xi} \in \Xi(X)$ we have

$$\Psi[(1-\alpha)M(\xi)+\alpha M(\bar{\xi})] = \Psi[M(\xi)] + \alpha \int_X \psi(x,\xi)\bar{\xi}(\mathrm{d}x) + o(\alpha;\xi,\bar{\xi}),$$

where the function $o(\,\cdot\,;\xi,\bar{\xi})$ satisfies $\lim_{\alpha\downarrow 0} o(\alpha;\xi,\bar{\xi})/\alpha = 0$.

The requirements (A3) and (A4) are rather obvious and natural as they refer to the properties which raise the possibility of applying methods of convex optimization. The assertion (A5) guarantees the existence of designs with finite values of Ψ. As the most restrictive one seems the assumption (A6) since it requires the existence of a specific form of the directional derivative. Note, however, that we have

$$\begin{aligned}
&\left.\frac{\partial \Psi[(1-\alpha)M(\xi)+\alpha M(\bar{\xi})]}{\partial \alpha}\right|_{\alpha=0^+} \\
&\quad= \lim_{\alpha\to 0^+} \frac{\Psi[(1-\alpha)M(\xi)+\alpha M(\bar{\xi})]-\Psi[M(\xi)]}{\alpha} \\
&\quad= \operatorname{trace}\left[\mathring{\Psi}[M(\xi)](M(\bar{\xi})-M(\xi))\right] \\
&\quad= \operatorname{trace}\left[\mathring{\Psi}[M(\xi)]\int_X \Upsilon(x)\,\bar{\xi}(\mathrm{d}x)\right] - \operatorname{trace}\left[\mathring{\Psi}[M(\xi)]M(\xi))\right] \\
&\quad= \int_X \left\{\operatorname{trace}\left[\mathring{\Psi}[M(\xi)]\Upsilon(x)\right] - \operatorname{trace}\left[\mathring{\Psi}[M(\xi)]M(\xi)\right]\right\}\bar{\xi}(\mathrm{d}x),
\end{aligned} \tag{2.35}$$

where

$$\mathring{\Psi}[M(\xi)] = \left.\frac{\partial \Psi(M)}{\partial M}\right|_{M=M(\xi)}. \tag{2.36}$$

From (2.35) it comes that, if Ψ is differentiable with respect to the elements of its matrix argument, the assumption (A6) is satisfied. Indeed, introduce the following representation of $\psi(x,\xi)$:

$$\psi(x,\xi) = \varsigma(\xi) - \phi(x,\xi), \tag{2.37}$$

where

$$\varsigma(\xi) = -\operatorname{trace}\left[\mathring{\Psi}[M(\xi)]M(\xi)\right] \tag{2.38}$$

$$\begin{aligned}
\phi(x,\xi) &= -\operatorname{trace}\left[\mathring{\Psi}[M(\xi)]\Upsilon(x)\right] \\
&= -\operatorname{trace}\left\{\frac{1}{t_f}\int_0^{t_f} \sigma^{-2}(x,t)F^{\mathsf{T}}(x,t)\mathring{\Psi}[M(\xi)]F(x,t)\Lambda^{-1}(t)\,\mathrm{d}t\right\}.
\end{aligned} \tag{2.39}$$

Armed with such properties of design criteria, we are ready to derive further characterizations of optimal designs.

Theorem 2.2. *Suppose that the the assumptions (A1)–(A6) hold. Then the optimal design $\xi^\star$ exists comprising no more than $m(m+1)/2$ support points. Moreover, the set of optimal designs is convex.*

Proof. The existence of an optimal design $\xi^\star$ follows from the compactness of $\mathbb{M}(X)$ (see Lemma 2.2) and the existence of designs with finite value of Ψ, cf. (A5). From the monotonicity of the criterion Ψ in (A3) it follows that $M(\xi^\star)$ has to be a boundary point of $\mathbb{M}(X)$. Indeed, if we assume that $M(\xi^\star)$ is an interior point of $\mathbb{M}(X)$, then there exists $\alpha > 1$ such that $\alpha M(\xi^\star) \in \mathbb{M}(X)$. Consequently, there exists some design ξ whose information matrix is given by $\alpha M(\xi^\star)$ (cf. Theorem 2.1). But, then we have

$$\Psi[M(\xi^\star)] > \Psi[M(\alpha\xi^\star)] = \Psi[M(\xi)],$$

and this contradicts the optimality of the design $\xi^\star$. Thus, if $M(\xi^\star)$ is a boundary point of $\mathbb{M}(X)$, from the second part of Theorem 2.1 we obtain the existence of a design with no more than $m(m+1)/2$ support points.

For the second part of the assertion, assume that $\xi_1^\star$ and $\xi_2^\star$ are optimal and $\xi^\star = \alpha\xi_1^\star + (1-\alpha)\xi_2^\star$. From the convexity of $\Psi(\,\cdot\,)$ (A4) and the set $\mathbb{M}(X)$, we have

$$\begin{aligned}\Psi[M(\xi^\star)] - \Psi[\alpha M(\xi_1^\star) + (1-\alpha)M(\xi_2^\star)] &\le \alpha\Psi[M(\xi_1^\star)] + (1-\alpha)\Psi[M(\xi_2^\star)]\\ = \alpha \min_{\xi\in\Xi(X)} \Psi[M(\xi)] + (1-\alpha) \min_{\xi\in\Xi(X)} \Psi[M(\xi)] &= \min_{\xi\in\Xi(X)} \Psi[M(\xi)],\end{aligned}$$

that is to say, $\xi^\star$ is an optimal design. ■

The next theorem presents necessary and sufficient conditions for the optimality of the designs.

Theorem 2.3. *Assume that (A1)–(A6) are satisfied. Then*

(i) A design $\xi^\star$ is optimal iff

$$\min_{x\in X} \psi(x,\xi^\star) = 0. \tag{2.40}$$

(ii) The function $\psi(x,\xi^\star)$ has zero value almost everywhere in $\operatorname{supp}\xi^\star$.

Proof. (*i*) From the convexity of Ψ it follows that the necessary and sufficient condition for the optimality of $\xi^\star$ over $\Xi(X)$ amounts to the nonnegativity of the directional derivative at $M(\xi^\star)$. By the assumption (A6), this is equivalent to the condition

$$\int_X \psi(x,\xi^\star)\, \xi(\mathrm{d}x) \ge 0, \quad \forall \xi \in \Xi(X). \tag{2.41}$$

Setting $\xi = \delta_x$, being a Dirac measure at point x, i.e., $\delta_x = \left\{ \begin{smallmatrix} x \\ 1 \end{smallmatrix} \right\}$, we get

$$\psi(x, \xi^\star) \geq 0, \quad \forall x \in X. \tag{2.42}$$

Clearly, (2.42) implies also (2.41). Consequently,

$$\min_{x \in X} \psi(x, \xi^\star) \geq 0. \tag{2.43}$$

However, from (A6) it follows that for any design measure $\xi \in \widetilde{\Xi}$,

$$\min_{x \in X} \psi(x, \xi^\star) \leq \int_X \psi(x, \xi^\star)\ \xi(\mathrm{d}x) = 0, \tag{2.44}$$

which combined with (2.43), establishes the formula.

(*ii*) The assumption (A6) implies the equality

$$\int_X \psi(x, \xi^\star)\, \xi^\star(\mathrm{d}x) = 0. \tag{2.45}$$

If $\xi^\star$ has finite cardinality of its support, i.e., it is given by (2.22), then from (2.45) it follows that

$$\sum_{i=1}^{\ell} p_i^\star \psi(x^{i\star}, \xi^\star) = 0. \tag{2.46}$$

However, (2.40) claims that

$$\psi(x^{i\star}, \xi^\star) \geq 0, \quad i = 1, \dots, \ell. \tag{2.47}$$

Since the weights $p^\star$ are also nonnegative, this proves the claim, otherwise the condition (2.46) will not be fulfilled. ■

The last theorem highlights the great decisive meaning of the function $\psi(x, \xi)$ in convex design theory, as it completely determines the location of the support points for the optimal solution. Additionally, its local minima for any arbitrary design ξ indicate points when the measurements provide the greatest amount of information about the parameters being the subject of our interest (of course, in the sense of a chosen criterion). Using the above result, it is possible to construct a simple test for the optimality of designs. In particular,

1. if the sensitivity function $\psi(x, \xi)$ is less than or equal to 0 for all $x \in X$, then ξ is optimal;
2. if the sensitivity function $\psi(x, \xi)$ exceeds 0, then ξ is not optimal.

Applications of analytical methods for finding optimal designs are dedicated rather for simple cases only. In general situations, there is a need for some iterative numerical procedures. In order to state a useful theorem for checking the optimality of designs, which will be called the *equivalence theorem*, we have to prove some auxiliary results.

Lemma 2.3. *For any design* $\xi \in \Xi(X)$, *we have*

(i) $\int_X \phi(x,\xi)\,\xi(\mathrm{d}x) = \varsigma(\xi)$, *and*

(ii) $\max_{x\in X} \phi(x,\xi) \geq \varsigma(\xi)$.

Proof. Taking into account (2.39), we obtain

$$\begin{aligned}\int_X \phi(x,\xi)\,\xi(\mathrm{d}x) &= -\int_X \operatorname{trace}\left[\mathring{\Psi}[M(\xi)]\Upsilon(x)\right]\xi(\mathrm{d}x)\\ &= -\operatorname{trace}\left[\mathring{\Psi}[M(\xi)]\int_X \Upsilon(x)\,\xi(\mathrm{d}x)\right]\\ &= -\operatorname{trace}\left[\mathring{\Psi}[M(\xi)]M(\xi)\right] = \varsigma(\xi).\end{aligned} \tag{2.48}$$

This establishes *(i)*. Then *(ii)* is a direct consequence of (2.48). ■

Lemma 2.4. *If* $\xi \in \widetilde{\Xi}$, $\bar{\xi} \in \Xi(X)$ *and* $\xi_\alpha = (1-\alpha)\xi + \alpha\bar{\xi}$, *then*

$$\left.\frac{\partial \Psi[M(\xi_\alpha)]}{\partial \alpha}\right|_{\alpha=0^+} = \varsigma(\xi) - \int_X \phi(x,\xi)\,\bar{\xi}(\mathrm{d}x). \tag{2.49}$$

Proof. From (2.35), we have

$$\begin{aligned}\left.\frac{\partial \Psi[M(\xi_\alpha)]}{\partial \alpha}\right|_{\alpha=0^+} &= \operatorname{trace}\left[\mathring{\Psi}[M(\xi)]\int_X \Upsilon(x)\,\bar{\xi}(\mathrm{d}x)\right] - \operatorname{trace}\left[\mathring{\Psi}[M(\xi)]M(\xi))\right]\\ &= \int_X \left\{\operatorname{trace}\left[\mathring{\Psi}[M(\xi)]\Upsilon(x)\right]\right\}\bar{\xi}(\mathrm{d}x) + \varsigma(\xi)\\ &= \varsigma(\xi) - \int_X \phi(x,\xi)\,\bar{\xi}(\mathrm{d}x).\end{aligned} \tag{2.50}$$

■

Now, we are capable of deriving our main result.

Theorem 2.4 (Generalized equivalence theorem). *The following conditions are equivalent:*

(i) the design $\xi^\star$ *minimizes* $\Psi[M(\xi)]$,

(ii) the design $\xi^\star$ *minimizes* $\max_{x\in X} \phi(x,\xi) - \varsigma(\xi)$, *and*

(iii) $\max_{x\in X} \phi(x,\xi^\star) = \varsigma(\xi^\star)$.

All the designs which satisfy (i)–(iii) and their convex combinations have the same information matrices equal to $M(\xi^\star)$, *provided that the criterion* $\Psi[\,\cdot\,]$ *is strictly convex.*

Proof. First, define $\xi_\alpha = (1-\alpha)\xi^\star + \alpha\xi_1$, where $\xi^\star \in \widetilde{\Xi}$, and $\xi_1 \in \Xi(X)$.
(i) $\Rightarrow$ *(ii)* If the optimal design $\xi^\star$ minimizes $\Psi[M(\xi)]$, then $\Psi[M(\xi^\star)] \leq \Psi[M(\xi_\alpha)]$ for any $\xi_1 \in \Xi(X)$; therefore,

$$\left.\frac{\partial \Psi[M(\xi_\alpha)]}{\partial \alpha}\right|_{\alpha=0^+} \geq 0, \quad \forall \xi_1 \in \Xi(X). \tag{2.51}$$

In particular, substituting $\xi = \xi^\star$ and $\bar{\xi} = \delta_x = \left\{ \begin{smallmatrix} x \\ 1 \end{smallmatrix} \right\}$ into (2.49), we get

$$\left.\frac{\partial \Psi[M(\xi_\alpha)]}{\partial \alpha}\right|_{\alpha=0^+} = \varsigma(\xi^\star) - \phi(x,\xi^\star) \geq 0, \quad \forall x \in X. \tag{2.52}$$

In connection with the second part of Lemma 2.3 this establishes *(ii)*.
(ii) $\Rightarrow$ *(iii)* Lemma 2.3 implies that $\max_{x\in X} \phi(x,\xi) - \varsigma(\xi)$ is bounded from below by zero. From (2.52) it follows that this zero bound is achieved at any design minimizing $\Psi[M(\xi)]$ (the existence of such a design is guaranteed by Theorem 2.2). This means that if $\xi^\star$ is a design characterized in *(ii)*, then necessarily $\max_{x\in X} \phi(x,\xi^\star) - \varsigma(\xi^\star) = 0$, which is exactly *(iii)*.
(iii) $\Rightarrow$ *(i)* Let $\xi^\star \in \Xi(X)$ satisfy $\max_{x\in X} \phi(x,\xi^\star) = \varsigma(\xi^\star)$. Setting $\xi_\alpha = (1-\alpha)\xi^\star + \alpha\bar{\xi}$ for $\bar{\xi} \in \Xi(X)$, from Lemma 2.4 we obtain

$$\left.\frac{\partial \Psi[M(\xi_\alpha)]}{\partial \alpha}\right|_{\alpha=0^+} = \varsigma(\xi^\star) - \int_X \phi(x,\xi^\star)\,\bar{\xi}(\mathrm{d}x) \geq \varsigma(\xi^\star) - \max_{x\in X} \phi(x,\xi^\star) = 0, \tag{2.53}$$

which implies the optimality of $\xi^\star$. The uniqueness of the information matrix for each optimal design follows from the convexity of the set $\mathbb{M}(X)$ and the strict convexity of the function $\Psi : M \mapsto \Psi[M]$ (from classical optimization theory it is known that there exists at most one global minimum of a strictly convex function over a convex set). ∎

Substituting a particular design criterion to Theorem 2.4, we obtain a specialized version of the equivalence theorem. Undoubtedly, the most famous classical result is the equivalence theorem for D- and G-optimal designs formulated first in [118] for a static regression model. The forms of the appropriate sensitivity functions for most popular criteria are listed in Table 2.1.

2.1.5 Nonlinear Models

As of now, we shall discuss a more complicated situation when the system considered is not linear with respect to the parameters. This is a straightforward consequence of the attempt to adopt results from the preceding section in parameter estimation of DPSs, because even if a system of PDEs constituting the model is linear, the state depends on the parameters in a highly

Table 2.1 Components of the sensitivity function for the most common optimality criteria

$\Psi[M(\xi)]$	$\phi(x,\xi)$	$\varsigma(x,\xi)$
$\ln\det M^{-1}$	$\operatorname{trace}\left\{\frac{1}{t_f}\int_0^{t_f}\sigma^{-2}(x,t)F^{\mathsf{T}}(x,t)M^{-1}F(x,t)\Lambda^{-1}(t)\,\mathrm{d}t\right\}$	m
$\operatorname{trace} M^{-1}$	$\operatorname{trace}\left\{\frac{1}{t_f}\int_0^{t_f}\sigma^{-2}(x,t)F^{\mathsf{T}}(x,t)M^{-2}F(x,t)\Lambda^{-1}(t)\,\mathrm{d}t\right\}$	$\operatorname{trace} M^{-1}$
$\operatorname{trace} AM^{-1}$	$\operatorname{trace}\left\{\frac{1}{t_f}\int_0^{t_f}\sigma^{-2}(x,t)F^{\mathsf{T}}(x,t)M^{-1}AM^{-1}F(x,t)\Lambda^{-1}(t)\,\mathrm{d}t\right\}$	$\operatorname{trace} AM^{-1}$
$-\operatorname{trace} M$	$\operatorname{trace}\left\{\frac{1}{t_f}\int_0^{t_f}\sigma^{-2}(x,t)F^{\mathsf{T}}(x,t)F(x,t)\Lambda^{-1}(t)\,\mathrm{d}t\right\}$	$\operatorname{trace} M$

nonlinear manner. This creates serious difficulties and makes closed-form solutions almost unavailable in most practical situations.

The sought generalization in (2.2) can be achieved by replacing the term $G^{\mathsf{T}}(t)\theta$ by the nonlinear multi-output system response

$$y(t;\theta) = \left[y(x^1,t;\theta),\dots,y(x^N,t;\theta)\right]^{\mathsf{T}}, \tag{2.54}$$

where

$$y(x,t;\theta) = \left[y_1(x,t;\theta),\dots,y_n(x,t;\theta)\right]^{\mathsf{T}}. \tag{2.55}$$

Analogously to (2.6), the least-squares estimator is then defined by

$$\hat{\theta} = \arg\min_{\theta\in\Theta}\frac{1}{2}\int_0^{t_f}\left[z(t)-y(t,\theta)\right]^{\mathsf{T}}C^{-1}(t)\left[z(t)-y(t,\theta)\right]\,\mathrm{d}t. \tag{2.56}$$

In contrast to the linear case, such an estimator is generally biased and the analytical form of the parameter covariance matrix, due to its dependence on θ, is very difficult to obtain. Most of the results for this case have only asymptotic character [13, 74, 75, 76, 312].

To derive an expression for the dispersion matrix, it is customary to linearize the system response in the vicinity of a prior estimate θ^0 of the unknown parameter vector θ. This estimate is assumed to be close enough to the true value of θ. Then, the system state can be approximated by expanding the function $y(t;\theta)$ in the Taylor series and retaining only the linear terms:

$$y(t;\theta) \cong y(t;\theta^0) + \left.\frac{\partial y(t;\theta)}{\partial \theta}\right|_{\theta=\theta^0} (\theta - \theta^0), \tag{2.57}$$

where

$$\frac{\partial y(t;\theta)}{\partial \theta} = \left[\frac{\partial y(x^1,t;\theta)}{\partial \theta}, \dots, \frac{\partial y(x^N,t;\theta)}{\partial \theta}\right]^{\mathsf{T}}, \tag{2.58}$$

and

$$\frac{\partial y(x,t;\theta)}{\partial \theta} = \begin{bmatrix} \frac{\partial y_1(x,t;\theta)}{\partial \theta_1} & \cdots & \frac{\partial y_1(x,t;\theta)}{\partial \theta_m} \\ \vdots & \ddots & \vdots \\ \frac{\partial y_n(x,t;\theta)}{\partial \theta_1} & \cdots & \frac{\partial y_n(x,t;\theta)}{\partial \theta_m} \end{bmatrix} \tag{2.59}$$

is the Jacobian matrix of the system response y with respect to the vector θ. Substituting (2.57) into (2.2), we obtain the following observation equation:

$$z(t) \approx y(t;\theta^0) + \left.\frac{\partial y(t;\theta)}{\partial \theta}\right|_{\theta=\theta^0} (\theta - \theta^0) + \varepsilon(t) \tag{2.60}$$

or, equivalently, after some rearrangement,

$$z_{\mathrm{m}}(t) = z(t) - y(t;\theta^0) + \left.\frac{\partial y(t;\theta)}{\partial \theta}\right|_{\theta=\theta^0} \theta^0 \approx \left.\frac{\partial y(t;\theta)}{\partial \theta}\right|_{\theta=\theta^0} \theta + \varepsilon(t). \tag{2.61}$$

In this way, we get an equivalent of (2.2) for the nonlinear case. The respective form of the average FIM (approximation of the inverse of $\operatorname{cov}\hat{\theta}$ up to a constant multiplier) is then

$$M(\xi_N) = \sum_{i=1}^{\ell} p_i \frac{1}{t_f} \int_0^{t_f} \sigma^{-2}(x^i,t) F(x^i,t) \Lambda^{-1}(t) F^{\mathsf{T}}(x^i,t)\,\mathrm{d}t, \tag{2.62}$$

where

$$F(x^i,t) = \left(\frac{\partial y(x^i,t;\theta)}{\partial \theta}\right)^{\mathsf{T}}_{\theta=\theta^0}. \tag{2.63}$$

It is clearly seen that the information matrix (2.62) depends on the prior estimate around which the model is linearized and it is valid only when the approximation (2.57) is accurate. In this sense, the results obtained with such an approach have only local character. However, it can be shown that, under rather mild assumptions the estimator (2.56) is strongly consistent, i.e.,

$$a.s. \lim_{N\to\infty} \hat{\theta} = \theta,$$

where $a.s. \lim$ denotes the *almost sure* limit (convergence with probability one). For detailed consistency results, the reader can be referred to [13, 74, 75, 76].

If both $y(x,t)$ and $F(x,t)$ are continuous in $\bar{\Omega} \times T$, then all the results from Sections 2.1.3 and 2.1.4 are directly applicable without any changes. In particular, the analogue of the FIM (2.64) can be written as

$$M(\xi) = \int_X \Upsilon(x)\, \xi(\mathrm{d}x), \tag{2.64}$$

where $X \in \bar{\Omega}$ is the region of admissible sensor locations, and

$$\Upsilon(x) = \frac{1}{t_f}\int_0^{t_f} \sigma^{-2}(x,t) \left(\frac{\partial y(x^i,t;\theta)}{\partial \theta}\right)^{\mathsf{T}}_{\theta=\theta^0} \Lambda^{-1}(t) \left(\frac{\partial y(x^i,t;\theta)}{\partial \theta}\right)_{\theta=\theta^0} \mathrm{d}t. \tag{2.65}$$

2.2 Optimal Sensor Scheduling Problem

2.2.1 System Description

Here, a class of systems to be considered within the framework of this monograph will be discussed. Attention is especially focused on the DPSs whose mathematical models are described by the system of PDEs of the general form:

$$\frac{\partial y(x,t)}{\partial t} = \mathcal{F}\Big(x,t,y(x,t),\nabla y(x,t),\nabla^2 y(x,t);\theta\Big), \quad (x,t) \in \Omega \times T, \tag{2.66}$$

where

- $\Omega \subset \mathbb{R}^d$ is a bounded simply-connected open domain with sufficiently regular boundary $\partial\Omega$,
- t signifies time,
- $T = (0, t_f)$ means the observation interval ($t_f < \infty$ denotes a fixed observation horizon),
- $x = (x_1, \ldots, x_d)$ stands for a spatial point belonging to the set $\bar{\Omega} = \Omega \cup \partial\Omega$,
- $y = \big(y_1(x,t), \ldots, y_s(x,t)\big)$ is the state variable with values in $\mathbb{R}^s$, and
- $\mathcal{F}$ stands for some known function which map its arguments to $\mathbb{R}^s$.

Note that this general setting includes the case when the function $\mathcal{F}$ may contain terms accounting for a priori known forcing inputs. The system (2.66) is supplemented with a suitable set of boundary conditions,

$$\mathcal{B}\Big(x,t,y,\nabla y;\theta\Big) = 0, \quad (x,t) \in \partial\Omega \times T, \tag{2.67}$$

and initial conditions,

$$\mathcal{I}\Big(x, y, \nabla y; \theta\Big) = 0, \quad (x,t) \in \partial\Omega \times \{0\}, \tag{2.68}$$

where $\mathcal{B}$ and $\mathcal{I}$ are some known functions.

It is assumed that there exists a unique solution of sufficient regularity to the system above. The model built on (2.66)–(2.68) contains the unknown parameter vector θ whose values belong to a parameter space Θ_{ad}. From the practical point of view, the most significant and common situation corresponds to constant parameter values, i.e., $\Theta_{\mathrm{ad}} \subset \mathbb{R}^m$, and from now on, we shall also make the assumption that the estimated parameters are constant. Note that the case of space-varying parameters can be easily tailored to this framework through appropriate parametrization, e.g., based on splines.

In order to give a taste of the role of PDEs in modelling and identification of DPSs, let us discuss briefly some examples of real engineering processes, being also amongst potential applications of the design approaches described further in the book.

Example 2.1. As an example of the discussed class of DPSs, the signal propagation process in a long-distance transmission line (e.g., a long electrical supply line or a high-frequency integrated circuit) can be examined. The large scale of such systems, due to the wavelength of the propagated electromagnetic signal, produces considerable spatial effects which have to be taken into account. In the case of a one-dimensional power supply line, the propagation phenomenon is described by the set of the so-called telegraph equations [49, 216]

$$\begin{aligned} -\frac{\partial y_2}{\partial t} &= L(x,t;\theta)\frac{\partial y_1}{\partial x} + R(x,t;\theta)y_1, \\ -\frac{\partial y_1}{\partial t} &= C(x,t;\theta)\frac{\partial y_2}{\partial x} + G(x,t;\theta)y_2, \quad x \in (0,l), \quad t \in (0,t_f), \end{aligned} \tag{2.69}$$

where $y_1(x,t)$ and $y_2(x,t)$ denote respectively the current intensity and voltage along the line, l is the maximal length of the line, and R, G, L, C stand for the spatial density of resistance, conductance, inductance and capacitance,

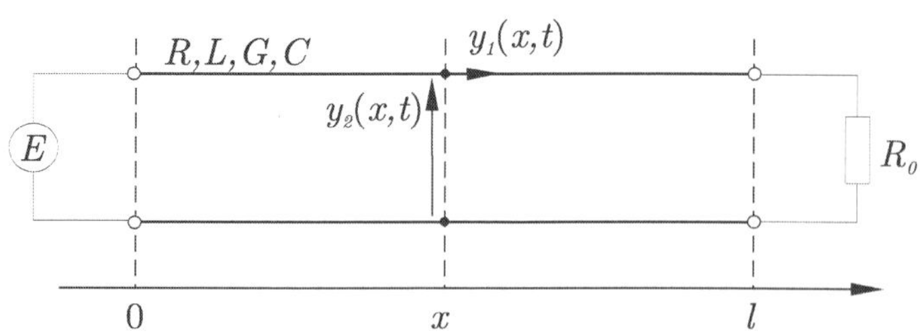

Fig. 2.1 Nonhomogeneous transmission line with a load of the resistance type

respectively. In the case when the line load is only of the resistance type (cf. Fig. 2.1), the initial conditions are defined as

$$y_1(x,0) = g_1(x), \quad y_2(x,0) = g_2(x), \quad x \in (0,l), \tag{2.70}$$

while the boundary conditions at the extremities of the line are

$$\begin{aligned} & y_1(0,t) = y_{10}(t), \quad y_2(0,t) = y_{20}(t), \\ & C(x,t;\theta)R_0\left(\frac{\partial y_1}{\partial t}\right)_{x=l} + \left(\frac{\partial y_1}{\partial x}\right)_{x=l} + G(x,t;\theta)R_0\, y_1(l,t) = 0, \\ & \frac{L(x,t;\theta)}{R_0}\left(\frac{\partial y_2}{\partial t}\right)_{x=l} + \left(\frac{\partial y_2}{\partial x}\right)_{x=l} + \frac{R(x,t;\theta)}{R_0}\, y_2(l,t) = 0, \quad t \in (0,t_f). \end{aligned} \tag{2.71}$$

Although the physical interpretation of the quantities which describe the properties of the transmission channel (i.e., R, G, L, C and R_0) is well established [49], their distributed nature give rise to severe difficulties in proper modelling. Some of these coefficients are known and available from direct measurements (e.g., R_0), yet others contain some components which are not measurable or such a measurement could be difficult to obtain. Therefore, precise calibration of model coefficients becomes a difficult but important task for accurate simulation and prediction of system behavior. ★

Example 2.2. Another interesting example is Electrical Impedance Tomography (EIT), which is an imaging technique for reconstruction of conductivity or permittivity for the examined object inferred from surface electrical measurements. Electrical impedance tomography has proved its usefulness in many areas including medical [102, 317] and industrial applications [174, 252, 313]. Impedance tomography is a two-step process [101, 174]. The experiment provides data which characterize the electrical response of a medium under the

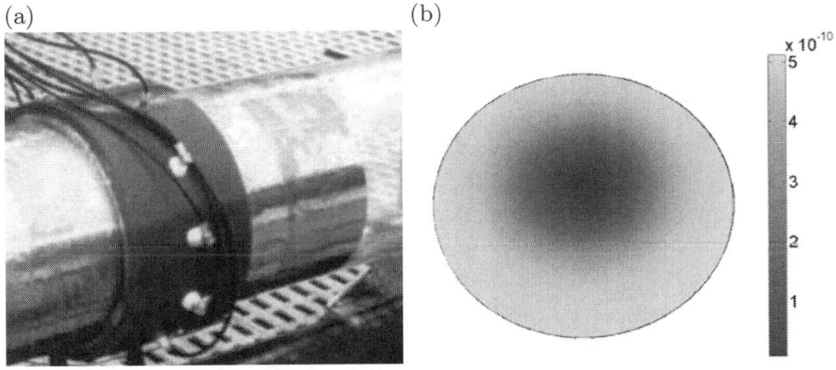

Fig. 2.2 Electrodes attached to the pipeline with oil and gas (a) and reconstruction of the conductivity map for the two-flow medium in the plane of the electrodes (bright and dark colors denotes oil and gas, respectively)

examination, and in the second step the distribution of the electrical properties is reconstructed from these data. Therefore, the forward problem is described in the simplest case by the Laplace equation [252, 313],

$$\nabla \cdot (\sigma(x)\nabla y(x)) = 0, \quad x \in \Omega, \tag{2.72}$$

where y is the electrical potential and σ is the conductivity coefficient. Equation (2.72) is supplemented by nonzero Dirichlet boundary conditions,

$$y(x) = y_0, \quad x \in \Gamma_1 \subset \partial\Omega, \tag{2.73}$$

at these points of the boundary which are contiguous with the supply electrodes and zero von Neumann boundary conditions,

$$\frac{\partial y(x)}{\partial n} = 0, \quad x \in \Gamma_2 = \partial\Omega \setminus \Gamma_1, \tag{2.74}$$

on the rest of $\partial\Omega$.

EIT inversion consist in addressing another question, the so-called *inverse* problem, defined for material parameters, (in this case the conductivity coefficient of distributed type). For a given potential distribution over the boundary, a suitable form of $\sigma(x)$ is constructed based on some lack-of-fit criterion defining the discrepancy between the modelled and measured responses. In Fig. 2.2, one of typical industrial applications of EIT is illustrated, namely, fluid-dynamic modelling (laminar and turbulent flows) [119, 174, 313]. However, it is well known that the inverse problem is usually ill-conditioned and regularization is required. In the presence of significant measurement noise, proper spatial configuration of both supply and measurement electrodes is of crucial importance so as to obtain the most informative data. ★

2.2.2 Observation Strategies

Given the model structure (2.66)–(2.68) up to a finite number of constant parameters θ, the main purpose of parameter estimation is to reconstruct θ in such a way that the model response imitates the behavior of the real state of the system $\tilde{y}$ as closely as possible. The crucial problem lies in the fact that usually not all the components of the state $\tilde{y}$ are observable, especially in real data acquisition systems. Measurements taking techniques can be split into the following four classes [40]:

1. observations at discrete spatial locations at discrete time instants,
2. observations at discrete spatial locations over a continuous time interval,
3. spatial observations at discrete time instants,
4. spatial observations continuously in a given time interval.

In general, 'ideal' spatial measurements are obviously not possible and, in fact, in the literature related to identification of DPSs a large number of contributions focus on pointwise observations, to mention a few. Note, however, that recent great developments in measurement techniques and instrumentation have begun changing this situation. Consequently, researchers also alter their point of view to the paradigms of distributed observations. In many scientific areas, techniques of the types 3 and 4 are successfully exploited and their outcomes are highly valuable (e.g., astrophysical observations of the universe in infrared or microwave ranges of electromagnetic radiation, thermography in medical and industrial diagnostics or meteorological observations of the atmosphere from satellites). On the other hand, in the majority of engineering applications, even if spatial observations are available, it is still impossible to measure the system state over the entire domain, and Methods 1 and 2 turn out to be most appropriate, and, consequently they dominate in the applications. This fact constitutes the main reason behind focusing our attention on these strategies of data acquisition.

In general, the observation strategy can be understood as any process of taking measurements, although for clarity of our further discussion it is necessary to define such a notion more precisely in terms of the standard language of statistics and metrology. Firstly, we should provide a more formal definition of the measurement result.

Definition 2.3 (ISO/IEC Guide 99:2007 [106]). The *result of a measurement* is only an approximation or an estimate of the value of a specific quantity subjected to the measurement, that is, the *measurand*, and thus the result is complete only when accompanied by a quantitative statement of its uncertainty.

Here, it is very important to emphasize that the class of the measurements studied in this work belongs to the so-called external observations (also known as black-box observations) made with noninertial measuring transducers whose influence on the system performance is negligible and may be omitted. Furthermore, characteristics of real sensors often depend on both spatial coordinates and time, and the measurement is a result of measurand averaging over some spatial domain. A common practice is, however, the assumption that this dependence can be approximated by a number of pointwise measurements. Thus, the resulting uncertainty about a measurement can be reduced to the components which are evaluated by statistical methods (i.e., an uncertainty of type A [106]). Otherwise, it is necessary to consider the components of the uncertainty evaluated by other means than the statistical analysis of a series of observations (the so-called type B evaluation [106]). With such a notion of observation being a result of the measurement, it is possible to introduce the definition of the measurement space and the strategy of observations which are used within the scope of this book.

Definition 2.4. The *measurement space* is the space induced by the subset of the Cartesian product $\bar{\Omega} \times T$ in which the observations are available.

Definition 2.5. The *strategy of observations* is understood as any subset of the measurement space.

In practice, the following main strategies of taking measurements can be distinguished:

- **Using stationary sensors.** In this case observations are made at a finite number of locations continuously or discretely in time. Because sensor positions are fixed, the resulting measurements are somewhat averaged in the time domain (of course, if the system state is time dependent). The main problem we are faced with here is the choice of optimal sensor locations in a given admissible spatial domain.
- **Using moving sensors.** Allowing the mobility of the sensors, we can increase their applicability and the capabilities in comparison with stationary sensors. In this way observations possess an additional degree of freedom regarding potential optimality. This is a direct consequence of the possibility of taking measurements at positions which are at given time moments most advantageous in the sense of a given performance index quantifying the information about the estimated parameters. Let us note that nonmobile sensors can be considered as movable sensors whose trajectories are reduced to fixed points. Thus the use of movable sensors constitutes a generalization including as a special case the use of stationary sensors. An observation strategy here is a set of sensor trajectories representing changes in the measurement locations in time. The problem of designing optimal trajectories becomes crucial for this case.
- **Scanning.** This strategy constitutes a combination of the previous two cases with a slight extension. In some situations, the observation system comprises multiple sensors whose positions are already specified and it is desired to activate only a subset of them during a given time interval while the other sensors remain dormant. Such a scanning strategy of taking measurements can be also interpreted in terms of several sensors which are mobile, but, since in general the number of activated sensors could vary in time, this technique is not a discrete-time version of the movable sensor method. This strategy is accompanied by the problem of determining the best spatio-temporal schedule of measurement taking.

Definition 2.5 may seem to be very abstract, especially in the case of moving observations when sensor trajectories should satisfy many constraints. However, such a formulation is convenient from the point of view of statistical analysis. Moreover, if necessary, in particular situations some additional restrictions can be made providing suitable regularization of the observation strategy.

2.2.3 Parameter Estimation

The objective of parameter estimation is to find the value of the unknown parameter vector $\hat{\theta} \in \Theta_{\mathrm{ad}}$ based on the appropriate set of process observations provided by N sensors such that the predicted response of the model (2.66)–(2.68) is close enough to the process observations (in the sense of some known quality measure). By analogy with Section 2.1, we focus our attention on the case of measurements taken continuously in time, as it is a more general situation and discrete time measurements may be covered in straightforward manner. Furthermore, to construct a proper statistical model it is assumed that the system state may not be available directly for measurement and the uncertainty of the system outputs can be described as additive disturbances. Different practical situations imply that the measurement process can be formally represented (similarly to (2.1)) by the following forms of the observation equation corresponding to presented observation strategies [180, 231, 276, 281]:

- stationary sensors,

$$z^j(t) = h\left(y(x^j, t; \theta), x^j, t\right) + \varepsilon(x^j, t), \quad t \in T, \quad j = 1, \dots, N; \tag{2.75}$$

- scanning sensors with switchings allowed at a finite number of time instants $t_0 < t_1 < \cdots < t_K = t_f$,

$$\begin{aligned} z^j(t) = h\Big(y\big(x_k^j, t; \theta\big), x_k^j(t), t\Big) + \varepsilon\big(x_k^j, t\big), \\ t \in T_k = [t_k - 1, t_k), \quad k = 1, \dots, K, \quad j = 1, \dots, N; \end{aligned} \tag{2.76}$$

- mobile sensors,

$$z^j(t) = h\big(y(x^j(t), t; \theta), x^j(t), t\big) + \varepsilon(x^j(t), t), \quad t \in T, \quad j = 1, \dots, N, \tag{2.77}$$

where $h(\,\cdot\,)$ is a given function mapping its arguments into $\mathbb{R}^{\tilde{s}}$, $z^j(t)$ is an $\tilde{s}$-dimensional output, x^j, x_k^j and $x^j(t) \in X$ stand respectively for the locations of the j-th stationary, scanning (during T_k) and movable sensor (at time instant t), X signifies the part of Ω where the measurements can be made, and $\varepsilon^j(\,\cdot\,)$ denotes the measurement noise. It is customary to assume that the noise is zero-mean, Gaussian, uncorrelated in both time and space [73, 138], i.e., its statistics are defined as

$$\mathrm{E}[\varepsilon(x^j, t)] = 0, \quad \mathrm{E}[\varepsilon(x^i, t)\varepsilon^{\mathsf{T}}(x^j, \tau)] = \delta_{ij}\delta(t - \tau)C(x^i, t), \tag{2.78}$$

$C(x^i, t) \in \mathbb{R}^{n \times n}$ is a known positive-definite matrix, δ_{ij} and $\delta(\,\cdot\,)$ stand for Kronecker's and Dirac's delta symbols, respectively. Although white noise is a physically impossible process, it constitutes a reasonable approximation to a disturbance whose adjacent samples are uncorrelated at all time instants for which the time increment exceeds some value which is small compared with the time constants of the DPS. A more general situation when the measurement noise is correlated in space and time domains will be considered in more detail in Section 9.2.

The parameter identification problem is customarily converted into an optimization one depending on the particular estimation method applied (cf., Section 2.1.2). For instance, if the least-square technique is used the parameter estimate $\hat{\theta}$ is determined via minimization of a suitable weighted criterion taking the following forms for particular situations, respectively:

- stationary sensors,

$$\mathcal{J}(\theta) = \frac{1}{2}\sum_{j=1}^{N}\int_{T} \|z^j(t) - h(\hat{y}(x^j, t; \theta), x^j, t)\|^2_{C(x^j,t)}\,\mathrm{d}t; \tag{2.79}$$

- scanning sensors, measurements discrete in time,

$$\mathcal{J}(\theta) = \frac{1}{2}\sum_{j=1}^{N}\sum_{k=1}^{K}\int_{T_k} \|z^j(t) - h(\hat{y}(x^j_k, t; \theta), x^j_k, t)\|^2_{C(x^j_k,t)}\,\mathrm{d}t; \tag{2.80}$$

- mobile sensors,

$$\mathcal{J}(\theta) = \frac{1}{2}\sum_{j=1}^{N}\int_{T} \|z^j(t) - h(\hat{y}(x^j(t), t_k; \theta), x^j(t), t)\|^2_{C(x^j(t),t)}\,\mathrm{d}t, \tag{2.81}$$

where $\hat{y}(x, t; \theta)$ is the solution to (2.66)–(2.68) for a given value of θ, and

$$\|e(x,t)\|^2_A = e^{\mathsf{T}}A^{-1}e,$$

for any positive-definite symmetric matrix A. On the analogy to the suggestions given in Section 2.1 for lumped systems, the pros and cons of using LSE as well as other estimation techniques are in general the same here.

2.2.4 Problem Formulation

Evidently, in each case the parameter estimate $\hat{\theta} = \arg\min_{\theta\in\Theta}\mathcal{J}(\theta)$ depends strongly on the coordinates of the sites where the measurements are taken at a given time. This fact allows us to consider the optimality of sensor locations in the case of stationary sensors and spatio-temporal schedule of position changes in the case of mobile and scanning sensors. Assuming that the estimator is unbiased, when the observation horizon is sufficiently long or the noise magnitude tends to zero, the inverse of the FIM is an asymptotic covariance matrix of the estimates [307]. As for a particular statistical interpretation, the inverse of the FIM constitutes, by the Cramér–Rao inequality [16, 88],

$$\operatorname{cov}\hat{\theta} \succeq M^{-1}, \tag{2.82}$$

the maximum accuracy available for any unbiased estimator. The optimal sensor schedule should thus lead to possibly small values of the inverse of the FIM, which will guarantee small dispersion of the estimates around the vector of the true parameter values. In the case of a biased estimator it is still possible to make use of the FIM, but it is necessary to incorporate the bias terms into (2.82), cf. [211].

For the class of DPS considered here a fundamental difficulty in exploiting the FIM consists in its dependence on the vector of true parameters, which are unknown [9, 73, 180, 210, 211, 281, 307]. Therefore, by analogy to results for nonlinear lumped models (cf., Section 2.1.5), under the assumptions regarding the noise statistics (2.78), the following approximations are used in applications:

- stationary sensors,

$$M = \sum_{j=1}^{N} \int_T G^\mathsf{T}(x^j, t) C^{-1}(x^j, t) G(x^j, t)\, \mathrm{d}t; \tag{2.83}$$

- scanning sensors, measurements discrete in time,

$$M = \sum_{j=1}^{N} \sum_{k=1}^{K} \int_{T_k} G^\mathsf{T}(x_k^j, t) C^{-1}(x_k^j, t) G(x_k^j, t)\, \mathrm{d}t; \tag{2.84}$$

- movable sensors, measurements continuous in time,

$$M = \sum_{j=1}^{N} \int_T G^\mathsf{T}(x^j(t), t) C^{-1}(x^j(t), t) G(x^j(t), t)\, \mathrm{d}t, \tag{2.85}$$

where

$$G(x,t) = \left(\frac{\partial h(y,t;\theta)}{\partial y} \right)_{y=y(x,t;\theta)} \frac{\partial y(x,t;\theta)}{\partial \theta} \tag{2.86}$$

is the Jacobian matrix of the parametric sensitivities. As has already been mentioned, a direct comparison of matrices is not always possible due to the Loewner ordering of FIMs which is only a partial ordering of measurement schedules. As was discussed in Section 2.1 in the context of lumped systems, it is reasonable to convert the problem to the optimization one using some scalar real-valued function Ψ operating on the FIM as an optimality criterion. In the literature one can find various choices for such a function [9, 73, 210, 211, 281, 307]; however, the most popular choices used in practical applications can be divided into two groups. The first one covers a class of criteria focusing on the accuracy of the regression parameters. For the purpose of interpretation, let assume that measurement noise is normally distributed. Under such an assumption, the *confidence* or *uncertainty ellipsoid* for parameter estimates is given by

$$\mathcal{E}(\hat{\theta}) = \{\theta : (\hat{\theta} - \theta)^{\mathsf{T}} M(\hat{\theta} - \theta) \le c\}, \tag{2.87}$$

where c is a positive real constant dependent on the assumed confidence level β, the estimated noise variance, the number of observations and parameters. The ellipsoid center is located at $\hat{\theta}$ and the probability that it contains a vector of true parameters is equal to β. The 'smaller' the confidence ellipsoid, the better the accuracy of the parameter estimates. Different strategies of compressing the ellipsoid lead to different design criteria:

- the A-optimality criterion,

  $$\Psi(M) = \operatorname{trace}(M^{-1}).$$

  An A-optimal design suppresses the variance of the estimates (but the correlation between the estimates is neglected), which is equivalent to minimizing the average axis length of the confidence ellipsoid;
- the D-optimality criterion,

  $$\Psi(M) = \ln\det(M^{-1}).$$

  A D-optimal design minimizes the volume of the uncertainty ellipsoid for the parameters;
- the E-optimality criterion,

  $$\Psi(M) = \lambda_{\max}(M^{-1}),$$

  where $\lambda_{\max}(\,\cdot\,)$ stand for the maximal eigenvalue of its argument. An E-optimal design minimizes the largest width of $\mathcal{E}(\hat{\theta})$ along its principal directions;
- the L-optimality criterion,

  $$\Psi(M) = \operatorname{trace}(AM^{-1}), \quad A \succ 0.$$

  It is a generalization of the A-optimality one. It is equivalent to minimizing a linearly weighted average axis length of $\mathcal{E}(\hat{\theta})$. In such a way it is possible to take into account the off-diagonal elements of the FIM;
- the sensitivity criterion,

  $$\Psi(M) = -\operatorname{trace}(M).$$

  This criterion does not possess a strong statistical interpretation, but it is sometimes used due to its simple form and the fact that its minimization increases the sensitivity of the outputs with respect to parameter changes.

In addition to its relative simplicity, the D-optimality has an important advantage in contrast to the E- and A-optimality criteria, namely, it is invariant under linear output transformations and changes in the parameter scales

[307]. The simplest one is the sensitivity criterion, but in many cases it leads to a singular FIM and serious problems with identifiability [281].

All these criteria are related to the eigenvalues of the FIM (e.g., minimizing the D-optimality criterion amounts to maximizing the product of all eigenvalues of M, while the use of the A-optimality criterion leads to minimizing the sum of the reciprocals of the eigenvalues). Therefore, most of the performance indices mentioned here can be treated as members of a wide class of criteria [73, 307] defined by the following functional:

$$\Psi_\gamma(M) = \begin{cases} \left[\dfrac{1}{m}\mathrm{trace}(PM^{-1}P^\mathsf{T})^\gamma\right]^{1/\gamma} & \text{if } \det M \neq 0, \\ \infty & \text{otherwise}, \end{cases}$$

where $P \in \mathbb{R}^{m\times m}$ is a weighting matrix. For instance, substituting $P = I$ for $\gamma = 1$, $\gamma \to \infty$ and $\gamma \to 0$, we obtain respectively the A-, E- and D-optimum design criteria.

The second important class of design criteria is related to the quality of the prediction or estimation of the output value at points where measurements have not been taken. Among this group of criteria, definitely the most popular ones include

- the G-optimality criterion,

  $$\Psi(M) = \max_{x\in X} \mathrm{var}\,\big(\hat{y}(x,t)\big).$$

  A G-optimal strategy minimizes the maximal variance of outputs (the worst case approach). This criterion is closely related to the concept of D-optimality concept (cf., Theorem 2.4).
- the Q-optimality criterion,

  $$\Psi(M) = \int_X \mathrm{var}\big(\hat{y}(x,t)\big)\,\mathrm{d}x.$$

  Q-optimal strategy minimizes the mean variances of outputs.

Obviously, there exist many more other performance indices, which are proposed in various publications [9, 307].

2.2.5 Main Difficulties

After conversion of the original problem of choosing an optimal measurement strategy to that of minimizing the appropriate performance index, one might think that its solution is only a matter of applying some well-known nonlinear optimization algorithms. Nevertheless, practical problems clearly show that such an impression is extremely misleading. Severe difficulties are encountered, which make the problem highly nontrivial and explain the scarcity of publications on this subject in contrast to a similar problem for state estimation. In what follows, the main impediments in solving the prescribed problem will be indicated and discussed.

2.2.5.1 Loss of the Estimator Quality

First of all, it should be emphasized that the approximation of the parameter dispersion matrix by the inverse of the FIM, which is based on converting the Cramér–Rao inequality into the equality, is legitimate only in situations when measurement errors have small magnitudes and are independently distributed, the nonlinearity of the model with respect to the parameters is mild, and the time horizon is comparatively large [307]. Although the first assumption is rather justified in most real situations, the other two are rarely satisfied. In fact, in more precise terms, the observational time horizon should satisfy $t_f \to \infty$ [232, 307]. From the technical point of view, an infinite observational horizon is impossible to implement and the loss of the accuracy of the estimator should be observed. However, if the length of the observation time interval is sufficiently large, this effect is commonly neglected in applications.

Another great drawback which can be encountered in practice is a strong nonlinearity of the model, and consequently the deterioration in the quality of the estimator. This situation necessitates a more careful application of estimation techniques.

2.2.5.2 Large Scale of the Problem

In practical engineering systems, the number of allocated sensors in a given spatial region may range from dozens to several hundred. The London Air Quality Network or the Air Quality Network of Berlin are typical examples of monitoring networks constructed for measurements of air pollution, identification of pollution sources, verification of the effect of air pollution control and prevention from detecting excessive smog levels within the framework of smog alert plans. They contain more than 150 monitoring stations of both stationary and mobile types. Recent advances in the area of wireless sensor networks and robotics make it possible to enlarge this number to thousands of sensors used for environmental monitoring or military applications. In the area of sensor arrays, the number of detectors involved in observation of DPS may be even greater by several orders. Moreover, since real processes often concern two or three dimensional spatial domains, the position of each sensor is determined by at least two spatial coordinates and the optimization problem may be multi-modal (when the global optimum is hidden in many local optima), severe difficulties are usually experienced when trying to solve it as a classical nonlinear constrained programming task.

2.2.5.3 Phenomenon of Sensor Clusterization

Another acute problem is potential sensor clusterization, i.e., the tendency of different sensors to take measurements at the same location, which is rather unacceptable in real situations. This phenomenon is a direct consequence of the assumption of spatially independent measurement errors. As an illustration of the clusterization effect, the following example can be considered.

Example 2.3. In this example we examine again the transmission line of Example 2.1 with a specified length $l = \ln(3)$, but attention will be focused only on the voltage signal y. Assume that the line is homogeneous (densities of its electrical parameters are constant along the line) without leakage conductance and inductance, i.e., $G = 0$ and $L = 0$ (this corresponds to the so-called *Thomson cable*). For notational simplicity, write down the time constant of the line as $RC = \theta_1$. Thus the signal propagation is described by the following telegraph equation of the parabolic type [49]:

$$\theta_1 \frac{\partial y}{\partial t} = \frac{\partial^2 y}{\partial x^2}, \quad x \in (0, l), \quad t \in (0, t_f). \tag{2.88}$$

Consider the transient state of the transmission line over the observation horizon $t_f = 1$ in the case when the system at both ends is supplied by the voltage sources whose characteristics are defined by the following boundary conditions:

$$\begin{cases} y(0,t) = \theta_2 e^{\theta_1 t}, \\ y(\ln(3), t) = 2\theta_2 e^{\theta_1 t}, \quad t \in (0,1), \end{cases} \tag{2.89}$$

and initial voltage distribution,

$$y(x,0) = \frac{1}{8}\theta_2\big(3\sinh(x) + 2e^x\big), \quad x \in (0, \ln(3)). \tag{2.90}$$

The unknown values of the parameter vector $\theta = (\theta_1, \theta_2)$ have to be estimated with the use of two stationary sensors. To find optimal locations x^1 and x^2, the D-optimality criterion will be applied.

An exact analytical solution for this problem exists and has the form

$$y(x,t) = \frac{1}{4}\theta_2 e^{\theta_1 t}\sinh(x) + \theta_2 e^{\theta_1 t}\cosh(x). \tag{2.91}$$

Since the measurements are continuous in time, the FIM can be obtained from (2.85), in which $C(x,t)$ is assumed to be the identity matrix and

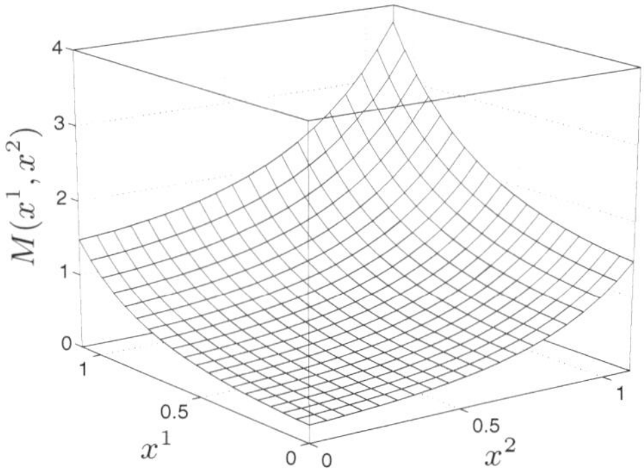

Fig. 2.3 Surface plot of $M(x^1, x^2)$ for Example 2.3 ($\theta = (0.05, 0.05)$)

$$G^{\mathsf{T}}(x,t) = \left[\frac{1}{4}\theta_2 t e^{\theta_1 t}\sinh(x) + \theta_2 t e^{\theta_1 t}\cosh(x) \frac{1}{4}e^{\theta_1 t}\sinh(x) + e^{\theta_1 t}\cosh(x)\right]. \tag{2.92}$$

Computation which can be performed using a computer-algebra system shows that

$$\det\left(M(x^1,x^2)\right) = \left(\frac{\theta_2}{16\theta_1^2}\right)^2 \left(e^{4\theta_1} - 2e^{2\theta_1} - 4e^{2\theta_1}\theta_1^2 + 1\right) \cdot \left(\left(\sinh(x^1) + 4\cosh(x^1)\right)^2 + \left(\sinh(x^2) + 4\cosh(x^2)\right)^2\right)^2. \tag{2.93}$$

The surface plot of $\det M(x^1, x^2)$ for $\theta = (0.05, 0.05)$ is shown in Fig. 2.3. Simple verification shows that the maximum value of the D-optimality criterion corresponds to the sensor locations,

$$x^{1\star} = x^{2\star} = \ln(3). \tag{2.94}$$

The conclusion that both sensors must be placed at the same spatial point, which illustrates the phenomenon of clusterization, is a direct consequence of the assumptions made while constructing the simplified mathematical model of the disturbances acting on the measurements. ★

From the statistical point of view, under the assumption of independent observations (satisfied at least approximately), a replication of measurement at the same spatial point may be advantageous and sometimes even necessary (e.g., if it is required to estimate noise characteristics). However, quite often this cannot be accepted because of basic experimental restrictions, such as the inability of repeating process trials or the impossibility of a close location of several sensors since during the measurements they can influence each other. In order to avoid clusterization tendency, the problem can usually be transformed to a combinatorial one [183, 200, 205, 291], when from among N fixed a priori positions the best $n < N$ ones are to be selected. In the case of movable sensors, additional constraints imposed on sensor trajectories are required, which highly complicates the approach. Another technique dedicated especially for the scanning strategy is to apply the idea of directly constrained design measures [73, 180, 188, 198, 281, 289], which will be presented in detail in Chapter 3.

2.2.5.4 Relationship between the Solution and Estimated Parameters

Perhaps the most serious obstacle which has to be overcome while trying to design an optimal measurement strategy is the dependence of the optimal solution on the estimated parameters. Since it is clear that these parameters are unknown prior to the experiment, in order to solve the problem, we need some preliminary estimate of their values. An illustrative example for this complication is given below.

Example 2.4. Reconsider the one-dimensional homogeneous transmission line of Example 2.1, but this time assume that the line is lossless, i.e., $G = 0$ and $R = 0$. Moreover, assume that $LC = \theta^2$, where the physical interpretation of θ is the velocity of the signal propagation along the line. The distribution of the voltage y is thus given by the hyperbolic 'wave' equation,

$$\theta^2 \frac{\partial^2 y}{\partial t^2} = \frac{\partial^2 y}{\partial x^2}, \quad x \in (0, \pi), \quad t \in (0, 2\pi). \tag{2.95}$$

The line is supplied at each end by sinusoidal electromotive forces, which correspond to the boundary conditions

$$y(0,t) = \sin(t), \quad y(\pi, t) = \cos(\pi\theta)\sin(t), \quad t \in (0, 2\pi). \tag{2.96}$$

The initial voltage distribution and the rate of changes along the line are respectively given by

$$y(x,0) = 0, \quad \left. \frac{\partial y(x,t)}{\partial t} \right|_{t-0} = \cos(\theta x), \quad x \in (0, \pi). \tag{2.97}$$

The analytical solution to the problem so defined is the function

$$y(x,0) = \cos(\theta x)\sin(t). \tag{2.98}$$

Our task here is to find the location x^1 of only one sensor in such a way as to obtain the most accurate estimate of θ. Because there is only one constant parameter, the FIM is a scalar and each criterion applied leads to the same

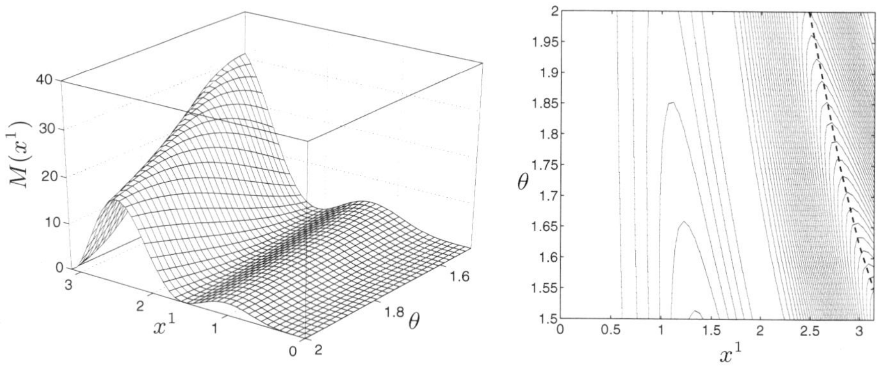

Fig. 2.4 Surface and contour plots of $M(x^1;\theta)$ in Example 2.4

solution. The FIM can be represented as

$$\begin{aligned} M(x^1) &= \int_0^{t_f} \left(\frac{\partial y(x^1,t;\theta)}{\partial \theta} \right)^2 \mathrm{d}t \\ &= \frac{1}{2} x^2 \sin(\theta x)^2 \cos(t_f) \left(t_f - \sin(t_f)\cos(t_f)\right). \end{aligned} \tag{2.99}$$

The surface and contour plots of (2.99) corresponding to $t_f = 2\pi$ are shown in Fig. 2.4. The first observation is the multi-modal character of the problem, as elucidated earlier. The second remark constitutes the essence of this example as it is clear that the optimal sensor position does depend on the value of the parameter θ. The optimal sensor position $x^{1\star}(\theta)$ is marked with a dashed line on the contour plot. ★

The dependence of the solutions on the unknown parameters is a significant drawback which implies the necessity of applying some a priori statistical knowledge about the parameter to be identified. Often some nominal values of physical parameters are available, otherwise there is a need for some pilot experiments to be conducted to obtain preliminary estimates of the parameters or for exploiting the so-called *sequential designs*, which comprises repeated consecutive experimentation and estimation steps. However, from the economical and technical points of view, such a procedure is often impractical and difficult to implement, especially in the case of on-line algorithms. Alternatively, there are some *robust-design* techniques, which minimize the amount of statistical information necessary to find an optimal solution on one hand, but on the other they significantly increase the complexity of the approach. This leads to the *average approach*, which takes into account information about statistical distributions of unknown parameters and *minimax designs* comprising the analysis in the worst conditions possible [180, 276, 281, 307].

2.2.5.5 Correlated Observations

One of the most characteristic properties of spatial data fusion is the fact that observations made at different sites are often determined by local correlations. This means that the sensor located in a close vicinity of another one may not provide essentially measurements which contain essentially new information about the DPS. As has been mentioned earlier, in addition to this, often there is no possibility of using replicated measurements because of technical experimental constraints. Obviously, correlation dependencies make the problem even more complex and challenging, therefore this issue is usually neglected in the literature on sensor design for parameter estimation of DPSs.

To deal with the problem, the very first idea is to exploit some well-known numerical techniques of constrained optimization where suitably defined additional constraints represent admissible distances between sensor nodes. However, this approach quickly leads to severe complexity of the problem as the number of sensors increases [180, 276, 281]. Another potentially more efficient approach consists in incorporating mutual correlations between measurements by proper construction of the information matrix [180, 191, 196, 281].

2.3 Summary

There is no doubt that preparation of experimental conditions in the sense of choosing a suitable observational strategy has great influence on the expected accuracy of parameter estimates, especially in the context of DPSs. The first part of the chapter contains a brief overview of the parameter identification problem and optimal experimental design for linear and nonlinear lumped dynamic systems. In particular, the notion of continuous designs is introduced, which dramatically reduces the problem dimensionality. Furthermore, generalization of some classical results from experimental design theory to multi-response systems with possibly correlated outputs is presented for the purpose of providing fundamental results which characterize the solutions and thus indicate possible numerical procedures. The inherent nonlinearity of DPSs with respect to the estimated parameters is certainly a problem of paramount importance and one of main complications in the delineated approach. However, under some assumptions, the results derived for systems linear in parameters can be extended to the class of nonlinear models based on suitable approximations of the FIM. Such a generalized classical theory paves the way to numerous applications of DPSs and, being crucial for specific problems, constitutes close connection with the chapters which follow.

In the second part of the chapter, an important class of models for spatio-temporal dynamic systems on which the attention of this work is focused is described along with the problem of choosing an optimal observation strategy for parameter estimation within this class. The problem, once formulated and transformed to an optimization one, is far form being trivial. In spite of the elegant formulation based on the use of a performance index defined on the FIM, serious difficulties inherent to the problem nature occur such as the clusterization effect or a large scale of the problem, not to mention the dependence of the solutions on the estimated parameters. Those complications make the problem really hard and the well-known nonlinear programming algorithms cannot be applied directly nor provide effective solutions for practical engineering applications. Bearing this in mind, the remainder of this book is dedicated to overcoming these difficulties, at least to some extent.

Chapter 3
Sensor Activation for Scanning Networks

In recent years, rapid advances in ad-hoc networking systems and sensor technology have begun a strong trend of measurement devices to become miniaturized and cheap, at the same time making the networks of sensors increasingly powerful. The reliability and robustness of DSNs have improved to the point where their wide applicability is unquestionable [29, 37, 98, 108, 110, 245, 309, 315]. This is especially important in the context of spatio-temporal systems where a large number of sensors can be used for the task of monitoring the dynamics of a system.

Definitely, a great number of the contributions to the measurement optimization problem based on the experimental design concern the choice of stationary sensor positions (for surveys, see [138, 180, 231, 276, 281]). However, the main drawback of using stationary sensors continuously monitoring the environment is the lack of flexibility regarding DPS dynamics. The measurements are taken during the whole observation horizon at the same positions even if the local data do not provide suitable information about the system dynamics. Moreover, for a large scale DSN, some serious operational and technical issues have to be overcome, such as the proper balance of energy consumption against the frequency of observation and node lifetime, the level of collaboration among the sensors and limitations with respect to communication range and channel overloads. Finally, the topology of network changes in time due to the failures of nodes and battery draining [1, 107, 304].

Those are the main reasons why it is worth paying closer attention to other strategies, such as scanning, which are more flexible and have more capabilities than nonmobile observations, since they exploit the time-measurement domain in a more effective manner, thereby offering additional degrees of freedom regarding optimality. The observation system then often comprises multiple sensors whose location is already specified, and it is desired to activate only a subset of them during a given time interval while the other sensors remain dormant [50, 56, 157]. In such a way, significant reduction in the observation system complexity and the costs of operation and maintenance can

M. Patan: Optimal Sensor Networks Scheduling in Identification of DPSs, LNCIS 425, pp. 47–95.
springerlink.com

be achieved [50]. Furthermore, since the sensor nodes can track the locations which give at a particular time instant the most significant knowledge about the estimated parameters of interest, the efficiency of the observational process can be increased. Simultaneously, valuable resources can be spared such as the energy of batteries, transmission bandwidth or/and computational cost of data fusion.

Another interesting interpretation of the scanning strategy can be given in terms of several sensors which are mobile, i.e., after taking measurements at fixed spatial positions on a time interval of nonzero length, they can change their locations to continue the measurements at more informative points, assuming that the time required for movement may be neglected compared to the scale of the observation horizon. Some attempts to the problem have been made in a related context of state estimation (see, e.g., [62, 150, 171]), although, as has been mentioned earlier, an extension to the parameter estimation problem is rather difficult, since in the latter case the current state usually depends nonlinearly on unknown parameters.

One of the major difficulties related to the scanning sensor scheduling problem is its combinatorial nature. It is even more difficult if sensor switchings are allowed to take place freely in continuous time. Thus, the main purpose of this chapter is to provide a class of methods, which instead of a laborious exhaustive search along the solution space, will construct solutions based on the proper reformulation of the scheduling problem in spirit of experimental design theory. In particular, apart from the extensions of the results adopting the concept of continuous designs for stationary sensors provided by Uciński [276, 281] and Rafajłowicz [231, 233] an extremely efficient approach based on directly constrained design measures is also presented [198, 289]. This technique was further refined by Uciński for optimization of the observability measure [281, 285] and generalized by Rafajłowicz to the context of so-called *sequences of designs* following dynamic changes in experimental environment [234].

In addition to this, the scanning measurement problem with free switching times is stated in terms of an optimal control one, creating a passage for effective use of common nonlinear programming algorithms [182, 281].

3.1 Fixed Switching Schedule

When the number of sensors is relatively high and the sensor switchings are arbitrarily fixed in time (i.e., the discretization of the time interval is known a priori, and the sensors are allowed to change their positions only according to this time partition), the first idea is to decompose the problem into a series of subproblems defined for given time instants and suitable adaptation of the continuous designs set forth in Section 2.1 [180, 234, 281]. Note that, in such a context, the strategy of stationary measurements can be considered a special case of scanning under the assumption that the number of subintervals the observation horizon is split to is equal to $K = 1$. Therefore, all results developed here are directly applicable to the simpler situation where the sensor positions are fixed along the whole time interval.

3.1.1 Problem Decomposition and Optimality Conditions

Let y be a state of the DPS defined by (2.66)–(2.68). In what follows, we form an arbitrary partition of the time interval T by choosing points $0 = t_0 < t_1 < t_2 < \cdots < t_K = t_f$ defining subintervals $T_k = [t_{k-1}, t_k)$, $k = 1, \dots, K$. We then consider N scanning sensors which possibly change their locations at the beginning of each time subinterval but then remain stationary till the end of this subinterval [281, 289]. In other words, the measurement process can be formally represented in the form of the output equation (2.76) or, assuming that the state is directly measured, by its simpler version,

$$z^j(t) = y(x_k^j, t; \theta) + \varepsilon(x_k^j, t), \quad t \in T_k, \tag{3.1}$$

for $j = 1, \dots, N$ and $k = 1, \dots, K$, and measurement noise $\varepsilon(\,\cdot\,,\,\cdot\,)$ satisfying the general characteristics defined by (2.78).

Sensor positions which guarantee the best accuracy of the least-squares estimates of θ are then found by choosing x_k^j for $j = 1, \dots, N$ and $k = 1, \dots, K$ so as to minimize a scalar measure of performance Ψ defined on the average FIM, which takes here the form [198, 289]

$$M = \frac{1}{N} \sum_{k=1}^{K} \sum_{j=1}^{N} \Upsilon_k(x_k^j), \tag{3.2}$$

where

$$\Upsilon_k(x) = \frac{1}{t_f} \int_{T_k} G^{\mathsf{T}}(x,t) C^{-1}(x,t) G(x,t)\, \mathrm{d}t, \tag{3.3}$$

$$G(x,t) = \left(\frac{\partial y(x,t;\theta)}{\partial \theta} \right)_{\theta = \theta^0}, \tag{3.4}$$

θ^0 stands for a prior estimate to the unknown parameter vector θ, G denotes the Jacobian matrix of the sensitivity coefficients defined analogously with (2.86), and C is the positive definite matrix responsible for the correlations between the outputs, which is constructed in the same manner as in (2.78).

Owing to the assumption of independent measurements made by different sensors (i.e., replicated observations), we may distinguish only different sensor locations in each time interval and introduce the continuous design of the experiment for subinterval T_k so that it consists of the following collection of variables:

$$\xi_k = \begin{Bmatrix} x_k^1, & x_k^2, & \dots, & x_k^{\ell(k)} \\ p_{k1}, & p_{k2}, & \dots, & p_{k\ell(k)} \end{Bmatrix}, \tag{3.5}$$

where $\ell(k)$ denotes the number of different locations on T_k. Obviously, for all k we have $p_{ki} > 0$ and $\sum_{i=1}^{\ell(k)} p_{ki} = 1, \quad i = 1, \dots, \ell(k)$. As for the relevant interpretation, p_{ki} can be understood in terms of the proportion of the total of N sensors measuring the system state at spatial point x_k^i during the time subinterval T_k. In such a way the weights define entirely the total proportion of the experimental effort spent at a given location during the particular observation interval.

On account of the above remarks, the FIM can be rewritten in the form

$$M(\xi) = \sum_{k=1}^{K} \sum_{i=1}^{\ell(k)} p_{ki} \Upsilon_k(x_k^i). \tag{3.6}$$

Relaxing the notion of the design even further (on the analogy of (2.24)), we obtain the equivalent notation in terms of the probability measures ξ_k,

$$M(\xi) = \sum_{k=1}^{K} M_k(\xi_k), \tag{3.7}$$

where

$$M_k(\xi_k) = \sum_{k=1}^{K} \int_X \Upsilon_k(x)\, \xi_k(\mathrm{d}x),$$

and

$$\xi = (\xi_1, \dots, \xi_K). \tag{3.8}$$

By definition, we have

$$\int_X \xi_k(\mathrm{d}x) = 1, \quad k = 1, \dots, K. \tag{3.9}$$

The integration in (3.9) and (3.7) is to be understood in the Lebesgue–Stieltjes sense.

Accordingly, we redefine the optimal design as a solution to the optimization problem.

Problem 3.1 (Optimal scheduling problem). Find $\xi^\star \in \Xi(X)$ to minimize

$$\mathcal{P}(\xi) = \Psi[M(\xi)],$$

where $\Xi(X)$ denotes the set of all designs of the form (3.8).

To provide suitable reasoning for the scanning approach, the validity of the conditions (A1)–(A5) from Section 2.1 is assumed, and instead of (A6) the following counterpart is taken, which will be used in the remainder of this section [281]:

(A7) $\forall \xi \in \tilde{\Xi} = \{\xi : \Psi[M(\xi)] \le q < \infty\} \neq \emptyset$, $\forall \bar{\xi} \in \Xi(X)$:

$$\begin{aligned}\Psi[(1-\alpha)M(\xi) + \alpha M(\bar{\xi})] = \\ \Psi[M(\xi)] + \alpha \sum_{k=1}^{K} \int_X \psi_k(x,\xi)\, \bar{\xi}_k(\mathrm{d}x) + o(\alpha;\xi,\bar{\xi}).\end{aligned} \tag{3.10}$$

where o is the Landau symbol, i.e., $\lim_{\lambda \downarrow 0} o(\lambda;\xi,\bar{\xi})/\lambda = 0$.

Just as in the previous deliberations regarding lumped systems, the assumption (A7) simply amounts to the existence of the directional derivative whose form must be on the one hand specific, while on the other, for most practical criteria such a condition is not particularly restrictive.

In fact, requiring Ψ to be differentiable with respect to the individual elements of its matrix argument, we obtain

$$\psi_k(x,\xi) = \underbrace{\operatorname{trace}\Big[\mathring{\Psi}[M(\xi)]\Upsilon_k(x)\Big]}_{-\phi_k(x,\xi)} - \underbrace{\operatorname{trace}\Big[\mathring{\Psi}[M(\xi)]M_k(\xi_k)\Big]}_{-\varsigma_k(\xi)}, \tag{3.11}$$

where

$$\mathring{\Psi}[M(\xi)] = \left.\frac{\partial \Psi(M)}{\partial M}\right|_{M=M(\xi)}.$$

In order to establish the necessary and sufficient conditions of optimality, we first give some auxiliary assertions being counterparts of Lemmas 2.3 and 2.4.

Lemma 3.1. *For any design $\xi = (\xi_1,\dots,\xi_K) \in \Xi(X)$ and all $k = 1,\dots,K$, we have*

(i) $\int_X \phi_k(x,\xi)\, \xi_k(\mathrm{d}x) = \varsigma_k(\xi)$, *and*
(ii) $\max_{x\in X} \phi_k(x,\xi) \ge \varsigma_k(\xi)$.

Proof. From (3.11), we obtain

$$\begin{aligned}\int_X \phi_k(x,\xi)\, \xi_k(\mathrm{d}x) &= -\int_X \operatorname{trace}\Big[\mathring{\Psi}[M(\xi)]\Upsilon_k(x)\Big]\, \xi_k(\mathrm{d}x) \\ &= -\operatorname{trace}\Big[\mathring{\Psi}[M(\xi)] \int_X \Upsilon_k(x)\, \xi_k(\mathrm{d}x)\Big] \\ &= -\operatorname{trace}\Big[\mathring{\Psi}[M(\xi)]M_k(\xi_k)\Big] = \varsigma_k(\xi).\end{aligned} \tag{3.12}$$

This establishes (i). Then

$$\varsigma_k(\xi) = \int_X \phi_k(x,\xi)\,\xi_k(\mathrm{d}x) \le \int_X \max_{x\in X} \phi_k(x,\xi)\,\xi_k(\mathrm{d}x) = \max_{x\in X} \phi_k(x,\xi), \quad (3.13)$$

which proves the second claim of the lemma. ■

Lemma 3.2. *If* $\xi = (\xi_1,\dots,\xi_K) \in \Xi_q$, $\bar{\xi} = (\bar{\xi}_1,\dots,\bar{\xi}_K) \in \Xi(X)$ *and* $\xi_\alpha = (1-\alpha)\xi + \alpha\bar{\xi}$, *then*

$$\left.\frac{\partial \Psi[M(\xi_\alpha)]}{\partial \alpha}\right|_{\alpha=0^+} = \varsigma(\xi) - \sum_{k=1}^{K} \int_X \phi_k(x,\xi)\,\bar{\xi}_k(\mathrm{d}x),$$

where

$$\varsigma(\xi) = \sum_{k=1}^{K} \varsigma_k(\xi).$$

Proof. Taking into account (3.10) and (3.11), we have

$$\begin{aligned}\left.\frac{\partial \Psi[M(\xi_\alpha)]}{\partial \alpha}\right|_{\alpha=0^+} &\\ &= \sum_{k=1}^{K} \int_X \left\{ \operatorname{trace}\left[\mathring{\Psi}[M(\xi)]\Upsilon_k(x)\right] - \operatorname{trace}\left[\mathring{\Psi}[M(\xi)]M_k(\xi_k)\right] \right\} \bar{\xi}(\mathrm{d}x) \\ &= -\sum_{k=1}^{K} \int_X \phi_k(x,\xi)\,\bar{\xi}_k(\mathrm{d}x) + \sum_{k=1}^{K} \varsigma_k(\xi).\end{aligned} \quad (3.14)$$

■

Now, it is possible to formulate the equivalence theorem for the scanning strategy.

Theorem 3.1. *The following statements are equivalent:*

(i) the design $\xi^\star = (\xi_1^\star, \xi_2^\star, \dots, \xi_K^\star)$ *minimizes* $\Psi[M(\xi)]$, *and*
(ii) $\max_{x\in X} \phi_k(x,\xi^\star) = \varsigma_k(\xi^\star), \quad k = 1,\dots,K.$

Proof. First, introduce the design $\xi_\alpha = (1-\alpha)\xi^\star + \alpha\bar{\xi}$, where $\xi^\star = (\xi_1^\star,\dots,\xi_K^\star) \in \Xi_q$, and $\bar{\xi} = (\bar{\xi}_1,\dots,\bar{\xi}_K) \in \Xi(X)$.
$(i) \Rightarrow (ii)$ If $\xi^\star = \arg\min_{\xi\in\Xi(X)} \Psi[M(\xi)]$, then

$$\left.\frac{\partial \Psi[M(\xi_\alpha)]}{\partial \alpha}\right|_{\alpha=0^+} \ge 0, \quad \forall \bar{\xi} \in \Xi(X). \quad (3.15)$$

From Lemma 3.2, setting $\xi = \xi^\star$ and $\bar{\xi}$ such that

$$\bar{\xi}_k = \begin{cases} \xi_k^\star & \text{if } k \neq i, \\ \left\{ \begin{smallmatrix} x \\ 1 \end{smallmatrix} \right\} & \text{if } k = i, \end{cases} \tag{3.16}$$

we obtain

$$\begin{aligned} \left.\frac{\partial \Psi[M(\xi_\alpha)]}{\partial \alpha}\right|_{\alpha=0^+} &= \sum_{k=1}^{K} \varsigma_k(\xi^\star) - \sum_{\substack{k=1 \\ k\neq i}}^{K} \int_X \phi_k(x,\xi^\star)\xi_k^\star(\mathrm{d}x) - \phi_i(x,\xi^\star) \\ &= \sum_{k=1}^{K} \varsigma_k(\xi^\star) - \sum_{k=1}^{K} \int_X \phi_k(x,\xi^\star)\xi_k^\star(\mathrm{d}x) - \phi_i(x,\xi^\star) \\ &\quad + \int_X \phi_i(x,\xi^\star)\xi_i^\star(\mathrm{d}x) = \varsigma_i(\xi^\star) - \phi_i(x,\xi^\star) \geq 0. \end{aligned} \tag{3.17}$$

The result is

$$\phi_i(x,\xi^\star) \leq \varsigma_i(\xi^\star), \tag{3.18}$$

which clearly forces

$$\max_{x\in X} \phi_i(x,\xi^\star) \leq \varsigma_i(\xi^\star). \tag{3.19}$$

Part (ii) of Lemma 3.1 now yields the desired conclusion.
$(ii) \Rightarrow (i)$ Let $\xi^\star \in \Xi(X)$ satisfy (ii). We have

$$\begin{aligned} \left.\frac{\partial \Psi[M(\xi_\alpha)]}{\partial \alpha}\right|_{\alpha=0^+} &= \sum_{k=1}^{K} \left[\varsigma_k(\xi^\star) - \int_X \phi_k(x,\xi^\star)\bar{\xi}_k(\mathrm{d}x) \right] \\ &\geq \sum_{k=1}^{K} \left[\varsigma_k(\xi^\star) - \max_{x\in X} \phi_k(x,\xi^\star) \right] = 0, \end{aligned} \tag{3.20}$$

which completes the proof. ■

Remark 3.1. The above result establishes direct applicability of the classical algorithms from the theory of optimum experimental design valid for stationary sensors, since the problem can be decomposed to a finite set of 'virtually independent' problems, where continuous-time observations are taken by stationary sensors over subintervals T_k. Since the only element joining the subproblems for consecutive time subintervals is the common global FIM, this also raises attractive possibilities for parallel implementations of the algorithm.

3.1.2 Optimization of the Experimental Effort

Our basic assumption in this section is that the set of admissible support points X, where the observations of the measurands are possible, is finite, and $|X| = \ell$. Because the number of locations from X is limited, any design

$\xi \in \Xi(X)$ is uniquely determined by the collection of the corresponding weights. Such a description is is very convenient as the problem is reduced to optimization of weights. As has already been mentioned, the weight assigned to a measurement point can be interpreted as the proportion of observations performed at this point, or the percentage of experimental effort spent at it. The potential solutions are of considerable interest while assessing which sensors are more informative than others at a given time instant and allow complexity reduction of the measurement system.

In the case under consideration, i.e., the design for fixed sensor locations, we focus our attention on the FIM given by (3.6), and Problem 3.1 can be rewritten as follows.

Problem 3.2. Find $P^\star = [p^\star_{ki}], \quad k = 1, \dots K, \quad i = 1, \dots, \ell$, to minimize

$$\mathcal{P}(P) = \Psi[M(\xi)] \tag{3.21}$$

subject to $P \in \mathbb{R}^{k \times \ell}$ being a stochastic matrix, i.e.,

$$P\mathbf{1} = \mathbf{1}, \quad P \geq 0, \tag{3.22}$$

where $\mathbf{1}$ is an ℓ-dimensional vector whose all components are equal to unity.

The constraint (3.22) simply means that each row $p_k = (p_{k1}, \dots, p_{k\ell})$ of P belongs to the canonical simplex or order ℓ,

$$p_k \in \mathbb{P}_\ell = \Big\{ p_k : \; p_{ki} \geq 0, \; i = 1, \dots, \ell; \quad \sum_{i=1}^{\ell} p_{ki} = 1. \Big\} \tag{3.23}$$

Therefore, Problem 3.2 is equivalent to a finite-dimensional optimization one over the product of canonical simplices $\mathbb{P}_\ell^K$.

Gradient Projection Method

At this point, it is possible to exploit some general constrained optimization routines in order to solve the formulated problem. However, taking advantage of the form of the constraints, a more straightforward procedure can be proposed, which boils down to using a gradient projection method [307]. The first step is to find a feasible direction, i.e., the one which guarantees a decrease in the value of the criterion Ψ, and then a step is taken along this line. The result is projected on $\mathbb{P}_\ell^K$, thereby obtaining a new feasible weight matrix P. Generation of a new feasible candidate point in the r iteration can be formalized as follows:

$$\tilde{P}^{(r)} = \Pi_{\mathbb{P}_\ell^K}[P^{(r)} + \beta_r \delta(P^{(r)})], \tag{3.24}$$

where $\delta(P^{(r)}) \in \mathbb{R}^{K \times \ell}$ is the matrix representing a feasible direction of weight modification, β_r is some coefficient which controls the correction process, and

$\Pi_{\mathbb{P}_\ell^K}[\,\cdot\,]$ stands for orthogonal projection onto the convex set of admissible weights $\mathbb{P}_\ell^K$. Because the derivatives

$$\frac{\partial \Psi[M(\xi)]}{\partial p_{ki}} = \phi_k(x^i, \xi), \quad i = 1, \ldots, \ell, \quad k = 1, \ldots, K, \tag{3.25}$$

are easy to calculate, the very first idea is to choose the direction $\delta(p^{(r)})$ determined by $-\nabla_P \Psi = \partial \Psi[M(\xi)]/\partial P$, being the counterpart of the negative gradient as in the steepest descent. Also note that there exist many possible choices of determining the step coefficient β_k. It can be taken as a suitable constant or adapted, e.g., according to the rule

$$\beta_{r+1} = \begin{cases} \gamma_1 \beta_r & \text{if } \Psi[M(\xi^{(r)})] \le \Psi[M(\xi^{(r-1)})], \\ \beta_r/\gamma_2 & \text{otherwise,} \end{cases} \tag{3.26}$$

where $\gamma_1 > 1$ and $\gamma_2 > 1$ are appropriate fixed parameters, or optimally, i.e.,

$$\beta_{r+1} = \arg\min_{\beta} \Psi\big[M(\tilde{\xi}^{(r)})\big], \tag{3.27}$$

where $\tilde{\xi}^{(r)}$ denotes the design with weights defined by (3.24). A detailed scheme of numerical procedure is provided in Algorithm 3.1.

Algorithm 3.1. Gradient projection weight optimization algorithm for a fixed finite set of measurement points

```
procedure GP_EFFORT_OPTIMIZATION
    Guess a starting set of weights P^(0) = [p_1^T, ..., p_K^T]^T, p_k ∈ P_ℓ, k = 1, ..., K
    Choose some positive tolerance ε ≪ 1.
    r ← 0.
    repeat
        Choose β_r according to (3.26) or (3.27)
        P̃^(r) ← SimplexProjection( P^(r) − β_r ∇_P Ψ[M(ξ^(r))] )
        α_r ← arg min_{α∈[0,1]} Ψ[αM(ξ̃^(r)) + (1 − α)M(ξ^(r))]
        P^(r+1) ← (1 − α_r)P^(r) + α_r P̃^(r)
        r ← r + 1
    until ||P^(r+1) − P^(r)|| < ε
end procedure
```

The gradient projection procedure delineated above is rather easy to implement due to its simplicity, but unfortunately it inherits all the drawbacks of stepeest-descent-like algorithms. For example, the convergence rate dramatically decreases in the vicinity of the minimum. This can be avoided to some extent with a suitable choice of the correction step β. It is also clear that the algorithm is useful as long as the projection operator is simple and effective. Fortunately, in the case considered, an algorithm can be developed which is almost as simple as a closed-form solution based on the very elegant algorithm for projection of a vector onto the canonical simplex proposed in [275]. This result can be generalized to the examined case of a product of canonical simplices by independent projection of each row of matrix P onto the canonical simplex. The projection operator $\Pi_{\mathbb{P}_\ell^K}$ is presented in the SIMPLEXPROJECTION routine embodied by Algorithm 3.2.

Algorithm 3.2. Orthogonal projection onto the product of canonical simplices; function returns matrix $\tilde{P}$ being a projected image of input matrix P, which is assumed to be sorted along the rows in a descent order, i.e., $p_{k1} \geq p_{k2} \geq \cdots \geq p_{k\ell}$ for any k

```
function SimplexProjection(P)
    for k ← 1, K do
        q ← 1 and s ← 0
        repeat
            q* ← q and s* ← s
            s ← s + q*(p_{kq*} − p_{kq})
            if s ≤ 1 then
                if m = ℓ then
                    q* ← q and s* ← s
                end if
            end if
        until m = ℓ or s > 1
        for i ← 1, ℓ do
            if i ≤ q* then
                p̃_{ki} ← p_{ki} + (1 − s*)/q* − p_{kq*}
            else
                p̃_{ki} ← 0
            end if
        end for
    end for
    return P̃
end function
```

Despite the simplicity of Algorithm 3.1, it still involves some additional numerical effort. For instance, the projection operator is not differentiable, so that the line search with respect to β in (3.27) requires a nonderivative algorithm.

Feasible-Direction Method

Careful analysis of the gradient projection method implies the conclusion that instead of a negative gradient it could be advisable to take under consideration other potential directions of weight modification providing better properties of the projection operator. One of such alternatives is the approach based on the mapping $\mathcal{T} : \Xi(X) \to \Xi(X)$ defined by multiplicative transformation,

$$\mathcal{T}\xi = \begin{Bmatrix} x_k^1, & \dots, & x_k^\ell \\ p_{k1}\phi(x_k^1,\xi)/\varsigma(\xi_k), & \dots, & p_{k\ell}\phi_k(x_k^\ell,\xi)/\varsigma_k(\xi) \end{Bmatrix}, \quad k = 1,\dots,K. \tag{3.28}$$

From Theorem 3.1 it follows that a design $\xi^\star$ is optimal if it is a fixed point of the mapping $\mathcal{T}$, i.e.,

$$\mathcal{T}\xi^\star = \xi^\star. \tag{3.29}$$

Certainly, from Lemma 3.1 it follows that the sum of weights is invariant with respect to $\mathcal{T}$. Thus the projection problem no longer exists as the proper weights projection is guaranteed by $\mathcal{T}$. As for the interpretation of this function, consider the situation when a design ξ is not optimal. Then the mapping $\mathcal{T}$ increases the weights of those support points of ξ at which the sensitivity function takes high values, i.e., $\phi_k(x,\xi) > \varsigma_k(\xi)$, in a way decreasing its maximal values. This is attained at the cost of decreasing the weights for support points with small values of $\phi_k(x,\xi)$ (where $\phi_k(x,\xi) < \varsigma_k(\xi)$). Obviously, this is possible as far as the weights are positive. Once the weight attains zero value, a multiplicative update forces it to remain zero, hence it is important to set all the weights to positive values in the initial design.

The scheme of a very simple procedure implementing the underlying idea is provided in Algorithm 3.3. Note that this procedure is nothing but a feasible

Algorithm 3.3. Feasible-direction weight optimization algorithm

procedure FD_EFFORT_OPTIMIZATION
  Choose some positive tolerance $\epsilon \ll 1$. $r \leftarrow 0$.
  Guess a starting design $\xi^{(0)} = (\xi_1^{(0)},\dots,\xi_K^{(0)})$ such that $p_{ki}^{(0)} > 0$ for $i = 1,\dots,\ell,\ k = 1,\dots,K$
  while $\max_{(k,i)} \left(\phi_k(x_k^i,\xi^{(r)})/\varsigma_k(\xi^{(r)})\right) \geq 1+\epsilon$ **do**
    for $k \leftarrow 1, K$ **do**
      for $i \leftarrow 1, \ell$ **do**
        $\tilde{p}_{ki}^{(r)} \leftarrow p_{ki}^{(r)} \dfrac{\phi_k(x_k^i,\xi^{(r)})}{\varsigma_k(\xi^{(r)})}$
      end for
    end for
    Choose α_r according to the one of the rules (3.30)–(3.32)
    $\xi^{(r+1)} \leftarrow (1-\alpha_r)\xi^{(r)} + \alpha_r\tilde{\xi}^{(r)}$
  end while
end procedure

direction method, cf. [22]. The convergence of the proposed iterative routine to the optimal design depend on suitable choice of the sequence $\{\alpha_r\}$. The possible selections are as follows:

- the sequence α_r satisfies the conditions

$$\lim_{r\to\infty} \alpha_r = 0, \quad \sum_{r=0}^{\infty} \alpha_r = \infty; \tag{3.30}$$

- α_r is the solution of the problem

$$\alpha_r = \arg\min_{\alpha\in[0,1]} \Psi\big[(1-\alpha)M(\xi^{(r)}) + \alpha M(\tilde{\xi}^{(r)})\big], \tag{3.31}$$

  where $\tilde{\xi}^{(r)}$ denotes the design with weights defined by $\tilde{P}^{(r)}$;
- α_r is taken as the value of α_{r-1} divided repeatedly by $\gamma < 1$ until it satisfies

$$\Psi[M(\xi^{(r-1)})] > \Psi[M(\xi^{(r)})]. \tag{3.32}$$

We have the following convergence result.

Theorem 3.2. *Assume that the optimality criterion Ψ is strictly monotonous in the sense that $\Psi[M_1] < \Psi[M_2]$ provided that $M_1 \succeq M_2$ for $M_1 \neq M_2$. If $\{\xi^{(r)}\}$ is a sequence of designs obtained according to Algorithm 3.3 and the rule* (3.31)*, then the sequence $\{\Psi[M(\xi^{(r)})]\}$ is nonincreasing, and it converges to* $\min\{\Psi[M(\xi)] : \xi \in \Xi(X)\}$.

Proof. Since we deal with a feasible-direction method, the convergence analysis proceeds in much the same way as in the classical framework of this category of methods, cf. [22, p. 213]. We must only show that the direction, i.e., $M(\tilde{\xi}^{(r)}) - M(\xi^{(r)})$, is also a descent direction,

$$\frac{\partial}{\partial\alpha}\Psi\left\{M(\xi^{(r)}) + \alpha\big[M(\xi_+^{(r)}) - M(\xi^{(r)})\big]\right\}\bigg|_{\alpha=0^+} < 0. \tag{3.33}$$

To this end, let us notice that for any design $\xi \in \Xi(X)$ such that $M(\xi_k) \succ 0,\ k = 1,\dots,K$, we have

$$\varsigma_k(\xi) > 0.$$

Indeed, the function $\Psi[\,\cdot\,]$ is convex if and only if [22, Prop. 3.3, p. 675]

$$\Psi[M(\xi) + A] \geq \Psi[M(\xi)] + \operatorname{trace}\left\{\mathring{\Psi}[M(\xi)]A\right\}, \quad \forall A \succeq 0. \tag{3.34}$$

This gives

$$-\operatorname{trace}\left\{\mathring{\Psi}[M(\xi)]A\right\} \geq \Psi[M(\xi)] - \Psi[M(\xi) + A] > 0, \tag{3.35}$$

for all nonzero $A \succeq 0$, which is valid due to the assumed monotonicity of $\Psi[\,\cdot\,]$. According to Lemma in [222, p. 9], this is equivalent to the condition

$$-\mathring{\Psi}[M(\xi)] \succ 0. \tag{3.36}$$

This clearly forces

$$\varsigma_k(\xi) = -\operatorname{trace}\left\{\mathring{\Psi}[M(\xi)]M(\xi_k)\right\} > 0. \tag{3.37}$$

We are now in a position to show our claim. It follows that

$$\begin{aligned}
&\frac{\partial}{\partial\alpha}\,\Psi[(1-\alpha)M(\xi^{(r)}) + \alpha M(\tilde{\xi}^{(r)})]\Big|_{\alpha=0^+} \\
&\quad = \operatorname{trace}\left\{\mathring{\Psi}[M(\xi^{(r)})][M(\tilde{\xi}^{(r)}) - M(\xi^{(r)})]\right\} \\
&\quad = \operatorname{trace}\left\{\mathring{\Psi}[M(\xi^{(r)})]\sum_{k=1}^{K}\sum_{i=1}^{\ell} p_{ki}^{(r)}\frac{\phi_k(x_k^i,\xi^{(r)}) - \varsigma_k(\xi^{(r)})}{\varsigma_k(\xi^{(r)})}M(\xi^{(r)})\right\} \\
&\quad = -\sum_{k=1}^{K}\sum_{i=1}^{\ell} p_{ki}^{(r)}\frac{\phi_k(x_k^i,\xi^{(r)}) - \varsigma_k(\xi^{(r)})}{\varsigma_k(\xi^{(r)})}\phi_k(x_k^i,\xi^{(r)}) \\
&\quad = -\sum_{k=1}^{K}\mathrm{E}\left[\frac{\phi_k(x,\xi^{(r)}) - \varsigma_k(\xi^{(r)})}{\varsigma_k(\xi^{(r)})}\phi_k(x,\xi^{(r)})\right] \\
&\quad = -\sum_{k=1}^{K}\frac{\mathrm{E}\left[\phi_k^2(x,\xi^{(r)})\right] - \varsigma_k(\xi^{(r)})\,\mathrm{E}\left[\phi_k(x,\xi^{(r)})\right]}{\varsigma_k(\xi^{(r)})} \\
&\quad = -\sum_{k=1}^{K}\frac{\mathrm{E}\left[\phi_k^2(x,\xi^{(r)})\right] - \mathrm{E}^2\left[\phi_k(x,\xi^{(r)})\right]}{\varsigma_k(\xi^{(r)})} \\
&\quad = -\sum_{k=1}^{K}\frac{\operatorname{var}\left(\phi_k(x,\xi^{(r)})\right)}{\varsigma_k(\xi^{(r)})} < 0,
\end{aligned} \tag{3.38}$$

provided that, there exists k such that not all $M\big(\xi(x_k^i)\big)$, $i = 1,\dots,\ell$, are equal. Here we have treated the weights $p_{ki}^{(r)}$, $i = 1,\dots,\ell$ as a probability distribution on the set of support points x_k^i, $i = 1,\dots,\ell$. ■

Remark 3.2. After some rearrangement of the weights update from line 7 of this fixed-point based algorithm, we get

$$\tilde{p}_{ki}^{(r)} = p_{ki}^{(r)} + \underbrace{p_{ki}^{(r)}\left(\frac{1}{\varsigma_k(\xi^{(r)})} - \frac{1}{\phi_k(x_k^i,\xi^{(r)})}\right)}_{\beta_k}\phi_k(x_k^i,\xi^{(r)}) \tag{3.39}$$

for any $i = 1,\dots,\ell$, $k = 1,\dots,K$. It is thus clear that the descent direction in this case is not certainly the negative gradient one. Nevertheless, a great advantage is that this direction automatically keeps all the designs for observation subintervals T_k, $k = 1,\dots,K$ in the simplex $\mathbb{P}_\ell$, essentially reducing the computational complexity of the algorithm. In spite of feasible direction not being an optimal one here, practical experiments show that the expected convergence rate can be in some situations comparable or even better than in the case of Algorithm 3.1, simultaneously providing great stability of performance.

Another interesting property is the independent update of the weights, which makes the algorithm advantageous from the point of view of potential parallel or even distributed implementation. Moreover the final design update in line 10, which guarantees the descent search direction, can be omitted or at least simplified for some particular criteria, including the D-optimum one, which will be discussed in the next paragraph. Finally, besides high efficiency coming from numerical simplicity, the ease of implementation should be noticed as another great benefit of the procedure.

From a practical point of view, the applicability of both techniques presented so far can be extended to a wide class of experimental design problems, in accordance with the following guidelines [233, 234]:

1. The set of admissible support points X is finite. In this case, the proposed routine converges to the optimal design as described above.
2. The set X is a bounded subset of $\mathbb{R}^d$. In such a framework, the following alternatives can be distinguished:
   a. If we know a small finite set $X_d \subset X$, which contains the optimal locations of the support points, the proposed routine can be used to find the optimal weights. The optimal design obtained for the set X_d constitutes simultaneously the optimal one for the set X. The main impediment is selection of a suitable subset X_d. For this purpose, other algorithms can be successfully applied, cf. [9, 73, 180, 233, 234, 276, 281, 307].
   b. We can choose a finite set $X_d \subset X$ with sufficiently distributed elements over the set X. For this set the algorithm can be applied, in the same manner as in Case 1. It guarantees that the resultant design is close enough to the optimal one (for the set X_d, of course), but for the quality assessment of the obtained solution, additional analysis is necessary.

From this point, one special case requires an additional comment as it is important in the framework of this work. Algorithm 3.3 is a generalization of the classical approach to the D-optimum design construction developed in the context of convex experimental design theory for nonlinear regression models and widely exploited in applications [178, 179, 210, 232, 253, 254, 269, 271, 272]. Hence, it can be improved when applied to a particular situation of the D-optimality criterion.

D-Optimum Experimental Effort

For the D-optimum criterion it can be shown that the line search in Algorithm 3.3 is not necessary for convergence. Therefore, it can be omitted in order to simplify the procedure. This leads to the simplified scheme presented in Algorithm 3.4, whose convergence in the case of only one system response was proved in [210, 224, 270, 271] within the framework of the classical optimum experimental design problem. The generalization of convergence result for MIMO systems is provided by Uciński [281].

Algorithm 3.4. Optimization of D-optimum experimental effort

procedure D-OPTIMUM_EFFORT_OPTIMIZATION
    Choose some positive tolerance $\epsilon \ll 1$. $r \leftarrow 0$.
    Guess a starting design $\xi^{(0)} = (\xi_1^{(0)}, \dots, \xi_K^{(0)})$ such that $p_{ki}^{(0)} > 0$ for $i = 1, \dots, \ell,\ k = 1, \dots, K$
    while $\max_{(k,i)} \left(\phi_k(x_k^i, \xi^{(r)}) / \varsigma_k(\xi^{(r)}) \right) \geq 1 + \epsilon$ **do**
        for $k \leftarrow 1, K$ **do**
            for $i \leftarrow 1, \ell$ **do**
                $\tilde{p}_{ki}^{(r)} \leftarrow p_{ki}^{(r)} \dfrac{\phi_k(x_k^i, \xi^{(r)})}{\varsigma_k(\xi^{(r)})}$
            end for
        end for
    end while
end procedure

In such a way, the efficient and extremely simple procedure known in optimum experimental design for lumped systems can be easily adopted for sensor scheduling in DPSs. It is important to notice that, although the convex combination of designs which appears in line 11 of Algorithm 3.3 is no longer required for the convergence, it could still be exploited to improve the convergence rate since the transformation $\mathcal{T}$ may be far from being the best choice of the correction of weights.

Reduction of the Measurement Space

It is clear that the numerical complexity of the discussed class of algorithms depends linearly on the number of admissible support points belonging to the set X, i.e., on the cardinality of this set. Note that the choice of the initial weights for Algorithm 3.3 is not crucial for the convergence, but no weights can be equal to zero because then it would be impossible to change them. This simple observation may lead to a significant improvement since, if during the run of the procedure a weight achieves a value close to zero, it is rather impossible that it will be increased in next iterations. Deleting such points with negligible weights significantly decreases the number of admissible locations in consecutive steps of the algorithm and thus increases the efficiency. To assure that the weight values for each time subinterval T_k sum up to unity, we modify them, e.g., according to the formula

$$p_{ki}^{\text{new}} = p_{ki}^{\text{old}} + \frac{S_{\text{rem}}}{|X_{\text{new}}|}, \qquad i \in X_{\text{new}}, \tag{3.40}$$

where X_{new} is the set containing the indices of the points which remain in the design at particular T_k, and $S_{\text{rem}} = \sum_{X \setminus X_{\text{new}}} p_{ki}$ is the sum of the weights of deleted points. Since this is associated with a very low computational cost, the general effect is very advantageous for the algorithm performance.

The essential question is what the threshold ϵ_{rem} for weights values to reject the points from X should be. In general, there is no proven result for such a threshold, which would be suitable to a wide class of design criteria, but experimental experience gives some indications for heuristic choice of ϵ_{rem}. In most practical situations, the number of $1/N$ where N is a total number of measurements is a reasonable choice [234]. This means that, if ϵ_{rem} is fixed at less than $1/N$, no measurements will be taken at points with lower weights.

In the case of the D-optimality criterion there exists a mathematical explanation of such empirical rule. Indeed, based on some matrix algebra dependencies, it can be shown that for D-optimality the following elegant result can be helpful.

Proposition 3.1. *Let X be some finite design space, $X = \{x^i \in \mathbb{R}^d,\ i = 1, \dots, \ell\}$, s denote the number of system outputs and $\xi^{(r)}$ be any design measure on X, with*

$$\epsilon_r = \max_{x_k^i \in X} \phi_k(x_k^i, \xi^{(r)}) - \varsigma_k(\xi). \tag{3.41}$$

Then any point x_k^i such that

$$\phi_k(x^i, \xi^{(r)}) < \varsigma_k(\xi) d(\epsilon_k) \tag{3.42}$$

with

$$d(\epsilon) = 1 + \frac{s\epsilon}{2} - \frac{\sqrt{s\epsilon(4 + s\epsilon)}}{2} \tag{3.43}$$

cannot be a support point of a D-optimum design measure on X.

Proof. The proof is a simple generalization of the result given by Pronzato [218] for a single output regression model. Since for multi-output scanning observations the main idea is to separately apply Lemma 1 from [218] for consecutive time intervals T_k, the proof is rather straightforward and of merely technical character. Therefore it is omitted. ∎

Armed with such a result, when using any algorithm of the D-optimum design, one can remove all the points satisfying (3.42) from the design space X in each iteration. Clearly, the acceleration that can be expected depends on the employed algorithm and the cardinality of X. Removing support points based on Proposition 3.1 implies some additional computations. Consequently, the best results can be obtained for the high power of the set X; otherwise the gain might not compensate the additional computational effort and a deceleration of the procedure may occur.

Semidefinite Programming Formulation

One more approach is proposed here since it makes it possible to employ very powerful algorithms for convex optimization based on Linear Matrix Inequalities (LMIs) or, more generally, on Semidefinite Programming (SDP), which has recently become a dynamically expanding research area. The SDP problem can be regarded as an extension of linear programming where the component-wise inequalities between vectors are replaced by matrix inequalities or, equivalently, the first orthant is replaced by the cone of positive semidefinite matrices. Most interior-point methods for linear programming have been generalized to semidefinite programs [32, 262, 297]. As in linear programming, these methods have polynomial worst-case complexity and perform very well in practice. SDP has been successfully applied in engineering (from control theory to structural design) and combinatorial optimization [28, 30, 32, 59, 61, 262]. Moreover, numerous public-domain solvers and numerical libraries dedicated for SDP are available, e.g., the SeDuMi package [262] or CVX [90].

Although potential applications of SDP in optimum experimental design were indicated in [32], the idea has not been pursued in the optimum experimental design community. In what follows, it is presented how to generalize Vandenberghe and Boyd's concept in the context of sensor scheduling for a scanning network on finite support sets. Its decided advantage, contrary to the standard experiment design techniques, is the possibility of solving problems for a wide class of design criteria, even for those which are nondifferentiable (e.g., the E-optimality criterion). In such a way this convenient formulation creates a very attractive alternative for classical approaches to optimization of the experimental effort.

As in general it is not possible to convert the general scheduling problem (3.21) (3.22) in terms of SDP, in the following it is shown how to deal with some of the most popular design criteria.

A-optimal design. The A-optimal experimental design problem is equivalent to that of minimizing $\operatorname{trace}(M^{-1}(\xi))$ over all weights at the k-th time interval belonging to the canonical simplex $\mathbb{P}_\ell$. But, equivalently,

$$\operatorname{trace}(M^{-1}(\xi)) = \sum_{j=1}^{m} e_j^\mathsf{T} M^{-1}(\xi) e_j, \tag{3.44}$$

where e_j is the usual unit vector along the j-th coordinate of $\mathbb{R}^m$. Introducing the vector of variables $q = [q_1, \dots, q_m]^\mathsf{T}$ and making use of the Schur complement [45], the A-optimum design problem can be cast as an SDP one.

Problem 3.3 (A-optimal SDP). Determine $q \in \mathbb{R}^m$ and $P \in \mathbb{R}^{K\times\ell}$ which minimize

$$\mathcal{P}(P, q) = \mathbf{1}^\mathsf{T} q \tag{3.45}$$

subject to the LMI

$$\left[\begin{array}{c:c} M(\xi) & I \\ \hdashline I & \mathrm{Diag}(q) \end{array}\right] \succeq 0 \tag{3.46}$$

and

$$P\mathbf{1} = \mathbf{1}, \quad P \geq 0, \tag{3.47}$$

where I and $\mathbf{1}$ are the identity matrix and vector with all elements equal to unity, respectively.

In the situation when the number of parameters is small compared to the cardinality of X (i.e., $m \ll \ell$), it is convenient and just simpler to solve the problem operating on the convex space of admissible information matrices $\mathbb{M}$, rather than on the design space $\Xi(X)$ itself. More precisely, it can be shown that the Problem 3.3 can be converted to the equivalent dual one:

Problem 3.4. Find $M = M(\xi) \in \mathbb{S}^m_{++}$ to maximize

$$\Psi(M) = \big(\mathrm{trace}(M^{1/2})\big)^2 \tag{3.48}$$

subject to

$$\mathrm{trace}\left(M\Upsilon_k(x^i_k)\right) \leq 1,\ i = 1, \dots, \ell, \quad , k = 1, \dots, K. \tag{3.49}$$

The Problem 3.4 is a convex one and the return to the primal space can be performed by complementary slackness,

$$p^\star_{ki}\left(1 - \mathrm{trace}\left(M^\star\Upsilon_k(x^i_k)\right)\right) = 0, \quad i = 1, \dots, \ell, \quad k = 1, \dots, K, \tag{3.50}$$

that is, the optimal experiment strategy contains only those locations which lie on the surface of the uncertainty ellipsoid.

E-optimal design. For E-optimal designs, the norm of the parameter covariance matrix estimate is minimized, i.e., the maximum eigenvalue of the FIM inverse. Since the diameter of the parameter confidence ellipsoid is proportional to $\|M^{-1}(\xi)\|_2^{1/2}$, the primal problem can be stated as follows.

Problem 3.5 (E-optimal SDP). Determine $q \in \mathbb{R}$ and $P \in \mathbb{R}^{K\times\ell}$ which minimize

$$\mathcal{P}(P, q) = -q \tag{3.51}$$

subject to the LMI

$$M(\xi) \succeq qI \tag{3.52}$$

and

$$P\mathbf{1} = \mathbf{1}, \quad P \geq 0. \tag{3.53}$$

Then, the dual problem takes the following form.

Problem 3.6. Find $M \in \mathbb{S}_+^m$ maximizing

$$\Psi(M) = \operatorname{trace}(M) \tag{3.54}$$

subject to

$$\operatorname{trace}\big(M\Upsilon_k(x_k^i)\big) \le 1,\ i = 1, \dots, \ell,\ \ k = 1, \dots, K. \tag{3.55}$$

The dual problem is also an SDP problem. This formulation of the E-optimum experimental design enables us to employ efficient numerical algorithms for solving convex optimization problems over LMIs (e.g., interior-point algorithms of the polynomial-time complexity). This is an essential advantage because, due to the nondifferentiability of this criterion, no other effective algorithms have been proposed in the literature so far. As for the A-optimal design, the quantities $P^\star$ and $M^\star$ also satisfy the condition (3.50), so the dual problem can be easily converted to the primal one.

Sensitivity criterion. For this very simple case the original problem constitutes the SDP formulation per se (this is because it is a linear programming task).

Problem 3.7. Determine $P \in \mathbb{R}^{K\times\ell}$ which minimize

$$\mathcal{P}(P) = -\operatorname{trace}\big(M(\xi)\big) \tag{3.56}$$

subject to

$$P\mathbf{1} = \mathbf{1}, \quad P \ge 0. \tag{3.57}$$

The corresponding dual problem is as follows.

Problem 3.8. Find $q \in \mathbb{R}^K$ minimizing

$$\mathcal{P}(q) = \mathbf{1}^\mathsf{T} q \tag{3.58}$$

subject to

$$\operatorname{trace}\big(\Upsilon_k(x_k^i)\big) \le v_k,\ i = 1, \dots, \ell,\ k = 1, \dots, K. \tag{3.59}$$

D-optimal design. For the determinant criterion, the discussed design problem can be formulated as the so-called *maxdet* one [298]:

Problem 3.9. Determine $P \in \mathbb{R}^{K\times\ell}$ which minimize

$$\mathcal{P}(P) = \ln\det M^{-1}(\xi) \tag{3.60}$$

subject to the LMI

$$P\mathbf{1} = \mathbf{1}, \quad P \ge 0. \tag{3.61}$$

The corresponding dual problem is of the following form.

Problem 3.10. Find $M \in \mathbb{S}^m_{++}$ maximizing

$$\Psi(M) = \ln\det M \tag{3.62}$$

subject to

$$\operatorname{trace}\big(M\Upsilon_k(x_k^i)\big) \leq 1,\ i = 1, \dots, \ell,\ \ k = 1, \dots, K. \tag{3.63}$$

Although in the case of D-optimality the problem is not a classical SDP one, it remains a convex optimization one with linear constraints (the same corresponds to dual formulation). Therefore, it is still possible to solve it using powerful algorithms of convex programming. However, it can also be solved in a similar manner to SDP problems defined for the other criteria. Consequently, an interior-point based method was proposed by Vandenberghe *et al.* [90, 298] for solving the maxdet problem subject to general LMI constraints preserving the quadratic convergence rate of this class of algorithms.

3.1.3 Optimization over the Continuous Design Space

In spite of the efficiency of the algorithms presented so far, they operate only on a finite set of available support points, and for some problems this approach may be too restrictive. This is especially related to the case when the measurement domain is continuous and multidimensional, therefore sometimes it may be difficult or simply impossible to provide a reasonable discretization of the domain which will offer a sufficient approximation of the design space. To increase the quality of the approximated solutions, some systematic algorithms are necessary which operate on the continuous set of spatial points where measurements may be taken and treat those locations as an additional design factor.

Unfortunately, because of the difficulties discussed earlier, including the large scale of practical issues combined with multi-modal character of the problem and the complications related to a simple description of the set $\mathbb{M}$, the number of contributed numerical algorithms for a continuous case is not so rich. Fortunately, the characterization constituting the general equivalence theorem from Section 2.1.4 highlights some ideas useful in construction of sequential numerical algorithms, which become a foundation for the whole class of numerical approaches within the framework of experimental design theory. The underlying reasoning relies on a correction of a nonoptimal design $\xi^{(r)}$ obtained after r iterations by convex combination with another design $\tilde{\xi}^{(r)}$, which hopefully improves the current solution, i.e.,

$$\xi^{(r+1)} = (1 - \alpha_r)\xi^{(r)} + \alpha_r\tilde{\xi}^{(r)} \tag{3.64}$$

for some convenient α_r. In this manner the experimental effort related to the design $\xi^{(r)}$ is reduced and, instead, the measurements at some locations corresponding to the design $\tilde{\xi}^{(r)}$ are taken. The crucial issue is how to determine a suitable design $\tilde{\xi}^{(r)}$ which would guarantee a proper decrease in the value of the optimality criterion. The answer is indicated by Theorem 2.4. First, recall that the points of the optimum design $\xi^\star$ coincide with the maxima of the function $\psi_k(x, \xi^\star)$. In such a way, by adding an appropriate design measure to the maxima of $\psi_k(x, \xi)$, an improvement in the current design can be expected. Indeed, if we take into account one-point designs $\tilde{\xi}^{(r)} = \{\begin{smallmatrix} x \\ 1 \end{smallmatrix}\}$, then, combining Lemmas 2.4 and 2.3, we have that the directional derivative is negative, i.e.,

$$\left.\frac{\partial \Psi[M(\xi^{(r+1)})]}{\partial \alpha_r}\right|_{\alpha_r=0+} = \varsigma(\xi^{(r)}) - \phi(x, \xi^{(r)}) < 0, \qquad (3.65)$$

which yields a decrease in the value of $\Psi[M(\xi^{(r+1)})]$ for a sufficiently small α_r.

3.1.3.1 First-Order Algorithm for a General Case

Sequential numerical design algorithms based on the ideas indicated above, which have been continually refined since the early 1960s, can be employed with some modifications to the sensor locations problem for parameter estimation. The general form of the one-point correction version of the delineated procedure can be expressed through the following scheme [68, 70, 73, 180, 233, 234, 276, 281, 307, 310]:

Algorithm 3.5. General first-order algorithm

procedure FIRST_ORDER_DESIGN_OPTIMIZATION
    Choose some positive tolerance $\epsilon \ll 1$. $r \leftarrow 0$
    Guess an initial nondegenerate design measure $\xi^{(0)}$ ▷ Design with finite criterion value
    repeat
        for $k \leftarrow 1, K$ **do**
            $x^k \leftarrow \arg\max_{x \in X} \phi_k(x, \xi^{(r)})$ ▷ Searching for a new sensor location at T_k
            Choose α_r according to one of the rules (3.66)–(3.68)
            for $k \leftarrow 1, K$ **do**
                $\xi_k^{(r+1)} \leftarrow (1 - \alpha_r)\xi_k^{(r)} + \alpha_r \delta(x^k)$ ▷ Correction of weights
            end for
            $r \leftarrow r + 1$
        end for
    until $\max_k \left(\phi_k(x^k, \xi^{(r)}) - \varsigma_k(\xi^{(r)})\right) \le \epsilon$
end procedure

On the analogy of Algorithm 3.3, dedicated to finite measurement spaces, it can be shown that a suitable choice of the sequence $\{\alpha_r\}$ guarantees the convergence of Algorithm 3.5. Common versions of the algorithm depending on the choice of $\{\alpha_r\}$ are [73, 180, 281] as follows:

- Wynn's rule,

$$\lim_{k\to\infty} \alpha_k = 0, \quad \sum_{k=0}^{\infty} \alpha_k = \infty; \tag{3.66}$$

- Fedorov's rule,

$$\alpha_k = \arg\min_{\alpha} \Psi\big[(1-\alpha)M(\xi^{(r)}) + \alpha M(\xi(x^k))\big]; \tag{3.67}$$

- adaptation rule,

$$\alpha_k = \begin{cases} \alpha_{k-1} & \text{if } \Psi[(1-\alpha_{k-1})M(\xi^{(r)}) + \alpha_{k-1}M(\xi(x^k))], \\ \gamma\alpha_{k-1} & \text{otherwise,} \end{cases} \tag{3.68}$$

where $\gamma < 1$ is suitably chosen.

Note that Algorithm 3.5 makes use only of information about the gradient of the performance index, and the rule (3.67) results in the steepest-descent algorithm. The conclusion is that the convergence rate of the presented numerical technique is comparable with its gradient counterparts from mathematical programming. That is, typically, a significant decrease in the performance index measure in the first few iterations is observed, and then serious moderation of the convergence rate occurs as the minimum is approached. Some second-order generalizations of Algorithm 3.5 are possible for specific criteria [69], but they involve high complexity of implementation and are connected with an improvement in the design weights rather than the support points, and in this context the characteristics of the first-order algorithm are satisfactory as the most significant support points are usually found in just few iterations.

A great advantage of Algorithm 3.5 is the possibility of extending it to the field of optimum experimental design for a wide class of DPSs. However, such an adaptation and further improvements require a little more detailed discussion of the implementation issues.

3.1.3.2 Implementation Details

Despite the fact that the above procedure offers a higher level of freedom regarding optimality in comparison with the algorithms dedicated to the discrete case, it still suffers from disadvantageous effects such as high numerical complexity and clusterization of support points. Therefore, the general first-order algorithm does not solve all the theoretical and technical problems which can be encountered in the area of DPSs and its applicability is far from being trivial. While implementing the algorithm, various problems should be addressed. The main are listed below.

1. In the literature some improvements of the first-order algorithm can be found, whose intention is to increase the convergence rate [234, 281].
   a. One of those is the idea of adding in each iteration all points which coincide with the maxima of $\phi(x, \xi^{(r)})$. Such an approach in the context of DPSs demands global maximization in multi-dimensional spaces. Such an adaptation is connected with dramatically increased numerical complexity. Because of this, multi-point correction is rather unavailable and not recommended.
   b. Inverting the underlying idea of Algorithm 3.5, it is possible to achieve an additional decrease in the performance index by removing some design measure from noninformative support points of the design $\xi^{(r)}$ and distributing this measure among the other design points. This leads to an accelerated version of the algorithm consisting in simultaneous addition of potentially significant support points and elimination of worthless ones from the design.
   c. In order to increase the efficiency of deleting redundant support points and to improve the quality of the update in line 9, the weight optimization algorithm for the discrete case can be exploited as a complement for one of the rules (3.66)–(3.68), since in line 9 the design measure for a newly selected point is uniformly subtracted from the current support points. This leads to a procedure that is a direct generalization of the two-phase design for the D-optimum criterion in the case of one-dimensional regression.
2. While implementing various versions of Algorithm 3.5, one may encounter numerous technical problems. The main include the following.
   a. The added support points tend to cluster in the vicinity of optimal ones. One of possible simple solutions is to represent all sites in a given cluster as one support point placed in the centre of the cluster with the weight equal to the sum of the weights for all clustered sites.
   b. Line 9 requires a comment concerning the choice of an optimal factor α in the variant of the stepeest descent rule. In a general case, when multi-output (or dynamic) systems are taken into account, the situation is slightly different from the static case of linear regression for which it is possible to determine a closed-form solution. Here an optimal α has to be found numerically and, because the searching range is fixed, golden-section or Fibbonacci methods will cope quite fine with this task.
   c. The removal of points with negligible weights is suggested in order to obtain designs with relatively small numbers of support points [233].
3. The crucial bottleneck of Algorithm 3.5 is a support candidate search in line 6, both from computational and implementation point of view.

First of all, complications are related to the necessity of calculating a global maximum of $\phi_k(x, \xi)$ over all points $x \in X$. Since this function has multi-modal nature and convergence can be guaranteed only if the added support point lies close to the global optimum, realization of this task becomes very difficult. Therefore, an effective global optimization procedure is required, which is crucial for efficient implementation. A major impediment while using PDE systems as DPS descriptions is that this involves a high cost of gradient evaluation, or approximations of the gradient may fail to be satisfactory (e.g., there may occur some scaling problems or insufficient smoothness of the underlying functions). Based on numerous computer experiments it has been found that, in order to overcome those difficulties, some procedures of the stochastic type are especially well suited for many practical problems. One of the examples which was successfully exploited is the extremely simple Adaptive Random Search (ARS) strategy proposed in [307]. It is dedicated to situations where the set of admissible measurement points X is a hypercube, i.e., $X = \{x = (x_1, \dots, x_d) : x_{i\,\min} \leq x_i \leq x_{i\,\max}\}$; however, it can be extended to more complex situations, as indicated by various authors [180, 281]. Another possibility is to use a normalized function $\phi_k(x, \xi)$ as a probability distribution of optimal support points. A very interesting approach based on such an idea and called the Selective Random Search (SRS), proposed by Rafajłowicz [234], constitutes the potential of the first-order approach.

Rounding procedures. Note that the discussed types of algorithms are oriented to calculating only approximate (continuous) designs. Namely, in the concept of approximated design measures we neglected the fact that in real experiments the number of measurements r_k^i to be taken at each support point x_k^i is integer-valued and the value of the weight $p_{ki} = r_k^i/N$ is rational. Nevertheless, the number of potential support points in the spatial setting is usually quite large and the set of candidate points is continuous (in the case of Algorithm 3.5) so that we can expect that some rounding procedure [223, 233] of the approximate designs calculated by the proposed algorithms will yield sufficiently good exact designs. In such circumstances, satisfactory results can be achieved using the simple numerical rounding

$$r_i^\star = \begin{cases} \lceil p_{ki} N \rceil & \text{if } \lceil p_{ki} N \rceil - p_{ki} N \leq 0.5, \\ \lfloor p_{ki} N \rfloor & \text{otherwise,} \end{cases} \tag{3.69}$$

where $\lceil a \rceil$ denotes the least integer greater than or equal to a and $\lfloor a \rfloor$ stands for the greatest integer less than or equal to a. If $\sum_{i=1}^{\ell} r_k^{i\,\star} = N_{\text{rep}} \neq N$, then the correction $(N - N_{\text{rep}})/N$ can be made to the weight of an arbitrary chosen support point or, alternatively, $|N - N_{\text{rep}}|$ points can be randomly selected and their weights suitably updated.

3.1.4 Clusterization-Free Designs for Scanning Sensors

In real engineering problems the scanning observation strategy is usually realized with the use of dedicated sensor networks or sometimes even with mobile monitoring stations. The assumption of independent observations, which is advantageous from a theoretical point of view, cannot be justified when clusterization of measurements should be avoided, as usually spatial data acquisition techniques do not raise the possibility of replicated observations at a single site. In addition to this, classical optimum experimental design techniques do not take into account local correlations which determine spatial locations of measurements. As far as noise characteristics are not the primal object of interest, several sensors situated in the close vicinity of one another usually do not provide significantly more information about the system behavior than a single sensor.

This has become a motivation and attracted researchers' attention to the so-called *clusterization-free* designs, where the idea is to properly distribute the sensors so as to assure minimal distances between their spatial locations, guaranteeing elimination of mutually dependent measurements. A first treatment of such a concept is related to applications of spatial statistics (cf. [170],[72],[34]) in the context of designs with no replications allowed. Especially the approach to constructing such replication-free designs based on directly constrained design measures proposed by Fedorov [44, 71, 72, 73] turns out to be very attractive. In spite of its somewhat abstract assumptions, the resulting algorithm of the exchange type is very easy to implement. It turns out that Fedorov's approach can be generalized to the problems of sensor scheduling considered in this monograph. The first and very successful adaptation of this approach to the task of sensor network design for estimation of DPSs was developed by Uciński [276] in the context of stationary sensors. Then, a very efficient extension to the scanning strategy was proposed by Uciński and Patan [289] and extensive theoretical substantiation provided in monograph [281]. The approach proved to be flexible to further refinements [187, 188, 198, 281, 295] which will be presented in what follows.

3.1.4.1 Problem of Optimal Sensor Densities

In order to avoid clustered sensor configurations, the idea of operating on the density of sensors (i.e., the number of sensors per unit area) instead of direct sensor locations can be adopted [44, 71, 72, 73]. This is justified when the total number of sensor nodes in the network is sufficiently large. In contrast to the designs discussed in the previous section, here, we impose the crucial restriction that the density of sensor allocation must not exceed some prescribed level, which in general can be defined separately for each observation subinterval T_k. This can be expressed with the condition

$$\xi_k(\mathrm{d}x) \le \omega_k(\mathrm{d}x), \quad k = 1, \dots, K, \tag{3.70}$$

where $\omega_k(\mathrm{d}x)$ signifies the maximal possible 'number' of sensors per $\mathrm{d}x$ [73] such that

$$\int_X \omega_k(\mathrm{d}x) \ge 1. \tag{3.71}$$

Note that, in the spirit of such a general formulation, the number of activated sensors may vary in time. A constant number of active sensors can be achieved by imposing a simple constraint that for any two observation intervals T_{k_1} and T_{k_2} the corresponding maximal densities be equal, i.e., $\omega_{k_1}(dx) = \omega_{k_2}(dx)$.

Introducing the constraints of the form (3.70), we face the following design problem, being the appropriate modification of Problem 3.1.

Problem 3.11. Find $\xi^\star = (\xi_1^\star, \dots, \xi_K^\star) \in \Xi(X)$ minimizing

$$\mathcal{P}(\xi) = \Psi[M(\xi)], \tag{3.72}$$

subject to

$$\xi_k(\mathrm{d}x) \le \omega_k(\mathrm{d}x), \quad k = 1, \dots, K. \tag{3.73}$$

Definition 3.1. The design $\xi^\star$ corresponding to the solution of Problem 3.11 is said to be a (Ψ, ω)-*optimal design* [289].

Apart from the assumptions (A1)–(A5) and (A7), proper mathematical characterization of the optimal solutions requires an additional proviso:

(A8) $\omega_k(\mathrm{d}x)$ is atomless, i.e., for any $\Delta X \subset X$ there exists a $\Delta X' \subset \Delta X$ such that

$$\int_{\Delta X'} \omega_k(\mathrm{d}x) < \int_{\Delta X} \omega_k(\mathrm{d}x), \quad k = 1, \dots, K. \tag{3.74}$$

In what follows, we write $\bar{\Xi}(X) \subset \Xi(X)$ for the collection of all the design measures (3.8) which satisfy the requirement

$$\xi_k(\Delta X) = \begin{cases} \omega_k(\Delta X) & \text{for } \Delta X \subset \operatorname{supp} \xi_k, \\ 0 & \text{for } \Delta X \subset X \setminus \operatorname{supp} \xi_k, \end{cases} \tag{3.75}$$

for $k = 1, \dots, K$.

Definition 3.2. For any given design ξ, the function $\psi_k(\,\cdot\,, \xi)$ defined by (3.11) *separates* sets X_1 and X_2 with respect to $\omega_k(\mathrm{d}x)$ if for any two sets $\Delta X_1 \subset X_1$ and $\Delta X_2 \subset X_2$ with equal nonzero measures we have

$$\int_{\Delta X_1} \psi_k(x, \xi)\, \omega_k(\mathrm{d}x) \le \int_{\Delta X_2} \psi_k(x, \xi)\, \omega_k(\mathrm{d}x). \tag{3.76}$$

Then a fundamental result which characterizes the (Ψ, ω)-optimal designs takes the following form.

Theorem 3.3. *Let the assumptions (A1)–(A5),(A7) and (A8) hold. Then*

(i) there exists an optimal design $\xi^\star \in \bar{\Xi}(X)$*, and*
(ii) a necessary and sufficient condition for $\xi^\star \in \bar{\Xi}$ *to be* (Ψ, ω)*-optimal is that* $\psi_k(\,\cdot\,, \xi^\star)$ *separates* $X_k^\star = \operatorname{supp} \xi_k^\star$ *and its complement* $X \setminus X_k^\star$ *with respect to the measure* $\omega_k(\mathrm{d}x)$ *for* $k = 1, \dots, K$*.*

Proof. The theorem is a minor generalization of Theorems 4.1 and 4.2 given by Uciński [281]. The only difference is that the measure ω_k may vary depending on the distinct observation subintervals. Since this is not essential for the proof, the results from [281] may be adopted in a straightforward manner. ■

3.1.4.2 Exchange Algorithm for Sensor Scheduling

A practical interpretation of Theorem 3.3 is that at all support points of an optimal design component $\xi_k^\star$ the function $\psi_k(\,\cdot\,, \xi^\star)$ should take lower values than at other points. This amounts to allocating observations in the vicinity of the points where the least is known about the system response (automatically, measurements at such locations will be most informative).

One of the interpretations of the resultant optimal designs is obtained after partitioning the domain X into subdomains ΔX_i (with relatively small areas). Then, on the subinterval T_k, we allocate to each of them the number

$$N_k^\star(\Delta X_i) = \left\lceil N \int_{\Delta X_i} \xi_k^\star(\mathrm{d}x) \right\rceil \tag{3.77}$$

of sensors whose positions may coincide with nodes of some grid which further could represent the possible locations of the scanning sensors.

Making use of the above properties, there is a possibility to develop a numerical algorithm of constructing approximated solutions to the analysed problem. Since $\xi_k^\star(\mathrm{d}x)$ should be nonzero in the areas where $\psi_k(\,\cdot\,, \xi^\star)$ takes on smaller values, the main idea is to move some measure from the areas with higher values of $\psi_k(\,\cdot\,, \xi)$ to those with smaller values, as we expect that such a procedure will improve ξ. This is embodied by the iterative procedure represented in Algorithm 3.6, being a natural extension of the clusterization-free strategy to scanning observations [289].

As for the convergence of this exchange-type procedure, it can be proven proceeding analogously to standard reasoning in the convergence analysis of feasible direction methods [22, 71]. For instance, the convergence of the algorithm is guaranteed for the sequence $\{\alpha_r\}_{r=0}^{\infty}$ satisfying the conditions

$$\lim_{r \to \infty} \alpha_r = 0, \quad \sum_{r=0}^{\infty} \alpha_r = \infty. \tag{3.78}$$

Algorithm 3.6. Clusterization-free scanning strategy algorithm

procedure CLUSTERIZATION_FREE_SCANNING_STRATEGY
    Guess an initial design $\xi^{(0)} \in \bar{\Xi}(X)$
    Choose some positive tolerance $\epsilon \ll 1$. $r \leftarrow 0$
    while true **do**
        for $k \leftarrow 1, K$ **do**
            $X_1^k(r) \leftarrow \operatorname{supp} \xi_k^{(r)}$
            $X_2^k(r) \leftarrow X \setminus X_1^k(r)$
            $x_1^k(r) \leftarrow \arg\max_{x \in X_1^k(r)} \psi_k(x, \xi^{(r)})$
            $x_2^k(r) \leftarrow \arg\min_{x \in X_2^k(r)} \psi_k(x, \xi^{(r)})$
        end for
        if $\max_k \left(\psi_k(x_1^k(r), \xi^{(r)}) - \psi_k(x_2^k(r), \xi^{(r)}) \right) < \epsilon$ **then return**
        end if
        for $k \leftarrow 1, K$ **do**
            if $\psi_k(x_1^k(r), \xi^{(r)}) > \psi_k(x_2^k(r), \xi^{(r)}) + \epsilon$ **then**
                Find two sets $S_1^k(r) \subset X_1^k(r)$ and $S_2^k(r) \subset X_2^k(r)$ such that
                $x_1^k(r) \in S_1^k(r), \quad x_2^k(r) \in S_2^k(r)$ and
                $\int_{S_1^k(r)} \omega_k(\mathrm{d}x) = \int_{S_2^k(r)} \omega_k(\mathrm{d}x) = \alpha_r$
            else
                $S_1^k(r) \leftarrow \emptyset$
                $S_2^k(r) \leftarrow \emptyset$
            end if
            $X_1^k(r+1) \leftarrow (X_1^k(r) \setminus S_1^k(r)) \cup S_2^k(r)$
            $\operatorname{supp} \xi_k^{(r+1)} \leftarrow X_1^k(r+1)$
        end for
        $r \leftarrow r + 1$
    end while
end procedure

The maximal numbers of sensors allocated to the spatial element $\mathrm{d}x$ can be usually expressed as $\omega_k(\mathrm{d}x) = \varrho_k(x)\mathrm{d}x$, $k = 1, \dots, K$, where the ϱ_ks play the roles of density functions. But then it is always possible to propose an appropriate change of coordinates which allows us to restrict our attention to constant ϱ_ks.

The proposed approach inherits all of the more and less valuable characteristics of the adopted clusterization-free strategy. For instance, a computer implementation forces the replacement of all integration operators by summing over some suitable regular grid elements. Furthermore, the sets X, $X_1^k(r)$, $X_2^k(r)$, $S_1^k(r)$ and $S_2^k(r)$ then simply consist of grid elements (in this case these are the potential sensor locations). As a result, a rather abstract form of the above iterative procedure is reduced to an exchange-type algorithm with the additional constraint that every grid element must not contain more than one support point in each time subinterval and the weights of all supporting points are equal. An additional assumption is that the α_rs are fixed and one-point exchanges are most often adopted, i.e., $S_1^k(r) = \left\{x_1^k(r)\right\}$

and $S_2^k(r) = \{x_2^k(r)\}$, which leads directly to an extremely simple and efficient implementation (see [281] for the detailed analysis). On the other hand, the convergence to an optimal design is assured only for a properly decreasing sequence $\{\alpha_r\}_{r=0}^{\infty}$ and, since it is not generally true for fixed α_rs, some minor oscillations in $\Psi[M(\xi^{(r)})]$ may sometimes be observed. A denser spatial grid usually constitutes a remedy for this predicament [170].

3.2 Optimal Switching Schedule

Although the approach based on directly-constrained design measures proposed in the preceding section turns out to be extremely efficient in practice, its main limitation is that it can be used only when the number of sensors is relatively high [198, 289]. Another important restriction is that the methods discussed so far assume that the switching instants are arbitrarily fixed along the time axis, which may not be the case in many applications. Obviously, if the sensor switchings are allowed to take place freely in continuous time, i.e., they are an additional design factor, the complexity of the scheduling problem significantly grows. In [150], a similar problem was considered for state estimation, where the proposed solution was to make use of some recently obtained results on discrete-valued optimal control problems. By introducing the transformation of the time scale described in [150], it was shown that the original discrete-valued control problem with variable switching times can be transformed into an equivalent continuous-valued optimal control problem which can then be solved using readily available optimal-control techniques. The aim of the deliberations presented in what follows and originally developed in [195] for a scalar case and further refined in [182] is to outline how this approach can be adopted to calculating optimal switching schedules for parameter estimation of DPSs.

3.2.1 Optimal Sensor Scheduling Problem

The examined class of MIMO spatio-temporal systems is described just like in the previous section, i.e., it is defined by (2.66)–(2.68) with the same notation and assumptions regarding spatial and temporal domains. In what follows, we suppose that there are N stationary sensors located at given points $x^1, \dots, x^N$ of $\Omega \cup \partial\Omega$. Additionally, in order to simplify the presentation and clarify the analysis, the approach is presented for the case that at a given time moment only one sensor may be active while others remain dormant. Then, the generalization for the case of multiple active sensors becomes straightforward. In the case considered, a sensor activation schedule can be represented by an integer-valued function,

$$u : T \to \Lambda = \{1, \dots, N\}. \tag{3.79}$$

In particular, $u(t) = j$ means that the j-th sensor is used at time t. The set of admissible sensor schedules is given by

$$\mathcal{U} = \{u : T \to \Lambda \mid u(\,\cdot\,) \text{ is Lebesgue measurable}\}. \tag{3.80}$$

In the general case, when the state may not be directly measurable, each sensor makes noisy observations of the system outputs continuously in time, which can be formally represented as

$$z(t) = \sum_{j=1}^{N} \chi_{\{u(t)=j\}}(t)\big[h\big(y(x^j,t;\theta),x^j,t\big) + \varepsilon(x^j,t)\big] \tag{3.81}$$

for $t \in T$, where

$$\chi_{\{u(t)=j\}}(t) = \begin{cases} 1 & \text{if } u(t) = j, \\ 0 & \text{otherwise,} \end{cases} \tag{3.82}$$

h being some known function responsible for mapping state components into measurements and $\varepsilon^j(\,\cdot\,)$ denoting the measurement noise assumed to be zero-mean, Gaussian and uncorrelated in both time and space.

Sensor positions which guarantee the best accuracy of the least-squares estimates of θ are then found by choosing $u(\,\cdot\,) \in \mathcal{U}$ so as to minimize some scalar measure of performance Ψ defined on the average FIM given in this situation by [227]

$$M(u) = \frac{1}{t_f}\sum_{j=1}^{N}\int_0^{t_f} \chi_{\{u(t)=j\}}(t) G^{\mathsf{T}}(x^j,t)C^{-1}(x^j,t)G(x^j,t)\,\mathrm{d}t, \tag{3.83}$$

with G given by (2.86).

3.2.2 Equivalent Mayer Problem

The sensor selection problem under consideration can be expressed as an optimal-control problem in Mayer canonical form [78]. Indeed, defining the quantity

$$\Pi(t) = \sum_{j=1}^{N}\int_0^{t} \chi_{\{u(\tau)=j\}}(\tau) G^{\mathsf{T}}(x^j,\tau)C^{-1}(x^j,\tau)G(x^j,\tau)\,\mathrm{d}\tau, \tag{3.84}$$

we get

$$M = \Pi(t_f). \tag{3.85}$$

Thus, to find $u \in \mathcal{U}$ minimizing $\Psi\big[M(u)\big]$ amounts to the following problem.

Problem 3.12. Choose $u^\star \in \mathcal{U}$ to minimize the performance index

$$J(u) = \Psi[\Pi(t_f)], \tag{3.86}$$

subject to the constraints in the form of nonlinear ODEs

$$\begin{cases} \dfrac{\mathrm{d}}{\mathrm{d}t}\Pi(t) = \displaystyle\sum_{j=1}^{N} \chi_{\{u(t)=j\}}(t) G^\mathsf{T}(x^j, t) C^{-1}(x^j, t) G(x^j, t), \\ \Pi(0) = 0. \end{cases} \tag{3.87}$$

Problem 3.12 constitutes an optimal control problem in which the main difficulty is that the range set of control is discrete and hence not convex. Furthermore, choosing the appropriate elements from the control set in a proper order is, in fact, a nonlinear combinatorial optimization problem.

3.2.3 Computational Scheme Based on the Control Parametrization Enhanced Technique

A first approach to handle Problem 3.12 is to view this optimal discrete-valued control problem as that of determining the switching points of the optimal discrete-valued control directly, but this may lead to serious numerical difficulties. This is caused by inherently combinatorial character of the problem on the one hand. On the other, the switchings may occur at any point of the interval $[0, t_f]$, which extremely complicates integration of the differential equation in the constraint (3.87) since the possible discontinuities produced by switchings may lead to instability of numerical solvers and therefore sophisticated robust integration methods [162] may be required. A novel problem transformation called the Control Parametrization Enhancing Technique (CPET) was thus proposed in [151] to address these difficulties. Under the CPET, the switching points are mapped onto the integers, and the transformed problem becomes just an ordinary optimal control problem with known and fixed switching points. It can then be readily solved numerically by commonplace optimization routines. Since the CPET transform was not developed directly in the context of sensor scheduling, in order to get more comprehensive background on the broad range of its applications the reader can be referred to [146, 147, 148, 149].

As in [150], where the CPET was employed to construct an optimal sensor schedule for finding optimal mean-square estimates of the system state, we set $Q = \gamma N^2$, where γ stands for the assumed maximum number of times any sensor $i \in \Lambda$ is selected. In spite of the already mentioned different nature of the corresponding problem of finding an optimal observation policy for parameter estimation, the CPET transform can still be adopted to improve the qualitative formulation of the Mayer problem (3.86)–(3.87). In such a way, we are able to build a new effective computational approach to be presented in what follows, in order to handle the wide class of optimal-control problems related to parameter estimation of DPSs.

Thus, we introduce a new time scale variable τ which varies from 0 to Q. Let $\mathcal{V}$ denote the class of nonnegative piecewise constant scalar functions defined on $[0, Q)$ with fixed interior knot points located at $\{1, 2, \dots, Q-1\}$. The CPET transformation from $t \in [0, t_f]$ to $\tau \in [0, Q]$ is defined by the differential equation

$$\frac{\mathrm{d}t}{\mathrm{d}\tau} = v(\tau), \quad t(0) = 0, \tag{3.88}$$

where the scalar function $v(\,\cdot\,) \in \mathcal{V}$ is called the enhancing control which satisfies

$$\int_0^Q v(\tau)\,\mathrm{d}\tau = t_f. \tag{3.89}$$

Furthermore, we introduce a fixed function $\mu : [0, Q] \to \Lambda$,

$$\mu(\tau) = (i \bmod N) + 1, \quad \tau \in [i, i+1), \tag{3.90}$$

for $i = 0, 1, \dots, Q-1$. The idea of this CPET transformation is to let any $u(\,\cdot\,) \in \mathcal{U}$ be naturally represented by a $v(\,\cdot\,) \in \mathcal{V}$ whenever this fixed $\mu(\tau)$ is defined. For instance, consider the simple example illustrated in Fig. 3.1, where the original control signal (cf. Fig. 3.1(a))

$$u(t) = \begin{cases} 2, & t \in [0, 0.2), \\ 1, & t \in [0.2, 0.5), \\ 3, & t \in [0.5, 1) \end{cases}$$

is used with $N = 3$ possible sensor configurations and only one ($\gamma = 1$) potential switching per every configuration is assumed. The new time scale τ is split into separate intervals which are designated for particular sensor configurations. In each interval, only a prescribed configuration can be activated. The duration of this activation is represented by the appropriate value of the function $v(\tau)$. In such a manner the interpretation of the function $\mu(\tau)$ becomes very clear, as its value represents the sensor configuration assigned to the interval containing a given value of τ (cf. Fig. 3.1(b)). Thus, the natural representation of the original control $u(t)$ in the $\mathcal{V}$ domain can be found as

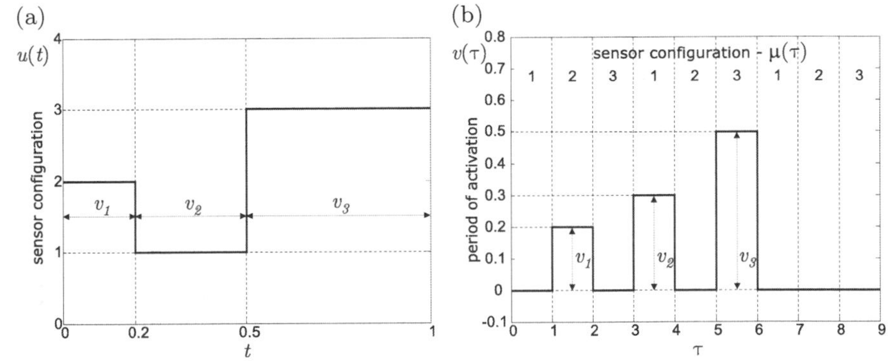

Fig. 3.1 Example of the control signal (a) and its natural representation in the transformed time scale τ (b)

$$
v(\tau) = \begin{cases} 0, & \tau \in [0,1), \\ 0.2, & \tau \in [1,2), \\ 0, & \tau \in [2,3), \\ 0.3, & \tau \in [3,4), \\ 0, & \tau \in [4,5), \\ 0.5, & \tau \in [5,6), \\ 0, & \tau \in [6,7), \\ 0, & \tau \in [7,8), \\ 0, & \tau \in [8,9), \end{cases}
$$

which is shown in Fig. 3.1(b).

Substituting (3.88) into (3.86)–(3.87) and setting $P(\tau) = \Pi(t(\tau))$, we obtain the following problem.

Problem 3.13. Find a $v \in \mathcal{V}$ such that the cost functional

$$
\mathcal{J}(v(\cdot)) = \Psi[P(Q)] \tag{3.91}
$$

is minimized, subject to the constraints

$$
\begin{aligned}
&\frac{\mathrm{d}}{\mathrm{d}\tau}P(\tau) - v(\tau)\left[\sum_{j=1}^{N} \chi_{\{\mu(\tau)=j\}}(t) G^{\mathsf{T}}(x^j, t(\tau)) C^{-1}(x^j, t(\tau)) G(x^j, t(\tau))\right], \\
&P(0) = 0
\end{aligned} \tag{3.92}
$$

and (3.89).

The sufficient conditions of the equivalence between Problems 3.12 and 3.13 are established by the following theorem.

Theorem 3.4. *Assume that the maximum number of selections (or activations) of any sensor $j \in \Lambda$ is finite and equal to γ. Then, Problems 3.12 and 3.13 are equivalent if $Q \geq \gamma N^2$.*

This result may be easily proved in much the same way as the corresponding result by Lee et al. [150, Th. 3]. Note that Problem 3.13 does not inherit any of the above-mentioned drawbacks of Problem 3.12 and can be solved with relative ease, as the switching points of the original control are mapped onto the set of integers in chronological order. Moreover, piecewise integration can now be easily performed since discontinuity points in the τ-domain are known and fixed.

As for a practical solution of this problem, note that controls v are piecewise constant, i.e., $v(\tau) = v_i$ if $\tau \in [i, i+1)$ for $i = 0, 1, \dots, Q-1$. Consequently, the only unknown variables to be determined are the values of parameters $v_0, v_1, \dots, v_{Q-1}$. These should be selected so that they are non-negative and sum up to t_f while simultaneously minimizing $\mathcal{J}(v)$ subject to (3.92). Note that, to determine the value of the performance index (3.91), the constraint (3.92) needs to be known only at point $\tau = Q$, and therefore from the point of view of implementation, it can be treated as part of the criterion $\mathcal{J}(v)$, i.e., equations (3.92) are solved during evaluation of the performance index (3.91). Thus, by the presented transformation, the optimal control problem becomes the following mathematical programming one.

Problem 3.14. Find scalars $v_0, \dots, v_{Q-1}$ that minimize $\mathcal{J}(v)$ subject to

$$v_i \geq 0, \quad i = 0, \dots, Q-1, \quad \sum_{i=0}^{Q-1} v_i = t_f.$$

It can be readily solved using any nonlinear programming routine for minimization of functions subject only to linear constraints. For instance, sequential quadratic programming [246] proved its efficiency in various numerical experiments related to the sensor location problem [180, 182, 281].

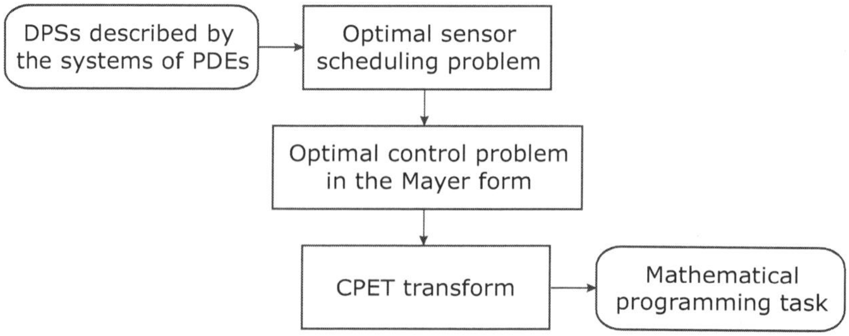

Fig. 3.2 General block scheme of the proposed approach

Although the delineated approach reveals new aspects of the CPET technique applied in order to build a novel methodology to manage the analyzed class of sensor location problems, there are still some technicalities which should be addressed. First of all, note that the method can be extended to the case of several scanning sensors with no major difficulties. The only change we have to make is definition of Λ. In particular, if ℓ sensors are to be used, there are altogether $N' = \binom{N}{\ell}$ ways of activating ℓ sensors from a total of N sensors. Then we can define $\Lambda = \{1, \dots, N'\}$ and the meaning of each $j \in \Lambda$ is to be understood as one of the sensor combinations.

Another question is determination of an optimal number of switchings, as in the above algorithm a fixed maximum number of switchings is assumed. A heuristic approach proposed in [151] to circumvent this problem is the following: Starting with a fixed γ, we solve Problem 3.12. We then increment the number of switchings and solve Problem 3.12 again. If there is no decrease in the optimal cost, we adopt the previous value of γ to be the optimal number of switchings, otherwise we further increase γ.

Finally, a problem of great importance is to avoid excessive numbers of sensor switchings or oscillations in switching, as it is clear that this involves costs related, e.g., to the potential wear and tear of the switching mechanism and transient responses which may decrease the quality of the data acquisition process. To overcome this impediment, the simplest idea is to impose additional constraints related to the lower bounds on v_is. Although such an approach is very simple to implement, it does not take account of the specific form of 'turn-on' characteristics for each sensor individually. A more systematic approach to this problem is presented by Lee et al. [150], where the appropriate penalization of the criterion is proposed, leading to exclusion of switchings with negligible values of $v(\tau)$ from the optimal solution.

3.3 Applications in Engineering

The proposed approach is motivated by a wide spectrum of practical applications, especially in the area of monitoring networks and sensor arrays for process quality control. At this point, some examples are provided which indicate a great potential of the methodology presented in this chapter.

3.3.1 Electrical Impedance Tomography

The first example is related to application of the proposed methodology in computer-assisted tomography of an electrolysis process [101, 174, 252]. Consider a cylindrical chamber of a reactor filled with an electrolyte. In the interior of the chamber, two metallic conductors are immersed in the liquid solution forming a conductive medium characterized by a spatially-varying conductivity coefficient. The three-dimensional domain of interest $\Omega = \{(x_1, x_2, x_3) : x_1^2 + x_2^2 \leq 1,\ 0 \leq x_3 \leq 1\}$ (scaled in meters) representing the reactor chamber with metallic plates is schematically shown in Fig. 3.3(a). Such an experimental setting is a typical configuration example in the galvanization or metal electrorefining processes. Metallic conductors are cuboids of size $1 \times 0.7 \times 0.05$ (in meters) placed symmetrically 5 cm below the upper head of the chamber at a distance of 1 m from each other.

An electric potential is applied to both the metallic plates playing a role of the cathodes and to the bottom of the chamber being the anode. Then, a steady current flow through the plates in connection with the absence of external current sources results in a physical model for this problem expressed

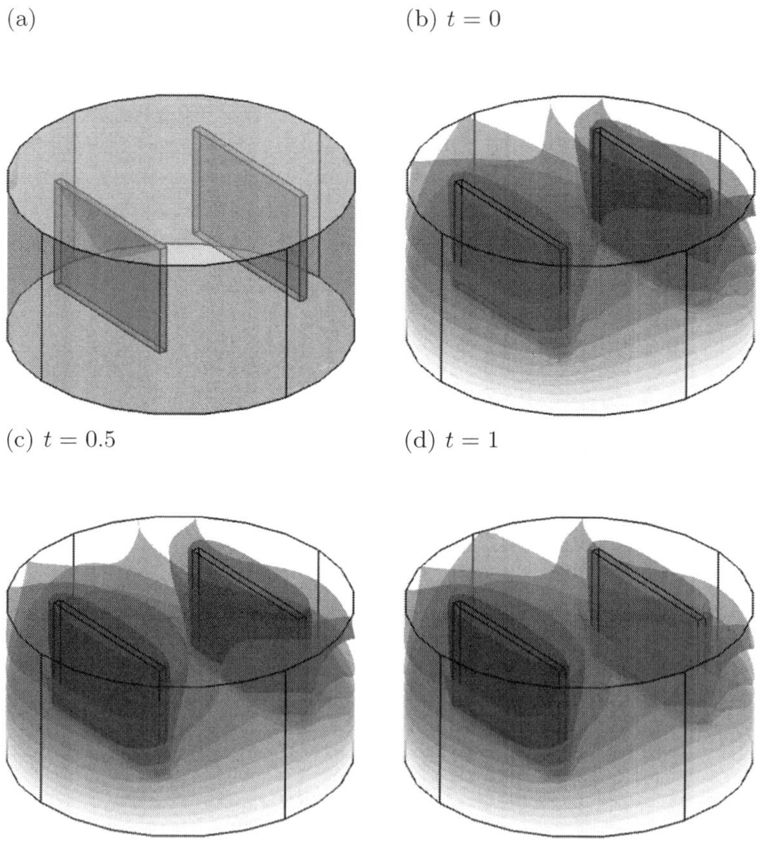

Fig. 3.3 Schematic view of the electrolytic chamber (a) and isosurface plots of evolution of the electric potential (b)–(d) (dark and bright colors denotes low and high voltage, respectively)

in the form of the Laplace equation over the observation interval $T = [0, 1]$ (in hours),

$$\nabla \cdot \big(\sigma_{\mathrm{dc}}(x,t)\nabla y(x,t)\big) = 0, \quad (x,t) \in \Omega \times T, \tag{3.93}$$

where y is the potential of the electric field in the conductive medium. Due to the reactions, the concentration of an electrolyte in the solution may change in time and space. Also, in order to control the rate of the electrolysis, a time-varying electric potential is applied to the plates. This is expressed in the respective boundary conditions,

$$\begin{cases} \dfrac{\partial y(x,t)}{\partial n} = 0 & \text{if } x \in \Gamma_1 \quad \text{(for insulated lateral and upper surfaces of chamber)}, \\ y(x,t) = -10(0.9 + 0.2t) & \text{if } x \in \Gamma_2 \quad \text{(for the left metallic conductor)}, \\ y(x,t) = -10(1.1 - 0.2t) & \text{if } x \in \Gamma_3 \quad \text{(for the right metallic conductor)}, \\ y(x,t) = 10 & \text{if } x \in \Gamma_4 \quad \text{(for the bottom of the chamber)}, \end{cases} \tag{3.94}$$

where $\partial y/\partial n$ denotes the derivative in the direction of the unit outward normal to Γ.

Our task is to find an optimum schedule of activation for scanning sensors to reconstruct the spatial distribution of the conductivity coefficient modelled in the following form:

$$\sigma_{\mathrm{dc}}(x,t) = 10(1 - \theta_1 x_3)(1 + \theta_2 x_1^2 + \theta_3 x_2^2)^{0.5(1+t)}, \tag{3.95}$$

which expresses the changes in the density of the electrolyte solution over the spatial area due to the reaction process. Thus, the aim of the parameter estimation is to optimally estimate the vector $\theta = (\theta_1, \theta_2, \theta_3)$. The state equation (3.93) is thus supplemented by three sensitivity equations, which results in a system of elliptic PDEs that have to be solved numerically, and the solution is to be stored in memory for further design. To this end, finite-element method solvers from COMSOL 3.5 [43] were employed with the time interval partitioned into 20 time steps, a spatial mesh composed of 8259 nodes and 50944 elements, and the initial estimate of $\theta^0 = (0.1, 0.05, 0.05)$ taken as a nominal one. The respective slow evolution of the spatial distribution of the electric potential over time is illustrated with isosurface plots in Figs 3.3(b)–(d). All covariance matrices $C(t)$ were then set to unity (a scalar system).

The requirement of a nondestructive experiment in computer-assisted tomography forces the placement of the measurement electrodes at the only accessible upper surface of the chamber, which thus becomes the design space X. Then, the observation interval is evenly partitioned into four subintervals, so the switching of sensor configurations are allowed at times $t_k = 0.25k, k = 1, \dots, 3$.

Optimization of experimental effort. In order to obtain an approximation of the D- and A-optimal designs, Algorithms 3.1 for the discrete case

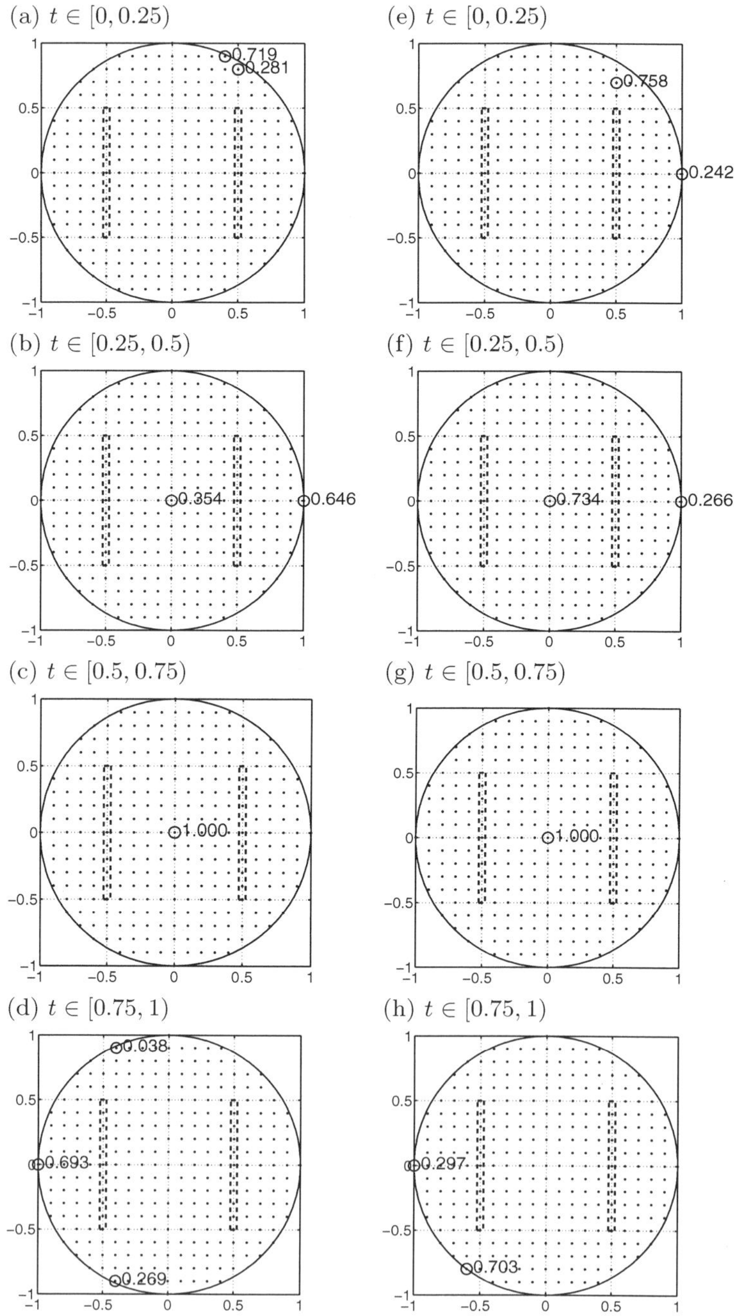

Fig. 3.4 D-optimal locations of sensors on the upper plane of the reaction chamber with the corresponding proportions of measurements (a)–(d) compared with the A-optimal scanning strategy (e)–(h)

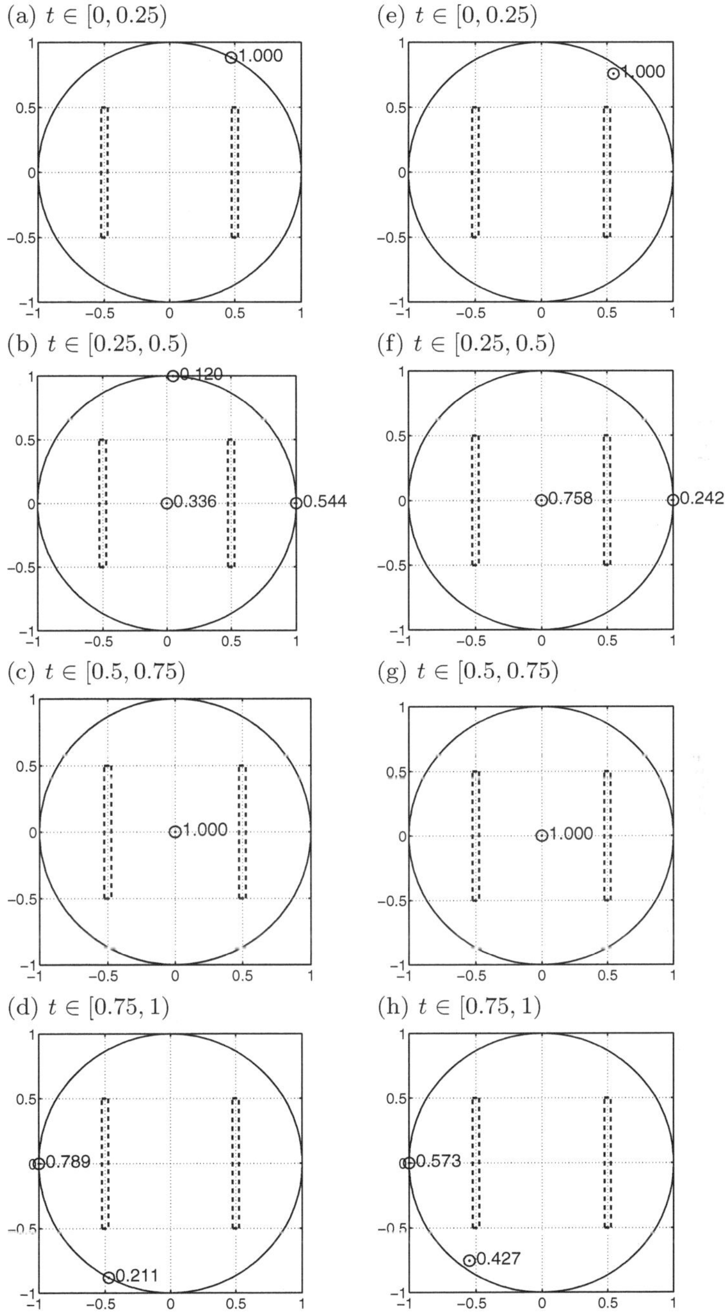

Fig. 3.5 D-optimal continuous locations of sensors with the corresponding proportions of measurements (a)–(d) compared with the A-optimal scanning strategy (e)–(h)

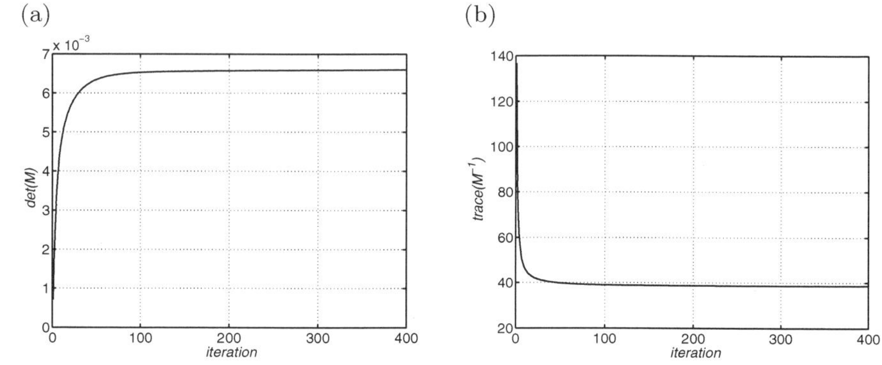

Fig. 3.6 Convergence of the multiplicative effort optimization algorithm for D-optimal (a) and A-optimal (b) criteria

and 3.5 for the continuous domain were implemented as a computer program with the Matlab 2009b environment on a PC (Intel Centrino 2.5GHz) running Windows 7.

For the discrete case, the set X_{d} of possible locations for measurement electrodes was proposed as a uniform grid of size 10 cm along the surface of the upper boundary (cf. Fig. 3.4, where the admissible locations are indicated with dots),

$$X_{\mathrm{d}} = \Big\{x : (x_1, x_2, x_3) = \big(\tfrac{i-11}{10}, \tfrac{j-11}{10}, 1\big),\ i, j = 1, \dots, 21,\ x_1^2 + x_2^2 \le 1\Big\},$$

Those support points ($|X| = 313$) with the uniform distribution of weights $p_{ki}{}^{(0)} = 1/313,\ i = 1, \dots, 313,\ k = 1, \dots, 4$ formed the starting design for weight optimization procedures.

Both algorithms produced very similar results, which proves the efficiency of the multiplicative weight optimization procedure. Given the accuracy of $\eta = 10^{-6}$, Algorithm 3.3 converged after 1021 and 1370 iterations (about 8 s of operational time) to the D- and A-optimal scheduling strategies, respectively. In turn, Algorithm 3.5 with the ARS adopted for searching for new candidates started from the randomly generated initial design attained the optimal solution using only nine additions of new support points (with $\eta = 10^{-6}$), which took about 20 seconds. The results for the A- and D-optimal scheduling criteria are compared in Figs 3.4 and 3.5. Points denote the possible sensor positions and open circles indicate the activated locations where the measurements should be taken. The numbers near the activated locations denote appropriate proportions of observations reallocated to particular points. As for the convergence of the multiplicative procedure, it is shown in Fig. 3.6, where it is clear that, after a short initial phase of very fast convergence, the algorithm slows down and the convergence rate becomes linear. Clearly, the A-optimal and D-optimal strategies focus on similar

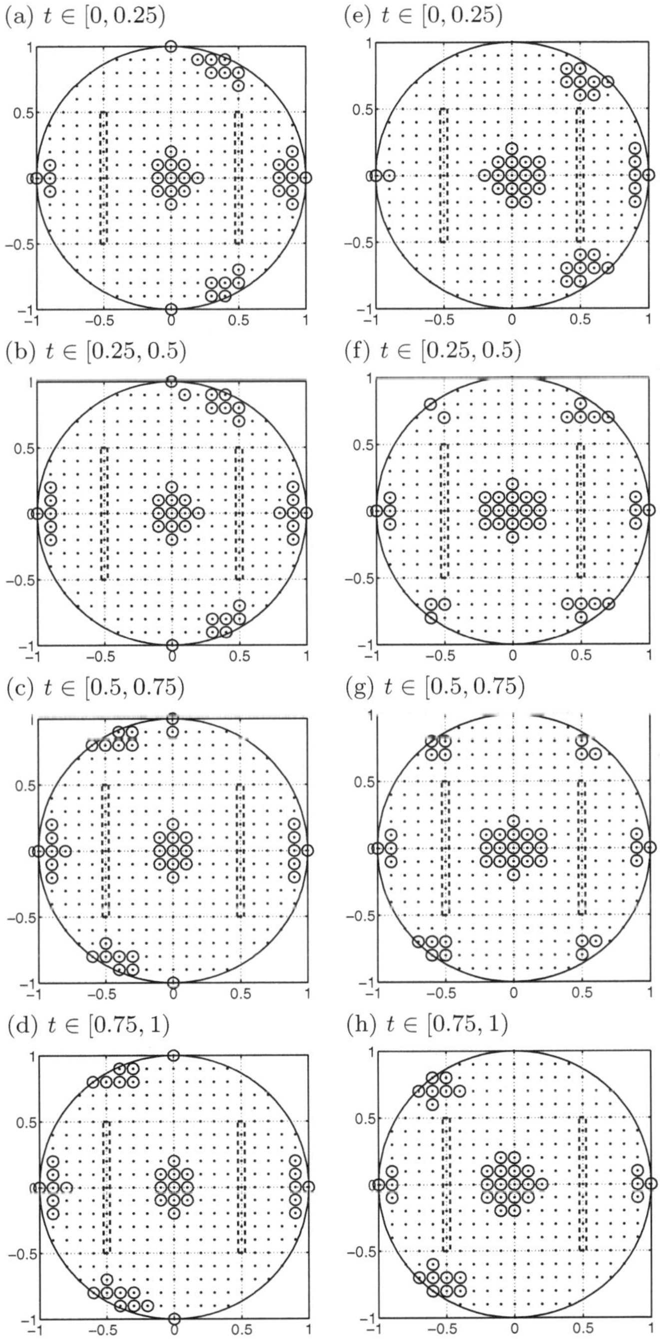

Fig. 3.7 D-optimal (a)–(d) and A-optimal (e)–(h) clusterization-free scanning schedule for 40 activated sensor nodes

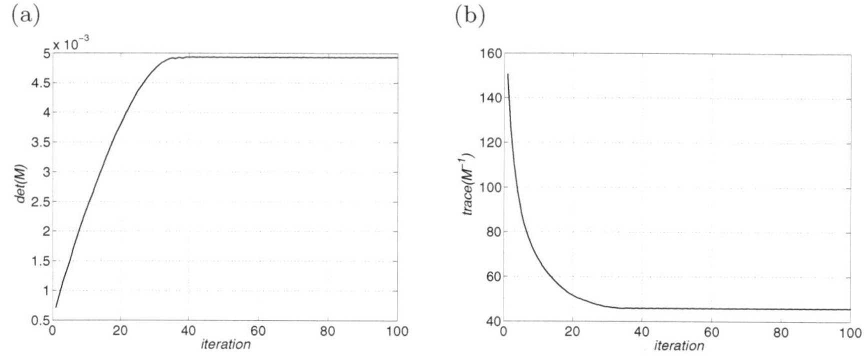

Fig. 3.8 Convergence of the clusterization-free exchange algorithm for D-optimal (a) and A-optimal (b) criteria

spatial positions of activated sensors, although there are significant differences in weights, which expresses different ideas adopted for compression of the parameter uncertainty. Moreover, the corresponding spatially continuous sensor scheduling turns out to be practically the same as for the discrete case. This means that the discretization of the boundary was good enough to obtain reasonably high quality of the solution.

Replication-free optimization. As the last approach applied in our experiment, the clusterization-free Algorithm 3.6 was used in the simplest form of a sequential one-point exchange procedure. In this case, the aim was to choose 40 locations of measurement electrodes from the set X_{d} defined at the beginning of the example. An initial design was created by randomly selecting support points. The algorithm calculated the solution very quickly (in both cases under 200 iterations for $\varepsilon \leq 10^{-6}$ and below 1 s of computation time). The final distributions of the optimal support points are shown in Fig. 3.7 and the convergence rates are depicted in Fig. 3.8. Since the replications of measurements are avoided (which is an additional constraint), it is clear that the degree of optimality is a little bit lower in comparison with optimization of the experimental effort as far as the values of the D- and A-optimality criteria are considered as a measure of the difference between them.

3.3.2 Monitoring of Groundwater Resources

Consider a confined aquifer lying on a river bank which is depicted in Fig. 3.9. The boundaries which are not adjacent to the river are assumed to be of the impermeable type or the flow throughout them is negligible. There are ten possible locations for the observation wells, O_1 to O_{10}.

The river is a source of pollution which spreads over the aquifer due to the hydrodynamic transport and dispersion. For illustrative convenience, a two-dimensional model will be considered as the description of the contamination process. Similar problems can be found in [130, 264] together with more sophisticated practical applications. In contrast to atmospheric pollution, the transport of substances in porous media takes place in a much larger time scale. The duration of such processes may cover months and even years, so pollution effects are also more permanent. In such a way, from a numerical point of view, it is convenient to suitably normalize the time scale.

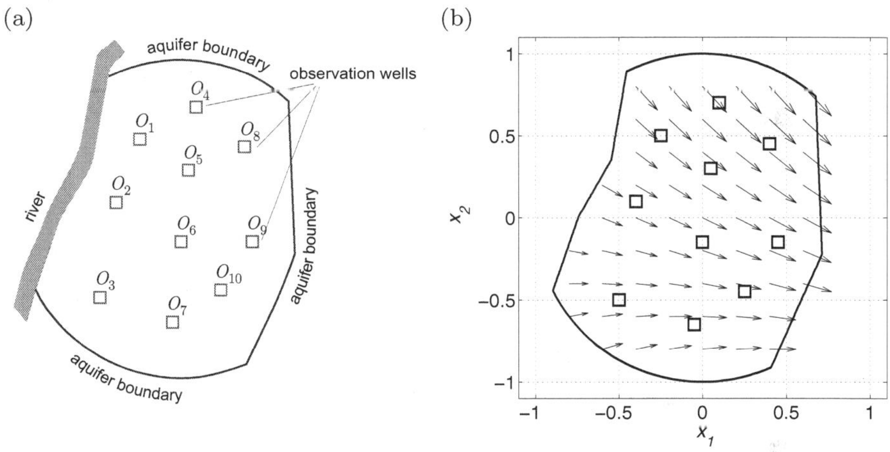

Fig. 3.9 General overview of a confined aquifer (a) and the velocity field with locations of observation wells (b)

This leads to the model for changes in the pollutant concentration $y(x,t)$ over the domain Ω in a unit time interval T, given by the advection-diffusion equation [180, 265]

$$\frac{\partial y(x,t)}{\partial t} + \nabla \cdot \big(v(x)y(x,t)\big) = \nabla \cdot \big(d(x)\nabla y(x,t)\big), \quad x \in \Omega, \ \ t \in T = [0,1], \tag{3.96}$$

where $v(x)$ is the vector describing the velocity field over the considered area, which can be found rather easily based on additional hydraulic equations and measurement data. The symbol $d(x)$ denotes the distributed coefficient of hydrodynamic dispersion, whose proper modelling is a rather challenging task.

From (3.96) it can be seen that the pollutant is assumed not to affect the density of groundwater, i.e., it is approximately constant in time. An additional assumption is that the bank of the river is contaminated all along its length adjacent to the domain Ω with the same constant rate. Moreover, initial contamination of the aquifer with the substance considered can be neglected. Thus, (3.96) can be supplemented with the following boundary and initial conditions:

$$\begin{aligned} y(x,t) &= 10, \ (x,t) \in \varGamma_1 \times T, \ \text{(river bank)} \\ \frac{\partial y(x,t)}{\partial n} &= 0, \quad (x,t) \in \varGamma_2 \times T, \ \text{(aquifer boundary)} \\ y(x,0) &= 0, \quad x \in \varOmega, \end{aligned} \tag{3.97}$$

where $\partial y/\partial n$ denotes the partial derivative with respect to the outward normal of $\partial \varOmega$.

Since exploratory wells of observation and pumping types are very expensive, the monitoring of water quality should be optimized in order to reduce the number of data sources and to properly use the available data. Because in our case the possible number of observation wells is rather small, application of the scanning strategy with an optimal switching schedule seems to be the most appropriate approach.

The parametric form of the distributed hydrodynamic dispersion was assumed as

$$d(x) = \theta_1 + \theta_2 \tanh(\theta_3 x_2), \tag{3.98}$$

where the elements of the vector $\theta^0 = (0.15, -0.02, 2.00)$ were taken as the nominal parameter values. Such a form of coefficient $d(x)$ plays the role of a smooth approximation to the changes in the effective permeability in the earthen foundation of the aquifer (i.e., this results from a changing level of clay over the area).

The velocity of the transport medium was taken in the simple form

$$v = (v_{x_1}, v_{x_2}) = (0.1x_1 + 0.4, \ -0.3x_2 - 0.1x_1 - 0.2), \tag{3.99}$$

Table 3.1 Combinations of the activated observation wells

Active wells	Control u_c
$\{O_1, O_2, O_3\}$	1
$\{O_1, O_2, O_4\}$	2
...	...
$\{O_1, O_9, O_{10}\}$	36
$\{O_2, O_3, O_4\}$	37
...	...
$\{O_2, O_9, O_{10}\}$	64
$\{O_3, O_4, O_5\}$	65
...	...
$\{O_8, O_9, O_{10}\}$	120

which is illustrated in Fig. 3.9(b). The chief aim of the experiment was to find an optimal sensor activation policy for determining the most accurate estimates of the true parameters θ_1 to θ_3. As the number of the exploited wells should be minimal, a reasonable choice is to use at every time instant only three from among ten available locations (cf. Fig. 3.9). The set of ten combinations of active wells, which were coded as successive integers, which are the levels of the input control signal u_c, is given in Table 3.1.

An elementary step in the design of an optimal sensor switching schedule is to use an effective procedure for computation of the so-called sensitivity coefficients (2.86), which are necessary when determining FIM elements. For that purpose, some methods known from shape optimization can be employed. In the context of this work, the Direct-Differentiation Method (DDM) was used since it provides exact results for the numerical solution, and mainly for that reason it seems to be most appropriate for a wide range of applications in engineering [95]. In this technique, the system equations are differentiated with respect to the individual parameters, which gives, after some rearrangement, the desired elements of the sensitivity matrix (2.86). Applying DDM to the problem (3.96)–(3.97), we obtain the following system of sensitivity equations:

$$\begin{aligned}\frac{\partial}{\partial t}\left(\frac{\partial y(x,t)}{\partial \theta_i}\right) &+ \nabla\cdot\left(v(x)\left(\frac{\partial y(x,t)}{\partial \theta_i}\right)\right) \\ &= \nabla\cdot\left(d(x)\nabla\left(\frac{\partial y(x,t)}{\partial \theta_i}\right)\right) + \nabla\cdot\left(\left(\frac{\partial d(x,t)}{\partial \theta_i}\right)\nabla y(x,t)\right),\end{aligned} \tag{3.100}$$

supplemented by suitable boundary and initial conditions,

$$\begin{aligned}\frac{\partial y(x,t)}{\partial \theta_i} &= 0,\ (x,t)\in \Gamma_1\times T,\ \text{(river bank)},\\ \frac{\partial}{\partial n}\left(\frac{\partial y(x,t)}{\partial \theta_i}\right) &= 0,\ (x,t)\in \Gamma_2\times T,\ \text{(aquifer boundary)},\\ \frac{\partial y(x,t)}{\partial \theta_i} &= 0,\ x\in \Omega,\end{aligned} \tag{3.101}$$

where $x \in \Omega$, $t \in T = [0,1]$, $i = 1,2,3$. Solving (3.96)–(3.97) simultaneously with (3.100)–(3.101) leads to the required sensitivity coefficients. Let us note that in our case the system equation is linear, therefore the form of the sensitivity problem is exactly the same and we can use the same computer code to solve both the system and sensitivity equations.

A computer implementation was done entirely in the Matlab 7.5 environment using a standard PC machine (Intel Core2Duo, 2.5GHz, 3BG RAM) running Windows Vista. The system of state and sensitivity PDEs was first solved with the use of the COMSOL environment [43] with a finite-element approximation of the domain (483 nodes, 893 triangles) and 40 divisions

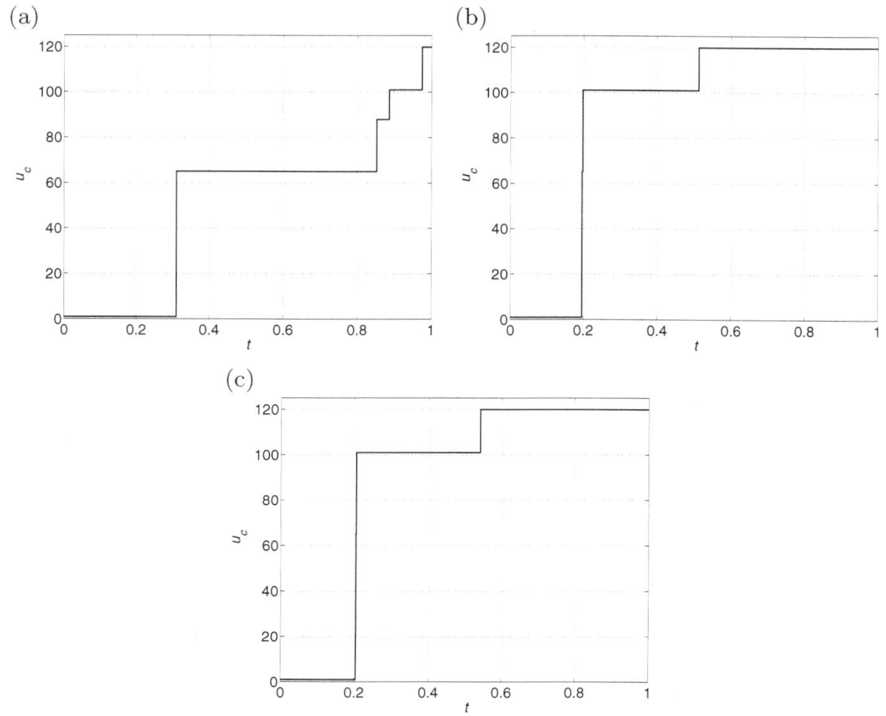

Fig. 3.10 Control signal for activation of observation wells for different criteria: D-optimality (a), A-optimality (b), and E-optimality (c)

of the time interval. The sensitivity coefficients were then interpolated and stored. Finally, the procedure based on the CPET approach with the standard Matlab function `fmincon` (based on the SQP solver) from the Matlab Optimization Toolbox [299] in the role of the optimizer produced the control signal describing the changes in the active well locations. For instance, for the D-optimality criterion, the control signal was (cf. Fig. 3.10(a))

$$u_c(t) = \begin{cases} 1 & \text{if } 0.0000 \le t < 0.3104, \\ 65 & \text{if } 0.3104 \le t < 0.8521, \\ 88 & \text{if } 0.8521 \le t < 0.8856, \\ 101 & \text{if } 0.8856 \le t < 0.9739, \\ 120 & \text{if } 0.9739 \le t \le 1.0000, \end{cases}$$

with the value of the criterion $\det(M) = 16.0445$. The corresponding solutions for the A-optimality and E-optimality criteria are also shown in Figs. 3.10(b) and (c) with the following values of the performance indices: $\mathrm{trace}(M^{-1}) = 2384.65$ and $\lambda_{\max}(M^{-1}) = 2384.27$, respectively. The constraints in the form of the ODE system (3.92) were implemented using the `ode45` routine from

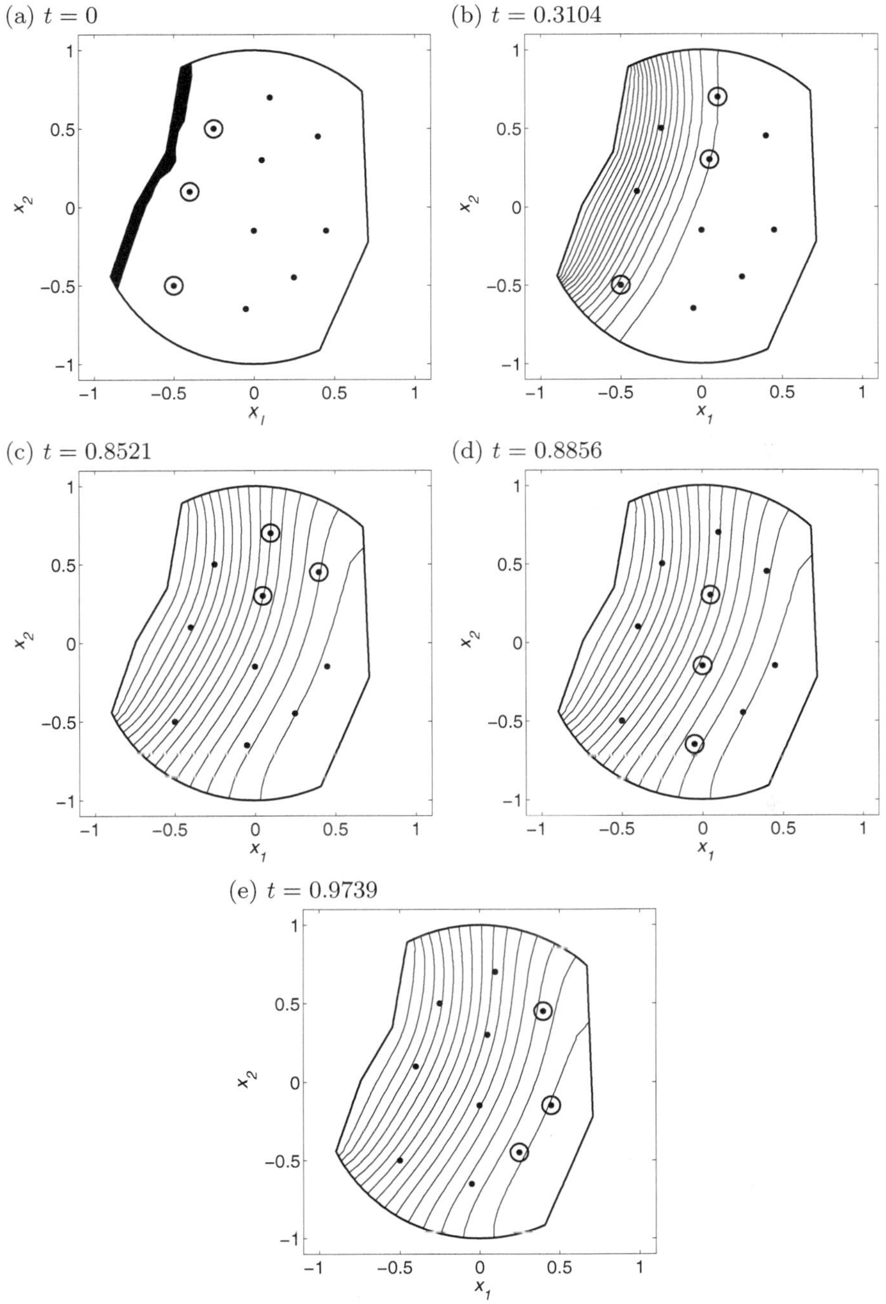

Fig. 3.11 Consecutive D-optimal switchings of the activated observational wells versus contour plots of the pollutant concentration

the Matlab ODE suite (piecewise integration over each interval of the s-domain where the solution is continuous). The maximal number of switchings was set *a prori* as equal to 2 and the time of calculations was about 8 minutes in each case.

Figure 3.11 illustrates the D-optimal sensor activation policy versus contour plots of the pollutant concentration, where open circles indicate the actually exploited observation wells. The sensor activation strategy can be interpreted based on the observation that the measurements are taken possibly close to the forehead of the 'pollutant wave' moving from the river bank to the right boundary of the aquifer and where we can expect significant changes in the pollutant concentration. Moreover, it is clearly seen from the assumed distributed model of the hydrodynamic dispersion that it takes greater values at the top of the aquifer than in its lower part. This means that the concentration changes are quicker while we move up along the x_2 axis. Those facts are reflected by the D-optimal sensor activation schedule as the sensors tend to make measurements in the vicinity of the indicated regions. In the case of the A- and E-optimality, the interpretation of results in terms of the velocity of pollutant concentration changes is not that obvious. This is a direct consequence of a more complicated dependence of the criterion upon the sensitivity coefficients. However, we can observe that the A- and E-optimal control signals are very similar (cf. Fig. 3.10(b)–(c)), which can be explained by a small size of the FIM, and this in turn leads to comparable local minima (the A-optimality criterion minimizes the sum of the FIM eigenvalues and E- the maximal eigenvalue). On the other hand, this perfectly illustrates how difficult it is to obtain an alternative solution with the use of the experimenter's experience as the only design tool.

3.4 Summary

Observation strategies for DPSs involving stationary sensors with time-continuous or time-discrete measurements are commonly exploited techniques in engineering applications due to their attractive properties such as low complexity of the observational system and a wide variety of methods. Nevertheless, such strategies are not always adequate to dynamic DPSs because of their low flexibility to fit into the system dynamics, which results in a significant decrease in the estimation quality.

A remedy is to employ more sophisticated measurement strategies, which may influence the experiment conditions operating directly on the time variable or indirectly through the dynamics of sensor motions. One of the possibilities is the scanning strategy, which can be split into two subcases, namely, the situation when the time schedule of sensor switchings is fixed a priori and when it also constitutes a design variable. Within the framework of this chapter, computationally attractive approaches were proposed for optimal placement of scanning sensors in both situations. In the former case, some ideas based on the concept of continuous experimental design can be effectively generalized to the setting considered here. In this chapter it was shown that construction of appropriate information matrices leads to a wide generalization of the methodology and algorithms developed for stationary sensor networks, including optimization of the experimental effort as well as clusterization-free designs. Accordingly, very efficient scanning measurement policies can be determined. In spite of their somewhat abstract assumptions, the resulting algorithms are easy to implement, which leads to extremely good performances. Bear in mind, however, that the clusterization-free approach should in principle be used if the number of sensors is relatively high. If this is not the case, we have to resort to standard discrete optimization routines which ensure that the constraints on the design measure are satisfied.

As for the situation of free switching moments, in order to select an optimal sensor activation strategy, a computational scheme based on the control parametrization enhancing technique was proposed in order to determine optimal schedules of scanning sensors which measure the state of a given DPS. The aim of this monitoring network is to accurately determine estimates of the unknown system parameters. Consequently, the highly combinatorial nature of the original problem is alleviated in a sense. The point is not in the computational effort, but in the possibility of applying widely available nonlinear programming algorithms to the resulting transformed optimal control problem.

Chapter 4
Resource Aware Mobile Sensor Routing

In modern observation systems, sensors can be located on various platforms which can be highly dynamic in motion. Each sensor node has a sensing capability, as well as limited energy supply, computing power, memory and communication ability. Endowing nodes in a sensor network with mobility drastically expands the spectrum of the network's capabilities. Moreover, assuming that each mobile node possesses a certain amount of decision making autonomy gives rise to a dynamic system with a considerable amount of flexibility, depending on the extent to which the nodes can cooperate in order to perform a mission. This flexibility, for example, allows us to handle a large number of data source targets with a much smaller number of nodes that can move comparing to the stationary or scanning sensing strategies. What is more, technological advances in communication systems and the growing ease in making small, low power and inexpensive mobile systems now make it feasible to deploy a group of networked vehicles in a number of environments, see [35, 37, 42, 163, 176, 255].

For instance, each networked sensor can be mounted on a mobile robotic platform usually called an *unmanned ground vehicle* (UGV) or an *unmanned aerial vehicle* (UAV), depending on the ability to explore two or three dimensional space. Each robot is actuated according to spatial and temporal information from sensors. Using all possible data, the data fusion system determines where sensor measurements should be taken and provides a suitable control driving the mobile sensors to appropriate sampling locations.

Clearly, mobility implies an increased level of complexity, see [37]. For example, if communication connectivity is to be maintained, we must ensure that each node remains within the range of at least some other nodes. We must also take into account that mobility consumes a considerable amount of energy, which amplifies the need for various forms of power control. However, the complexity of the resulting sensor management problem is compensated by a number of benefits. Specifically, sensors are not assigned to fixed spatial positions, but are capable of tracking points which provide at a given time

M. Patan: Optimal Sensor Networks Scheduling in Identification of DPSs, LNCIS 425, pp. 97–134.
springerlink.com

moment the best information about the contamination sources. In order to take advantage of these possibilities, sensors must be deployed and then guided so as to maximize the information extracted from the mission space while maintaining acceptable levels of communication and energy consumption. In addition to this, mobile sensing may significantly alter fundamental system properties, e.g., systems which are not observable under nonmoving observations become observable after transition to moving measurements [36].

4.1 Adaptation of Continuous Designs for Moving Sensors

When adapting the idea of continuous designs to the field of movable observations, it has to be emphasized that the connection between the classical experimental design theory and the problem of determining optimal sensor movements is, beyond doubt, rather complex. The main obstacle is the representation of the sensor motion trajectories in the form of generalized conditional distributions or, more precisely, Radon probability measures on the Borel sets of a given compact set of admissible locations. Despite this impediment, however, some relevant approaches can still be developed and successfully applied.

4.1.1 Time-Dependent Design Measures

The direct approach proposed by Rafajłowicz [232] for the scalar case has become one of the very scarce classical references on moving sensors. It was then expanded by Uciński [276] and further by Patan [180] for the case of MIMO systems.

As before, denote by X the compact set, in which the state $y \in \mathbb{R}^s$ of a DPS can be measured.

Definition 4.1. The mapping $T \ni t \mapsto x(t) \in X$, where $x(\,\cdot\,)$ is measurable in the Lebesgue sense, stands for the *trajectory* of the sensor motion.

The term 'trajectory' is used here to emphasize the dependence of $x(t)$ on time. Such trajectories are elements of the corresponding design space. For discrete-time observations, i.e., $T = \{t_k,\ k = 1,\dots,K : t_1 \leq t_1 \leq \dots \leq t_K\}$, the trajectory of the j-th sensor may be interpreted as a vector containing a series of observations that are taken at consecutive time moments t_k, i.e.,

$$x^j = [x^j(t_1),\dots,x^j(t_K)]^\mathsf{T}. \tag{4.1}$$

For the continuous observation interval $T = [0, t_f]$, each trajectory is determined by an observation curve for the j-th sensor,

$$x^j(t) \in X, \quad \text{a.e. on } T. \tag{4.2}$$

Now we can formulate our main objective, which is the design of an optimal measurement strategy for estimation of the unknown system parameter vector $\theta \in \mathbb{R}^m$, while the observations are made by N pointwise moving sensors. Thus, the observation equation is given by (2.77) with the measurement noise assumed to be a realization of a white Gaussian random field. According to (2.78), correlations between observations on the same trajectory may occur, but observations from different trajectories are not correlated. This fact is crucial for the approach as the additivity of the information matrices for measurements from different trajectories is assured similarly to the stationary and scanning sensor strategies.

Another important assumption is that there exists a neighbourhood of some known preliminary estimate θ^0 of the unknown parameter θ where the state $y(x, t; \cdot)$ is continuously differentiable with respect to θ. In such a way, the average Fisher information matrix is given by

$$M = \frac{1}{Nt_f} \int_0^{t_f} \left\{ \sum_{j=1}^{N} G^\mathsf{T}(x^j(t), t) C^{-1}(x^j(t), t) G(x^j(t), t) \right\} \mathrm{d}t, \tag{4.3}$$

where $G(x, t)$ defined by (2.86) is required to be continuous in $\bar{\Omega} \times T$.

On the other hand, the assumption of independent measurements leads directly to the clusterization effect for some time moments, i.e., at a given time moment more than one sensor may take measurements at a point $x^j(t)$. Taking this into account, by suitably relabeling the sensors (i.e., such that $i_1 \neq i_2 \Rightarrow x^{i_1}(t) \neq x^{i_2}(t)$, $1 \leq i_1, i_2 \leq \ell(t)$, where $\ell(t)$ is the number of different sensor locations at time t) the following collection of variables may be introduced by analogy to exact design for a specified time moment t:

$$\xi_N(t) = \begin{Bmatrix} x^1(t) & \dots & x^{\ell(t)}(t) \\ p_1(t) & \dots & p_{\ell(t)}(t) \end{Bmatrix}, \tag{4.4}$$

where $p_i(t) = r_i(t)/N$, $\sum_{i=1}^{\ell(t)} r_i(t) = 1$ and $r_i(t)$ denotes the number of sensors occupying the position $x^i(t)$. Hence the FIM may be rewritten as

$$M(\xi_N) = \frac{1}{t_f} \int_0^{t_f} \left\{ \sum_{i=1}^{\ell(t)} p_i(t) G^\mathsf{T}(x^i(t), t) C^{-1}(x^i(t), t) G(x^i(t), t) \right\} \mathrm{d}t. \tag{4.5}$$

Analogously to the case of stationary sensors, the notion of the exact design can be extended to the more general form of a probability measure ξ over X, which can be considered a function

$$\xi : T \ni t \mapsto \xi(\mathrm{d}x|t) \in \Xi_t(X) \tag{4.6}$$

where $\Xi_t(X)$ is the set of all probability measures on X. In our further deliberations, $\Xi(X)$ will be understood as the family of all such mappings ξ. This yields the following form of the FIM:

$$M(\xi) = \frac{1}{t_f} \int_0^{t_f} \left\{ \int_X G^\mathsf{T}(x,t) C^{-1}(x,t) G(x,t)\, \xi(\mathrm{d}x|t) \right\} \mathrm{d}t. \tag{4.7}$$

Introducing a fixed measure $\xi(\mathrm{d}t)$ of the observation effort in the interval T, we arrive at a further generalization, i.e.,

$$\begin{aligned} M(\xi) &= \int_T \left\{ \int_X G(x,t)^\mathsf{T} C^{-1}(x,t) G(x,t)\, \xi(\mathrm{d}x|t) \right\} \xi(\mathrm{d}t) \\ &= \iint_{X\times T} G^\mathsf{T}(x,t) C^{-1}(x,t) G(x,t)\, \xi(\mathrm{d}x,\mathrm{d}t), \end{aligned} \tag{4.8}$$

with the marginal distribution defined as

$$\xi(\mathrm{d}t) = \int_X \xi(\mathrm{d}x, \mathrm{d}t). \tag{4.9}$$

Notice that the identity $\xi(\mathrm{d}t) \equiv \mathrm{d}t/t_f$ corresponds to the uniform distribution of the experimental effort. In other words, $\xi(\mathrm{d}t)$ defines a 'density' of replications.

The task of finding an optimal observation strategy can be reformulated as determining

$$\xi^\star = \arg \min_{\xi\in\Xi(X)} \Psi[M(\xi)], \tag{4.10}$$

where $\xi(\mathrm{d}t)$ is assumed to be fixed a priori. Obviously, introducing the measure

$$\xi(\mathrm{d}t) = \frac{1}{t_f}\, \mathrm{d}t \tag{4.11}$$

brings us back to (4.7). The time dependent measure $\xi^\star(\mathrm{d}x|t)$ can be considered the optimal generalized trajectory.

To transfer the results from the stationary and scanning sensors, the conditions (A4)–(A5) on page 21 are assumed to be satisfied and instead of (A6) the following assumption is necessary:

(A9) $\forall \xi \in \Xi_q = \{\xi : \Psi[M(\xi)] \le q < \infty\}, \forall \bar{\xi} \in \Xi(X)$,

$$\Psi[(1-\alpha)M(\xi)+\alpha M(\bar{\xi})] = \Psi[M(\xi)] + \alpha \iint_{X\times T} \psi(x,t,\xi)\bar{\xi}\,(\mathrm{d}x,\mathrm{d}t) + o(\alpha;\xi,\bar{\xi}),$$

where the scalar q is chosen so that $\Xi_q \neq \emptyset$.

For differentiable criteria Ψ, we have

$$\begin{aligned}\delta_+\Psi\big(M(\xi),M(\bar{\xi})\big) &= \left.\frac{\partial\Psi[(1-\alpha)M(\xi)+\alpha M(\bar{\xi})]}{\partial\alpha}\right|_{\alpha=0^+}\\ &= \operatorname{trace}\left[\mathring{\Psi}[M(\xi)]\big(M(\bar{\xi})-M(\xi)\big)\right]\\ &= \iint_{X\times T}\Big\{\operatorname{trace}\left[\mathring{\Psi}[M(\xi)]G^{\mathsf{T}}(x,t)C^{-1}(x,t)G(x,t)\right]\\ &\quad -\operatorname{trace}\left[\mathring{\Psi}[M(\xi)]M(\xi)\right]\Big\}\,\bar{\xi}(\mathrm{d}x,\mathrm{d}t),\end{aligned} \tag{4.12}$$

where

$$\mathring{\Psi}[M(\xi)] = \left.\frac{\partial\Psi(M)}{\partial M}\right|_{M=M(\xi)}.$$

Hence, the function $\psi(x,t,\xi)$ takes the form

$$\psi(x,t,\xi) = \varsigma(\xi) - \phi(x,t,\xi), \tag{4.13}$$

where

$$\varsigma(\xi) = -\operatorname{trace}\left[\mathring{\Psi}[M(\xi)]M(\xi)\right] \tag{4.14}$$

and

$$\phi(x,t,\xi) = -\operatorname{trace}\left[G^{\mathsf{T}}(x,t)\mathring{\Psi}[M(\xi)]G(x,t)C^{-1}(x,t)\right]. \tag{4.15}$$

At this point, the following necessary and sufficient condition for optimality can be formulated.

Theorem 4.1. *A design $\xi^\star$ is optimal iff*

$$\int_T \min_{x\in X}\psi(x,t,\xi^\star)\,\xi(\mathrm{d}t) = 0. \tag{4.16}$$

Proof. Since the explicit form of the FIM is not crucial for the proof, the studied generalization to multi-response systems can be obtained by following the reasoning of the relevant result from [281, Th. 4.3, p. 125]. First, the convexity of Ψ yields a necessary and sufficient condition for the optimality of $\xi^\star$:

$$\begin{aligned}\inf_{\bar{\xi}\in\Xi(X)}\delta_+\Psi\big(M(\xi^\star),M(\bar{\xi})\big) &= \inf_{\bar{\xi}\in\Xi(X)}\operatorname{trace}\left[\mathring{\Psi}[M(\xi^\star)]\big(M(\bar{\xi})-M(\xi^\star)\big)\right]\\ &= \operatorname{trace}\left[\mathring{\Psi}[M(\xi^\star)]\big(M(\xi^\star)-M(\xi^\star)\big)\right] = 0.\end{aligned} \tag{4.17}$$

Setting $x^\star(t) = \arg\min_{x\in X} \psi(x,t,\xi^\star)$, we get

$$\begin{aligned}\int_T \psi(x^\star(t),t,\xi^\star)\,\xi(\mathrm{d}t) &= \int_T \left\{ \int_X \psi(x,t,\xi^\star)\delta\big(x-x^\star(t)\big)\,\mathrm{d}x \right\} \xi(\mathrm{d}t) \\ &\geq \inf_{\bar\xi} \int_T \left\{ \int_X \psi(x,t,\xi^\star)\,\bar\xi(\mathrm{d}x|t) \right\} \xi(\mathrm{d}t) \\ &\geq \inf_{\bar\xi} \int_T \left\{ \int_X \min_{x\in X} \psi(x,t,\xi^\star)\,\bar\xi(\mathrm{d}x|t) \right\} \xi(\mathrm{d}t) \qquad (4.18) \\ &= \inf_{\bar\xi} \int_T \psi(x^\star(t),t,\xi^\star) \left\{ \int_X \bar\xi(\mathrm{d}x|t) \right\} \xi(\mathrm{d}t) \\ &= \int_T \psi(x^\star(t),t,\xi^\star)\,\xi(\mathrm{d}t).\end{aligned}$$

Hence

$$\inf_{\bar\xi} \int_T \left\{ \int_X \psi(x,t,\xi^\star)\,\bar\xi(\mathrm{d}x|t) \right\} \xi(\mathrm{d}t) = \int_T \psi(x^\star(t),t,\xi^\star)\,\xi(\mathrm{d}t). \qquad (4.19)$$

Combining the last equation with (4.17) and (4.12) gives (4.16), which completes the proof. ■

In such a way, it is possible to formulate the following version of the general equivalence theorem.

Corollary 4.1. *The following statements are equivalent:*

(i) the design $\xi^\star$ *minimizes* $\Psi[M(\xi)]$,

(ii) the design $\xi^\star$ *minimizes* $\int_T \max_{x\in X} \phi(x,t,\xi)\,\bar\xi(\mathrm{d}t) - \varsigma(\xi)$, *and*

(iii) $\int_T \max_{x\in X} \phi(x,t,\xi^\star)\bar\xi(\mathrm{d}t) = \varsigma(\xi^\star)$.

Corollary 4.1 constitutes a generalization of Rafajowicz's result for D-optimum designs [232, Th. 1]. Moreover, in [232] some further sufficient optimality conditions are given, whose use allows us to reduce the problem to a series of optimization tasks for each time moment separately.

4.1.2 Parametrization of Trajectories

The main drawback of the approach outlined in the previous section is that the resulting trajectories are only guaranteed to be measurable in the Lebesgue sense according to the very abstract form of Definition 4.1. This may be insufficient in real applications, but can be avoided by appropriate parametrization of the trajectories. In such a manner, some additional conditions can be imposed on the regularity of the solutions and, additionally, a finite-dimensional form of the optimization problem can be provided.

To focus our attention, assume that the sensor trajectories can be approximated by parametric curves of the form

$$x^j(t) = \eta(t, \beta^j), \quad t \in T, \tag{4.20}$$

where β^j is a $\widetilde{m}$-dimensional constant parameter vector belonging to a compact set $\widetilde{B}$. Here η denotes a known function which is assumed to be sufficiently flexible to approximate the trajectory of the j-th sensor. Moreover, it is required that for any values of β^j and t, the functions $\eta(\,\cdot\,, \beta^j)$ and $\eta(t, \,\cdot\,)$ be continuous. Restricting our attention to the trajectories lying completely inside the set of admissible locations X (i.e., each point of the trajectory belongs to X), we assume the existence of a nonempty set

$$B = \{\beta \in \widetilde{B} : \eta(t, \beta) \in X, \forall t \in T)\}, \tag{4.21}$$

which is rather natural and not particularly restrictive in practice. It can be easily verified that B is also compact.

Taking into account that the independent measurements taken by different sensors lead potentially to replicated trajectories, it is easy to derive the form of the average FIM, being the counterpart of (4.5),

$$M(\xi_N) = \frac{1}{t_f} \int_0^{t_f} \left\{ \sum_{i=1}^{\ell} p_i(t) G^\mathsf{T}\big(\eta(t, \beta^i), t\big) C^{-1}\big(\eta(t, \beta^i), t\big) G\big(\eta(t, \beta^i), t\big) \right\} \mathrm{d}t, \tag{4.22}$$

with a new formulation of the continuous design,

$$\xi_N = \begin{Bmatrix} \beta^1 & \beta^2 & \dots & \beta^\ell \\ p_1 & p_2 & \dots & p_\ell \end{Bmatrix}, \tag{4.23}$$

$p_i = r_i/N$, r_i being the number of sensors moving along the i-th curve.

In terms of designs interpreted as generalized probability measures ξ for all Borel sets of B including single points, we can replace (4.22) with

$$M(\xi) = \int_B \Upsilon(\beta)\, \xi(\mathrm{d}\beta), \tag{4.24}$$

where

$$\Upsilon(\beta) = \frac{1}{t_f} \int_0^{t_f} G^\mathsf{T}\big(\eta(t, \beta), t\big) C^{-1}\big(\eta(t, \beta), t\big) G\big(\eta(t, \beta), t\big)\, \mathrm{d}t. \tag{4.25}$$

If the optimal design $\xi^\star$ is understood as the one which minimizes the scalar measure $\Psi[M(\xi)]$, it is clear that the form of the reformulated problem is exactly the same as that in Section 2.1.3 (p. 18) with β, B and $G^{\mathsf{T}}(\eta(t,\beta),t)$ substituted for x, X and $F(x,t)$, respectively. As a result, the corresponding theory, characterizations of solutions and other results remain valid in this context. Furthermore, note that the dimension of this optimization problem is $\widetilde{m}$, namely the dimension of vector β.

In addition to dimensionality reduction, the approximation (4.20) offers the possibility of imposing various constraints on the trajectories in a relatively easy manner, as will be shown in subsequent sections, where some limitations on the path length or/and velocity along the trajectories will be introduced. Moreover, proper parametrization clears the way for direct applicability of the algorithms presented in the previous chapter for scanning sensor scheduling. Especially, a Wynn–Fedorov procedure embodied in Algorithm 3.5 can be effectively exploited for this purpose [180] as will be shown further in the chapter.

However, in spite of the relative simplicity of the ideas presented above, the main drawback of the outlined approach still remains a high computational cost. To provide a sufficient approximation of the trajectories, the size of the parameter vector β has to be much larger than in the case of stationary sensors, where the size of the design variable was simply equal to the number of spatial coordinates. Consequently, the computational burden connected with the proposed approach usually remains quite heavy. Nevertheless, such a cost is unavoidable when trying to increase the degree of freedom regarding optimality. It is then necessary to search for and to develop alternative approaches which would minimize the numerical effort spent on finding a solution.

At this juncture, in order to illustrate the potential benefits of applying mobile sensors in implementation of the observational strategies, let us consider the following example.

Example 4.1. For the sake of comparison with the case of stationary sensors, consider a signal propagation process over an infinite one-dimensional transmission line over the time interval $T = [0,1]$ governed by the homogeneous wave equation of the form

$$\frac{\partial^2 y(x,t)}{\partial t^2} = \theta^2 \nabla^2 y(x,t), \quad -\infty \le x \le \infty, \quad 0 < t \le 1, \tag{4.26}$$

subject to the initial conditions

$$\begin{cases} y(x,0) = \tanh(x) & -\infty \le x \le \infty, \\ \left.\dfrac{\partial y(x,t)}{\partial t}\right|_{t=0} = \theta(\operatorname{sech}(x))^2, & -\infty \le x \le \infty, \end{cases} \tag{4.27}$$

where $\operatorname{sech}(\,\cdot\,)$ denotes a hyperbolic secant function (inverse of hyperbolic cosine). The initial Cauchy problem above describes the propagation of a voltage signal with amplitude y and velocity θ along the line. From a practical point of view, the assumption of an infinite length of the line is not too abstract as the distributed nature of such a system appears when its dimensions are much greater than the wavelength of the signal.

The problem considered possesses a closed-form solution of the fundamental form

$$y(x,t) = \tanh(x - \theta t). \tag{4.28}$$

Our task consists in finding the best allocation for stationary sensors and then trajectories of mobile sensors which assure the best estimate of the velocity θ, whose nominal value is assumed to be equal to $\theta^0 = 10.0$. The identification of velocity θ in this case can be performed using only one sensor. Hence, the FIM takes the form of the scalar

$$M(x^1(t)) = \int_0^{t_f} \left(\frac{\partial y(x^1(t), t; \theta)}{\partial \theta} \right)^2_{\theta=\theta^0} \mathrm{d}t = \int_0^1 t^2 \operatorname{sech}^4(x^1(t) - 10t)\,\mathrm{d}t. \tag{4.29}$$

Since the FIM is positive, the minimum of any optimality criterion defined on the FIM corresponds to its maximum value. To determine an optimal trajectory $x^1(\cdot)$, we have to solve the variational problem of maximizing the functional (4.29) with respect to the motion curve $x^1(\cdot)$. Fortunately, since the integrand is positive for any t, the maximum of (4.29) is achieved for a trajectory which follows the maximum of the integrand along time, and it is easy to verify that this is attained for the curve

$$x^{1\star}(t) = \theta t, \tag{4.30}$$

which constitutes the exact optimal solution. The value of the FIM in this case is $M(x^{1\star}(t)) = \frac{1}{3}$.

A solution for the case of a stationary sensor can be obtained by reducing the trajectory to a constant function independent of time, i.e., $x^1(t) = x^1$. Then we obtain the one-dimensional optimization problem which does not have a solution in a simple closed form, but it can be easily achieved numerically. Then a location

$$x^{1\star} \approx 9.08637$$

is an approximate solution with the criterion value $M(x^{1\star}) \approx 0.103134$ which is three times less than the corresponding value for the movable sensor strategy. This means that instead of one mobile sensor we have to use three stationary sensors in order to achieve the same identification accuracy. Such interpretation shows why it is worth paying attention to mobile sensors through more flexible observation strategies. ★

4.2 Optimal-Control Approach to Sensor Routing

In practical realizations of mobile DSNs based on different types of vehicles conveying the sensors (UGVs or UAVs), a trajectory planning task requires imposing numerous conditions describing both the technological limits and additional control goals for robotic platforms used. First of all, the dynamics of the vehicles must be taken into account as well as various geometric constraints on sensor movements (induced, e.g., by the admissible measurement regions and allowable distances between the sensors). Finally, in many situations, some additional control objectives have to be satisfied simultaneously to achieving high identification quality (related, e.g., to sensor distribution, power consumption or limited path lengths). From this point of view, a reasonable and systematic approach is to convert the problem to an optimal-control formulation for which solutions can be obtained numerically using some existing effective numerical solvers.

The main purpose of this section is to describe some original results concerning off-line determination of moving sensor positions which maximize parameter-identification accuracy subject to various constraints imposed on sensor motions. More precisely, refinements of the optimal-control approach developed by Uciński [276, 280, 281] are provided with further extensions to different experimental settings related to limited resources of sensor nodes [182, 184, 193, 273, 284].

4.2.1 Limitations on Sensor Movements

4.2.1.1 Node dynamics

With no loss of generality, we assume that all sensors are conveyed by vehicles whose motions are described by

$$\dot{x}^j(t) = f(x^j(t), u^j(t)) \quad \text{a.e. on } T, \quad x^j(0) = x_0^j, \tag{4.31}$$

where a given function $f : \mathbb{R}^d \times \mathbb{R}^b \to \mathbb{R}^d$ is required to be continuously differentiable, $x_0^j \in \mathbb{R}^d$ defines an initial sensor position, and $u^j : T \to \mathbb{R}^b$ is a measurable control function which satisfies

$$u_l \leq u^j(t) \leq u_u \quad \text{a.e. on } T \tag{4.32}$$

for some constant vectors u_l and u_u, $j = 1, \ldots, N$.

For simplicity of notation, let us write

$$q(t) = \begin{bmatrix} x^1(t) \\ \vdots \\ x^N(t) \end{bmatrix}, \quad u(t) = \begin{bmatrix} u^1(t) \\ \vdots \\ u^N(t) \end{bmatrix}, \quad q_0 = \begin{bmatrix} x_0^1 \\ \vdots \\ x_0^N \end{bmatrix}, \quad \forall t \in T. \tag{4.33}$$

Then, in lieu of (4.31), we shall subsequently use a vector system of ODEs

$$\dot{q}(t) = f(q(t), u(t)) \quad \text{a.e. on } T, \quad q(0) = q_0, \tag{4.34}$$

where

$$h(q(t), u(t)) = \begin{bmatrix} f(x^1(t), u^1(t)) \\ \vdots \\ f(x^N(t), u^N(t)) \end{bmatrix}. \tag{4.35}$$

Given any initial sensor configuration q_0 and any control function, there is a unique, absolutely continuous function $q : T \to \mathbb{R}^d$ which satisfies (4.34) a.e. on T. In what follows, we will call it the state trajectory corresponding to q_0 and u, and use the following notational convention: If q appears without a mention in a formula, it is always understood that a control u and an initial condition q_0 have been specified and q is the trajectory corresponding to u and q_0 through (4.34).

In particular, there exist various choices for the structure of (4.34) in the literature, including different forms of first and second order linear equations [281]. However, in the special case, where the sensor dynamics is not of paramount importance and attention is focused on the trajectories themselves, the following variant may be exploited:

$$\dot{q}(t) = u(t), \quad q(0) = q_0, \tag{4.36}$$

(i.e., control of the sensor velocities), which ensures the compromise between generality of description and flexibility in applications [207, 281].

4.2.1.2 Pathwise State Constraints

In reality, some restrictions on the motions are inevitably induced. First of all, all sensor network nodes should stay within the admissible region Ω_{ad} where measurements are allowed. We assume that it is a compact set defined as follows:

$$\Omega_{\text{ad}} = \left\{ x \in \bar{\Omega} : b_i(x) \le 0,\ i = 1, \dots, I \right\}, \tag{4.37}$$

where b_is are given continuously differentiable functions. Accordingly, the conditions

$$\lambda_{ij}(q(t)) = b_i(x^j(t)) \le 0, \quad \forall t \in T \tag{4.38}$$

must be fulfilled, where $1 \le i \le I$ and $1 \le j \le N$.

Optionally, to avoid collisions, we can directly restrict the admissible distances between the sensors by imposing the constraints

$$\mu_{ij}(q(t)) = D^2 - \|x^i(t) - x^j(t)\|^2 \leq 0, \quad \forall\, t \in T, \tag{4.39}$$

where $1 \leq i < j \leq N$ and D stands for a minimum allowable distance. The same constraints may be used to guarantee that the measurements taken by the sensors can be considered independent, thus providing a replication-free observation schedule.

To shorten the notation, after relabeling, we rewrite the constraints (4.38) and (4.39) in the form

$$\gamma_\ell(q(t)) \leq 0, \quad \forall\, t \in T, \tag{4.40}$$

where γ_ℓ, $\ell = 1, \dots, IN$ tally with (4.38), whereas γ_ℓ, $\ell = IN + 1, \dots, [I + (N-1)/2]N$ coincide with (4.39). In the sequel, $\bar{\nu}$ stands for the set of indices $\{1, \dots, [I + (N-1)/2]N\}$.

4.2.2 Optimal Control Formulation

The goal in the optimal measurement problem is to determine the forces (controls) applied to each vehicle conveying a sensor, which maximize the design criterion $\Psi[\,\cdot\,]$ defined on the FIM of the form (2.85), and are determined unequivocally by the corresponding trajectories, subject to the constraints (4.32) on the magnitude of the controls and induced state constraints (4.40). In general, apart from the control u, an initial sensor configuration q_0 can also be treated as a design variable, hence increasing the degree of optimality. Clearly, the correctness of such a formulation necessitates some additional restrictions on the smoothness of sensitivity coefficients G in (2.86). In what follows, we thus assume the continuity of G, and so $\partial G/\partial x$.

The above formulation can be interpreted as the following optimization problem.

Problem 4.1 (Optimal sensor routing). Find the pair (q_0, u) which minimizes

$$\mathcal{J}_1(q_0, u) = \Psi[M(q)] \tag{4.41}$$

over the set of feasible pairs

$$\mathcal{Q} = \big\{(q_0, u) \mid u : T \to \mathbb{R}^{bN} \text{ is measurable,} \\ u_l \leq u(t) \leq u_u \text{ a.e. on } T,\ q_0 \in \Omega_{\mathrm{ad}}^N\big\}, \tag{4.42}$$

subject to the pathwise state inequality constraints (4.40).

Evidently, the high nonlinearity of the problem above excludes any possibility of finding closed-form formulas for its solution. Accordingly, we must resort to numerical techniques. However, before trying to apply any optimization procedure directly, observe that, in spite of its apparently nonclassical form, the resulting optimal-control problem can be easily cast as a classical Mayer problem where the performance index is defined only via terminal values of state variables.

4.2.3 Equivalent Canonical Problem

The aim of this section is to convert our problem into a canonical optimal control one with an endpoint cost and inequality-constrained trajectories, see [214]. Such a transcription will make it possible to employ existing software packages for numerically solving dynamic optimization and optimal control problems.

For notational convenience, define the function svec : $\mathbb{S}^m \to \mathbb{R}^{m(m+1)/2}$ that takes the lower triangular part (elements only on the main diagonal and below) of a symmetric matrix A and stacks them columnwise into a vector a:

$$\begin{aligned} a &= \operatorname{svec}(A) \\ &= [A_{11}, A_{21}, \dots, A_{m1}, A_{22}, A_{32}, \dots, A_{m2}, \dots, A_{mm}]^{\mathsf{T}}. \end{aligned} \tag{4.43}$$

Similarly, let $A = \operatorname{Smat}(a)$ be a symmetric matrix such that $\operatorname{svec}(\operatorname{Smat}(a)) = a$ for any $a \in \mathbb{R}^{m(m+1)/2}$.

Consider the matrix valued function

$$\Pi(q(t), t) = \sum_{j=1}^{N} G^{\mathsf{T}}(x^j(t), t) C^{-1}(x^j(t), t) G(x^j(t), t). \tag{4.44}$$

Setting $r : T \to \mathbb{R}^{m(m+1)/2}$ as the solution of the differential equations

$$\dot{r}(t) = \operatorname{svec}(\Pi(q(t), t)), \quad r(0) = 0, \tag{4.45}$$

we have

$$M(q) = \operatorname{Smat}(r(t_f)), \tag{4.46}$$

i.e., maximization of $\Psi[M(q)]$ reduces to maximization of a function of the terminal value of the solution to (4.45).

Introducing the augmented state vector

$$w(t) = \begin{bmatrix} q(t) \\ r(t) \end{bmatrix}, \tag{4.47}$$

we obtain

$$w_0 = w(0) = \begin{bmatrix} q_0 \\ \mathbf{0} \end{bmatrix}. \tag{4.48}$$

Then the equivalent canonical optimal control problem can be cast as follows.

Problem 4.2 (Optimal sensor routing in Mayer form). Determine a pair $(w_0, u) \in \overline{\mathcal{Q}}$ which minimizes the performance index

$$\mathcal{J}_2(w_0, u) = \overline{\Psi}(w(t_f)) \tag{4.49}$$

subject to

$$\begin{cases} \dot{w}(t) = \overline{h}(w(t), u(t), t), \\ w(0) = w_0, \\ \overline{\gamma}_\ell(w(t)) \le 0, \quad \forall\, t \in T, \quad \ell \in \overline{\nu}, \end{cases} \tag{4.50}$$

where

$$\overline{\mathcal{Q}} = \big\{(w_0, u) \mid u : T \to \mathbb{R}^{bN} \text{ is measurable}, \\ u_l \le u(t) \le u_u \text{ a.e. on } T,\ q_0 \in \Omega_{\text{ad}}^N\big\}, \tag{4.51}$$

and

$$\overline{h}(w, u, t) = \begin{bmatrix} h(q(t), u(t)) \\ \operatorname{svec}(\Pi(q(t), t)) \end{bmatrix}, \tag{4.52}$$

$$\overline{\Psi}(w(t)) = \Psi[\operatorname{Smat}(r(t))], \tag{4.53}$$

$$\overline{\gamma}_\ell(w(t)) = \gamma_\ell(q(t)). \tag{4.54}$$

The above problem in canonical form is a standard problem studied extensively in most works on optimal control in the presence of state inequality constraints (see [78, 94, 226] for details regarding existence and optimality conditions) and can be readily solved using one of the existing packages for numerically solving dynamic optimization problems, such as RIOTS_95, cf. [248], DIRCOL, cf. [261], or MISER, cf. [111]. In the context of this work, the first package, i.e., RIOTS_95, is of particular importance since it has been successfully employed for wide spectrum of mobile sensors routing problems. It is designed as a MATLAB toolbox written mostly in C and runs under Windows 2000/XP/Vista and Linux. It provides an interactive environment for solving a very broad class of optimal control problems. The input problems can be prepared purely as M-files and no compiler is required to solve them. To speed up the solution process, the functions defining the problem can be coded in C and then compiled and linked with some pre-built linking libraries. The implemented numerical methods are supported by the theory outlined by [214], which uses the approach of consistent approximations. The system dynamics can be integrated with fixed step-size Runge–Kutta integration, a discrete-time solver or a variable step-size method. The software automatically computes gradients for all functions with respect to the controls and any free initial conditions. The controls are represented as splines, which allows a high-degree accuracy of function approximation without requiring a large number of control parameters. There are three main optimization routines, each suited for different levels of generality, and the most general one is based on sequential quadratic programming methods (it was also used in our computations reported at the end of the chapter).

4.2.4 Resource Aware Problem Extensions

From a practical point of view, apart from the monitoring of distributed system, a crucial issue in administration of any sensor network is to maintain it in a fully operational state. This can be achieved only through proper management of network resources. Therefore, Problem 4.2 requires further extensions toward the possibility of simultaneous assessment of additional resource-related objectives. Motivations come from technical limitations imposed on the vehicles conveying the measurement equipment. These are inherent to mobile platforms carrying sensors, which are supplied with power from batteries and characterized by a limited range of wireless communication. Unfortunately, researchers' attention is predominantly focused only on the achieved precision while neglecting the problem of realizability of the produced solutions. Thus, the main aim of this section is to produce results which can be useful when such factors come into consideration. In particular, generalizations of Problem 4.2 concerning efficient area coverage, limited distances travelled by the nodes and limited energy of motion are discussed.

4.2.4.1 Guaranteed Efficiency Constraints

Sometimes, resource limits raise the question of how to minimize their consumption while guaranteeing an acceptable quality level of measurement data. This leads to alternative formulation of our control problem based on the notion of the so-called *Ψ-efficiency*, which quantifies the suboptimality of given solution. Setting a reasonable positive threshold $c < 1$, we can introduce an efficiency constraint

$$E_{\Psi}(q) \geq c, \tag{4.55}$$

which guarantees a suboptimal yet reasonable solution. By analogy to the classical optimum experimental design (cf. [9, 307]) for some particular criteria this measure can be defined as follows.

D-efficiency. In the case of D-optimality, we have

$$E_D(q) = \left\{ \frac{\det(M(q))}{\det(M(\widehat{q}))} \right\}^{1/m}, \tag{4.56}$$

where $\widehat{q}$ stands for the D-optimal trajectories obtained for the observations with no constraints imposed on the resources of sensor nodes (hence they offer a greater degree of optimality). The value of $\det(M(\widehat{q}))$ can be determined in advance, therefore it can be easily seen that the efficiency constraint (4.55) is equivalent to the constraint

$$\Psi[M(q)] \leq \Psi_c, \tag{4.57}$$

where $\Psi_c = \Psi[M(\widehat{q})] - m\log(c)$.

Taking the m-th root of the ratio of the determinants in (4.56) results in an efficiency measure which has the dimensions of variance, irrespective of the dimension of the model [9]. So two replicates of an experiment with trajectories for which $E_D = 0.5$ would be as efficient as one replicate of optimum sensor routing.

A-optimality. In this case, the efficiency is defined by

$$E_A(q) = \left\{ \frac{\operatorname{trace}(M^{-1}(\widehat{q})}{\operatorname{trace}(M^{-1}(q))} \right\}. \tag{4.58}$$

Since here the efficiency has the dimension of variance, normalization with the m-th root is not necessary and we have accordingly that (4.55) is equivalent in this case to

$$c\Psi[M(q)] \geq \Psi[M(\widehat{q})]. \tag{4.59}$$

G-optimality. For this criterion, we have

$$E_G(q) = \left\{ \frac{\int_T \max_{x\in X} \phi(x,t,\widehat{g})\,\bar{\xi}(\mathrm{d}t)}{\int_T \max_{x\in X} \phi(x,t,q)\,\bar{\xi}(\mathrm{d}t)}, \right\} \tag{4.60}$$

and (4.55) is equivalent to

$$\Psi[M(q)] \leq \frac{1}{c}\Psi[M(\widehat{q})]. \tag{4.61}$$

4.2.4.2 Distribution of Sensors

Owing to the assumption of independent observations, a crucial problem which arises when trying to maximize the estimation accuracy is that the trajectories may coincide and sensors tend to be very close to each other. Therefore, it is required to provide a suitable distribution of sensors over the domain considered through an appropriate criterion, forcing the required behavior of sensor nodes. This task is also related to the problem of proper coverage of the domain with the trajectories so as to gather the data from possibly large number of different locations. Since in this framework we are particularly interested in avoiding the effect of sensor clusterization (which also can be understood in terms of collision avoidance), our primary control objective can be created based on the average minimal distance between movable sensors over the observation interval

$$\Phi(q) = \frac{1}{t_f} \int_0^{t_f} \min_{\substack{(i,\,j)\\ i \neq j}} \|x^i(t) - x^j(t)\|^2 \, dt. \tag{4.62}$$

Then, as an additional termination state constraint, the performance index based on the Ψ-efficiency can be applied in order to guarantee the required estimation accuracy according to (4.55). Consequently, this leads to the following formulation of the problem.

Problem 4.3. Determine the pair $(q_0, u) \in \mathcal{Q}$, which maximizes

$$\mathcal{J}_1(q_0, u) = \Phi(q), \tag{4.63}$$

subject to the constraints (4.55) and (4.40).

Although the criterion (4.63) representing an average minimal distance between sensors is quite suitable for our control objectives and is less conservative than the 'worst case' minimax criterion, it is clear that the nondifferentiability of the integrand leads to serious difficulties during the minimization, as common nonlinear programming packages have strong requirements regarding the smoothness of the cost functions (e.g., continuous differentiability). Obviously, it is possible to use some specialized procedures, but an alternative solution, which is very suitable here, consist in replacing the original nonsmooth problem by minimization of a smooth function being an approximation to the minimal distance. Being more precise, it is possible to exchange the integrand in (4.63) by a smooth symmetric exponential penalty function [214, p. 248]

$$f_\delta(s) = -\delta \log\Big(\sum_{\substack{(i,\,j)\\ i \neq j}} \exp^{-\|x^i(t) - x^j(t)\|^2/\delta} \Big), \tag{4.64}$$

where $\delta > 0$ is a parameter. It is a C^∞ convex function and it is easy to check that it possesses the uniform approximation property to the integrand of Φ (see [214, p.249] or [281, p.292] for further details).

Equivalent Lagrange problem. Note that the sensor routing defined by Problem 4.3 can be cast as an optimal-control problem in Lagrange form. Indeed, making use of (4.44) and (4.45), we obtain the following equivalent formulation.

Problem 4.4. Find the pair $(w_0, u) \in \overline{\mathcal{Q}}$, which minimizes

$$\mathcal{J}_2(w_0, u) = -\Phi(q), \tag{4.65}$$

subject to the constraints (4.55) and (4.50).

Thus, we have the classical control-optimal problem in Lagrange form, which from the qualitative point of view is of the same level of difficulty as Problem 4.2 and can be effectively solved using dedicated numerical solvers.

4.2.4.3 Limited Path Lengths

If the motion capabilities of the nodes are severely limited due to the limited battery power, one potential policy is to be primarily concerned with restrictions imposed on nodes' path lengths. Two formulations are thus proposed: the first for a specified limit for the longest path, and the other for the case in which the longest path length constitutes a decision variable which leads to minimax formulation. It is also shown that these formulations can be transformed into equivalent optimal control problems in canonical Mayer form.

First, note that the distance travelled by the j-th sensor over the interval $[0, t]$ can be expressed as

$$s_j(t) = \int_0^t \|\dot{x}^j(\tau)\| \, \mathrm{d}\tau = \int_0^t \|f(x^j(\tau), u^j(\tau))\| \, \mathrm{d}\tau, \tag{4.66}$$

where $\|\cdot\|$ signifies the Euclidean norm.

Trajectory design with hard constraints on path lengths. The distances travelled by sensor nodes are a critical factor especially in the context of a cooperative mobile sensor network formed from a number of wheeled mobile robots (e.g., differential drives, synchronous drives, etc.), if a major problem in the design is the power consumption by the robots. Then we demand that the lengths of the trajectories do not exceed a given limit $s_{\max}$, i.e., we require that

$$s_j(t_f) \le s_{\max}, \quad j = 1, \dots, N. \tag{4.67}$$

Since sensor trajectories q are unequivocally determined as solutions to the state equation (4.34), the above control problem can be interpreted as the following optimization problem.

Problem 4.5. Find the pair $(q_0, u) \in \mathcal{Q}$, which maximizes

$$\mathcal{J}_1(q_0, u) = \Psi[M(q)], \tag{4.68}$$

subject to the constraints (4.34), (4.40) and (4.67).

Minimum-length trajectories with guaranteed efficiency. Making use of the concept of Ψ-efficiency, we can formulate the following version of Problem 4.5.

Problem 4.6. Find the pair $(q_0, u) \in \mathcal{Q}$ which minimizes

$$\mathcal{J}_2(q_0, u) = \max_{j=1,\dots,N} s_j(t_f) \tag{4.69}$$

subject to the constraints (4.34), (4.40) and (4.55).

The above minimax problem is slightly more difficult than Problem 4.5. Therefore, in what follows, we shall fix our attention on Problem 4.6 only, as Problem 4.5 can be addressed in much the same way.

Equivalent canonical Mayer problem. Again, it is clear that Problems 4.5 and 4.6 are both highly nonlinear and we are not capable of finding closed-form formulae for their solutions. Accordingly, we must resort to numerical techniques.

The distances s_j travelled by sensor nodes, cf. (4.66), are easily incorporated into the usual optimal control formulation by augmenting the system dynamics with additional states,

$$s(t) = [s_1(t), \dots, s_N(t)]^\top, \tag{4.70}$$

being the solutions of the Cauchy problems

$$\dot{s}(t) = g(q(t), u(t)), \quad s(0) = 0, \tag{4.71}$$

where

$$g(q(t), u(t)) = \begin{bmatrix} \|f(x^1(t), u^1(t))\| \\ \vdots \\ \|f(x^N(t), u^N(t))\| \end{bmatrix}. \tag{4.72}$$

Defining the augmented state vector

$$w(t) = \begin{bmatrix} q(t) \\ r(t) \\ s(t) \end{bmatrix}, \tag{4.73}$$

we obtain

$$w_0 = w(0) = \begin{bmatrix} q_0 \\ \mathbf{0} \\ \mathbf{0} \end{bmatrix}. \tag{4.74}$$

Then the equivalent canonical optimal control problem is as follows.

Problem 4.7. Find a triple $(w_0, u, \varrho) \in \overline{\mathcal{Q}} \times \mathbb{R}_+$ which minimizes the performance index

$$\mathcal{J}(w_0, u, \varrho) = \varrho \tag{4.75}$$

subject to

$$\begin{cases} \dot{w}(t) = \overline{h}(w(t), u(t), t), \\ w(0) = w_0, \\ \overline{\Psi}(w(t_f)) \le \Psi_c, \\ \overline{\gamma}_\ell(w(t)) \le 0, \quad \forall t \in T, \quad \ell \in \bar{\nu}, \\ \overline{\omega}_j(w(t_f)) \le \varrho, \quad j = 1, \dots, N, \end{cases} \tag{4.76}$$

where

$$\overline{h}(w,u,t) = \begin{bmatrix} h(q(t),u(t)) \\ \operatorname{svec}(\Pi(q(t),t)) \\ g(q(t),u(t)) \end{bmatrix}, \tag{4.77}$$

$$\overline{\Psi}(w(t)) = \Psi[\operatorname{Smat}(q(t))], \tag{4.78}$$

$$\overline{\gamma}_\ell(w(t)) = \gamma_\ell(q(t)), \tag{4.79}$$

$$\overline{\omega}_j(w(t)) = s_j(t). \tag{4.80}$$

4.2.4.4 Constraints on Motion Energy

Another idea for controlling a limited power consumption related to a particular realization of sensor motions is to concentrate on the constraints imposed on the motion energy of individual nodes. Similarly as in the preceding section, two formulations are also proposed to cover this issue: one for a specified limit for the maximum energy per sensor node, and the other for the case in which the highest energy constitutes a decision variable which leads to the minimax formulation.

In the following, we shall focus on the restricted power supplies and assume that energy consumption can be expressed in terms of physical work put into sensor motion. The energy transfer related to the translational kinetic energy of the j-th sensor over the interval $[0,t]$ is proportional to the square velocity of motion,

$$e_j(t) = \int_0^t \|\dot{x}^j(\tau)\|^2 \, \mathrm{d}\tau = \int_0^t \|f(x^j(\tau),u^j(\tau))\|^2 \, \mathrm{d}\tau, \tag{4.81}$$

where $\|\cdot\|$ signifies the Euclidean norm. Note that, if the trajectories are of our primary interest and a simple form of sensor dynamics is applied as in (4.36), the energy can be expressed in terms of the energy of the control function,

$$e_j(t) = \int_0^t \|u^j(\tau)\|^2 \, \mathrm{d}\tau. \tag{4.82}$$

Trajectory design with direct constraints on energy. Since the energy transferred into sensor motions is a critical factor, we demand that the maximum kinetic energy level do not exceed a given limit $e_{\max}$, i.e., we require that

$$e_j(t_f) \le e_{\max}, \quad j = 1,\dots,N. \tag{4.83}$$

On the analogy to the preceding section, the control problem can be interpreted as an optimization problem.

Problem 4.8. Find the pair $(q_0,u) \in \mathcal{Q}$ which minimizes

$$\mathcal{J}_1(q_0,u) = \Psi[M(q)] \tag{4.84}$$

subject to the constraints (4.34), (4.40) and (4.83).

Minimum-energy trajectories with guaranteed efficiency. Again, making use of the Ψ-efficiency we are able to reformulate Problem 4.8.

Problem 4.9. Find the pair $(q_0, u) \in \mathcal{P}$ which minimizes

$$\mathcal{J}_2(q_0, u) = \max_{j=1,\dots,N} e_j(t_f) \tag{4.85}$$

subject to the constraints (4.34), (4.40) and (4.55).

Equivalent canonical optimal control problem. As in the case of path lengths, for the sake of clarity, we shall fix our attention on the more difficult Problem 4.9 only, as Problem 4.8 can be addressed analogously.

The motion energies e_j of sensor nodes, cf. (4.81), can be incorporated into the optimal control formulation by augmenting the system dynamics with additional states,

$$e(t) = [e_1(t), \dots, e_N(t)]^\mathsf{T}, \tag{4.86}$$

being the solutions of the Cauchy problems

$$\dot{e}(t) = g(q(t), u(t)), \quad e(0) = 0, \tag{4.87}$$

where

$$g(q(t), u(t)) = \begin{bmatrix} \|f(x^1(t), u^1(t))\|^2 \\ \vdots \\ \|f(x^N(t), u^N(t))\|^2 \end{bmatrix}. \tag{4.88}$$

Defining the augmented state vector and its initial value as

$$w(t) = \begin{bmatrix} q(t) \\ r(t) \\ e(t) \end{bmatrix}, \qquad w_0 = w(0) = \begin{bmatrix} q_0 \\ \mathbf{0} \\ \mathbf{0} \end{bmatrix}, \tag{4.89}$$

the equivalent Mayer optimal control problem has the same formulation as Problem 4.7, where instead of the condition 4.80 we have

$$\overline{\omega}_j(w(t)) = e_j(t).$$

4.3 Trajectory Design for Heterogeneous Mobile Sensors

In the literature on mobile sensors, it is most often assumed that the optimal measurement problem consists in the design of trajectories of a given number of identical sensors. In this section, the problem will be formulated in quite a different manner. First of all, apart from sensor controls and initial positions, the number of sensors constitutes an additional design variable. Additionally, we allow for different measurement accuracies of individual sensors, which are quantified by weights steering the corresponding measurement variances. This leads to a much more general formulation which most often produces an uneven allocation of experimental effort between different sensors [273]. The corresponding solutions could then be implemented on a sensor network with heterogeneous mobile nodes. It turns out that these solutions can be determined using convex optimization tools commonly applied in optimum experimental design. As a result, much better accuracies of the parameter estimates can be achieved.

4.3.1 Reinterpretation of Continuous Designs

There is another very important interpretation of sensor weights within the trajectory design (4.4) which has not been mentioned yet, although it is significant for movable observations and could be exploited here. Note that, instead of several mobile sensors whose accuracies are characterized by the same variance σ^2, we use sensors for which the variance of measurement errors is σ^2/p_j. This means that a large weight p_j indicates that the j-th sensor guarantees more precise measurements than sensors with lower weight values. With no loss of generality, we assume that the weights $p_j \in \mathbb{P}_N$, i.e., they belong to the probability simplex .

Based on the generalized weighted least-squares method (cf. Section 2.1.2), where the performance index

$$\mathcal{J}(\theta) = \frac{1}{2}\sum_{j=1}^{N}\int_T p_j[z^j(t) - \hat{y}(x^j(t), t; \theta)]^\mathsf{T} \Lambda^{-1}(t)[z^j(t) - \hat{y}(x^j(t), t; \theta)]\, \mathrm{d}t \tag{4.90}$$

is minimized, each weight can be interpreted as the reciprocal of the variance of the observation error along a given trajectory. Thus we may think of the weights as sensitivities of the measurement devices. Such an interpretation is not only very reasonable, but also very practical from an engineering point of view, as it allows us to include such a generalization into the analysis of the experiment without essential changes. Particularly, the direct approach and optimal control techniques discussed in the preceding sections can be effectively combined, resulting in a powerful approach to sensor routing capable to cope with real-world complex problems [273].

Control parametrization. Although measurement accuracies may vary from sensor to sensor, we assume that all sensors are conveyed by identical vehicles whose motions are described by (4.34) and are subject to the pathwise state constraints (4.40). Once the sensor dynamics is introduced, instead of parametrization of trajectories discussed in Section 4.2.2, it is convenient to introduce the proper parametrization of controls.

On the analogy to (4.20), we make the assumption that the controls of the available vehicles can be represented in the parametric form,

$$u^j(t) = \eta(t, \widetilde{\beta}^j), \quad t \in T, \tag{4.91}$$

where η denotes a given function such that $\eta(\,\cdot\,, \widetilde{\beta}^j)$ is continuous for each fixed $\widetilde{\beta}^j$ and $\eta(t, \,\cdot\,)$ is continuous for each fixed t, the constant parameter vector $\widetilde{\beta}^j$ ranging over a compact set $\widetilde{B} \subset \mathbb{R}^{\widetilde{m}}$. For instance, such parametrization can rely on using B-splines as employed in numerous optimal control solvers.

Based on this parametrization, we can define the mapping $\mathcal{X}$ which assigns every $\beta^j = (x_0^j, \widetilde{\beta}^j) \in \Omega_{\text{ad}} \times \widetilde{B}$ a trajectory $x^j = \mathcal{X}(\beta^j)$ through solving (4.31) for the initial position x_0^j and the control defined by (4.91).

Since only the controls and trajectories satisfying the imposed constraints are interesting, we introduce the set

$$\begin{aligned} B = \big\{\beta = (x_0, \widetilde{\beta}) \in \widetilde{B} \times \Omega_{\text{ad}} : \eta(\,\cdot\,, \widetilde{\beta}) \text{ satisfies (4.32)}, \\ \mathcal{X}(\beta) \text{ satisfies (4.38)}\big\}, \end{aligned} \tag{4.92}$$

and assume that it is nonempty. A trivial verification shows that B is also compact.

Given N sensors, we obtain trajectories x^j corresponding to vectors $\beta^j \in \mathbb{R}^{Nd+\widetilde{m}}$, $j = 1, \dots, N$. Applying the terminology of optimum experimental design, we can redefine the continuous design as

$$\xi_N = \begin{Bmatrix} \beta^1, & \beta^2, & \dots, & \beta^N \\ p_1, & p_2, & \dots, & p_N \end{Bmatrix}, \tag{4.93}$$

where $p_j \in \mathbb{P}_N$ and the FIM takes the form of (4.22). Analogously to Section 4.1.2, extending the design to a probability measure ξ for all Borel sets of B including single points, we can define the FIM in the same way as in (4.24).

4.3.2 Numerical Scheme for Heterogeneous Sensor Routing

Problem formulation and optimality results. For clarity, the following notational conventions are adopted: the symbol $\Xi(B)$ will be used to denote the set of all probability measures on B. Let us also introduce the notation $\mathbb{M}(B)$ for the set of all admissible information matrices, i.e.,

$$\mathbb{M}(B) = \big\{M(\xi) : \xi \in \Xi(B)\big\}. \tag{4.94}$$

Then we may redefine the optimal design as a solution to the optimization problem

$$\xi^\star = \arg \max_{\xi \in \Xi(B)} \Psi[M(\xi)]. \tag{4.95}$$

To characterize the solutions, the following counterparts of the results from Section 4.1.1 can be provided. Let us start with the sufficient and necessary conditions for optimality being a straightforward generalization of Theorem 4.1.

Corollary 4.2. *We have the following properties:*

(i) A design $\xi^\star$ is optimal iff

$$\max_{\beta \in B} \psi(\beta, \xi^\star) = 0, \tag{4.96}$$

where

$$\psi(\beta, \xi) = \varsigma(\xi) - \phi(\beta, \xi) \tag{4.97}$$

and

$$\phi(\beta, \xi) = -\operatorname{trace}\left[\mathring{\Psi}[M(\xi)]\Upsilon(\beta)\right], \qquad \varsigma(\xi) = -\operatorname{trace}\left[\mathring{\Psi}[M(\xi)]M(\xi)\right].$$

(ii) For any purely discrete optimal design $\xi^\star$, the function $\psi(\,\cdot\,, \xi^\star)$ has value zero at all support points.

Then, the following useful complement of the *equivalence theorem* can be given.

Corollary 4.3. *The following characterizations of an optimal design $\xi^\star$ are equivalent in the sense that each one implies the other two:*

(i) the design $\xi^\star$ minimizes $\Psi[M(\xi)]$,
(ii) the design $\xi^\star$ minimizes $\max_{\beta \in B} \phi(\beta), \xi) - \varsigma(\xi)$, and
(iii) $\max_{\beta \in B} \phi(\beta, \xi^\star) = \varsigma(\xi^\star)$.

The outlined results pave the way for direct adaptation of the first-order Wynn–Fedorov procedure derived by Algorithm 3.5 in the context of the scheduling of scanning sensors. Its appropriate modification for the case of mobile sensors is presented in a very simple scheme of Algorithm 4.1.

Computationally, line 5 is of crucial significance but at the same time it is the most time-consuming step in the algorithm. Complications arise, among other things, due to the necessity of calculating a global maximum of $\psi(\,\cdot\,, \xi)$, which is usually multimodal (getting stuck in one of local maxima leads to precocious termination of the algorithm). It turns out that this subproblem can be conveniently stated in the spirit of the optimal control approach presented in Section 4.2, which allows alleviating, in a sense, the difficulties mentioned above.

Algorithm 4.1. First-order Wynn–Fedorov algorithm for mobile routing

1: **procedure** FIRST_ORDER_SENSOR_ROUTING
2: Choose some positive tolerance $\epsilon \ll 1$. $r \leftarrow 0$
3: Guess an initial nondegenerate design measure $\xi^{(0)}$.
4: **repeat**
5: $\beta^{(r)} \leftarrow \arg\max_{\beta \in B} \psi(\beta, \xi^{(r)})$ ▷ Searching for a candidate path
6: Choose α_r according to the one of the rules (3.66)–(3.68)
7: $\xi^{(r+1)} \leftarrow (1-\alpha_r)\xi^{(r)} + \alpha_r \delta(\beta^{(r)})$ ▷ Correction of sensor accuracy
8: $r \leftarrow r+1$
9: **until** $\phi(\beta^{(r)}, \xi^{(r)}) \le \varsigma(\xi^{(r)}) + \epsilon$
10: **end procedure**

Optimal control formulation of the search for the candidate path. Line 5 of the delineated Wynn–Fedorov algorithm necessitates determination of $\arg\max_{\beta \in B} \psi(\beta, \xi^{(r)})$. This formulation can be interpreted as a finite-dimensional approximation to the following optimization problem.

Problem 4.10 (Search for the candidate path). Find the pair (x_0, u) which minimizes

$$\mathcal{J}(x_0, u) = \operatorname{trace}\left[\mathring{\Psi}[M(\xi^{(r)})] \left(\int_T G(x(t), t) \Lambda^{-1}(t) G^{\mathsf{T}}(x(t), t)\, \mathrm{d}t - M(\xi^{(r)}) \right) \right] \tag{4.98}$$

over the set of feasible pairs

$$\mathcal{Q} = \big\{ (x_0, u) \mid u : T \to \mathbb{R}^r \text{ is measurable}, \\ u_l \le u(t) \le u_u \text{ a.e. on } T,\ x_0 \in \Omega_{\mathrm{ad}} \big\}, \tag{4.99}$$

subject to the pathwise state inequality constraints (4.38).

Note that, at each iteration of Algorithm 4.1, Problem 4.10 is related to trajectory design for one sensor node only, which leads to solution of a sequence of relatively simple control problems. Consequently, this fact, combined with an easy implementation of the general scheme, results in an extremely effective approach to trajectory design compared with aggregated optimization of the controls for the whole sensor network at the same time.

4.4 Applications in Engineering

At this point, more examples of potential applications of the theory and methods proposed so far are provided. They have been selected so as to be relatively easy for interpretation (therefore the number of sensors nodes is quite low), but at the same time they are already far from triviality and may be used as prototypes for more sophisticated frameworks.

4.4.1 Heat Transfer

In this example the heat transfer process is considered over a two dimensional spatial domain. The phenomenon is governed by the diffusion equation

$$\frac{\partial y}{\partial t} = \nabla \cdot (\kappa \nabla y) + \zeta \tag{4.100}$$

for $x \in \Omega = (0,1)^2$ and $t \in [0,1]$, subject to homogeneous zero initial and Dirichlet boundary conditions, where $\zeta(x,t) = 20\exp(-50(x_1 - t)^2)$ plays the role of a moving source transferring heat to the domain. The spatial distribution of the diffusion coefficient is assumed to have the form

$$\kappa(x_1, x_2) = \theta_1 + \theta_2 x_1 + \theta_3 x_2. \tag{4.101}$$

In our example, we select the initial estimates of the parameter values as $\theta_1^0 = 0.1$, $\theta_2^0 = -0.05$ and $\theta_3^0 = 0.2$, which are assumed to be nominal and known prior to the experiment. The excitation function ζ in (4.100) simulates a source with a vertical line support along the x_2-axis, which moves like a plane wave with constant speed from the left to the right boundary of Ω within the observation interval $[0,1]$.

The determination of the Fisher information matrix for a given experiment requires the knowledge of the vector of the sensitivity coefficients $g = [g_1, g_2, g_3]^\mathsf{T}$ along sensor trajectories. The FIM can be obtained using the direct-differentiation method [281] by solving the following system of PDEs:

$$\begin{aligned}
\frac{\partial y}{\partial t} &= \nabla \cdot (\kappa \nabla y) + F, \\
\frac{\partial g_1}{\partial t} &= \nabla \cdot \nabla y + \nabla \cdot (\kappa \nabla g_1), \\
\frac{\partial g_2}{\partial t} &= \nabla \cdot (x_1 \nabla y) + \nabla \cdot (\kappa \nabla g_2), \\
\frac{\partial g_3}{\partial t} &= \nabla \cdot (x_2 \nabla y) + \nabla \cdot (\kappa \nabla g_3),
\end{aligned} \tag{4.102}$$

in which the first equation represents the original state equation and the other three equations are obtained from the differentiation of the first equation with respect to the parameters θ_1, θ_2 and θ_3, respectively. The initial and Dirichlet boundary conditions for all the four equations are homogeneous.

The system (4.102) has been solved numerically using routines from the MATLAB PDE toolbox and stored g_1, g_2 and g_3 interpolated at the nodes of a rectangular grid in a four-dimensional array (we applied uniform partitions using 21 grid points per each spatial dimension and 31 points in time), cf. Appendix I in [281] for details. Since the values of g_1, g_2 and g_3 may have been required at points which were not necessarily nodes of that grid, the relevant interpolation was thus performed using cubic splines in space (for this purpose, the MATLAB procedure `interp2` has been applied) and linear splines in time. Since, additionally, the derivatives of g with respect to spatial variables and time were required during the trajectory optimization process, they were approximated numerically using the central difference formula.

Next, we used RIOTS_95 to determine D-optimal sensor trajectories in accordance with the Wynn–Fedorov algorithm. The dynamics follow the simple model (4.36), and additional constraints,

$$|u_i^j(t)| \leq 0.7, \quad \forall t \in T, \quad i = 1, \ldots, 6, \tag{4.103}$$

restricting the maximum sensor velocity components, were imposed on the controls. Our goal is to design their trajectories so as to obtain possibly the best estimates of θ_1, θ_2 and θ_3.

A program was implemented using a low-cost PC (AMD Athlon 3800+, 2GB RAM) running Windows XP and MATLAB 7.1 (R2006a). We ran the program twice with 4 iterations and 200 randomly chosen initial positions for each iteration. Each run took between 10 and 45 seconds for each initial position. This is necessary if we wish to get an approximation to a global maximum in line 5 of the Wynn–Fedorov algorithm. This constitutes a trade-off between the computation time and the number of possible initial positions. Note that in RIOTS_95 the controls are internally approximated by linear, quadratic or cubic splines, and this immediately defines the parametrization (4.91).

Figures 4.1 and 4.2 present the results obtained for these two simulations. The initial sensor positions are marked with open circles, and the sensor positions at the consecutive points of the time grid are marked with dots. When available, weights are inserted inside the figures, each weight being positioned by its respective trajectory.

The first run gives two different trajectories with the weights of 0.54807 and 0.45193. Based on the generalized weighted LS criterion, each weight can be interpreted in terms of an experimental cost, which is inversely proportional to the variance of the observation error along a given trajectory. Thus we may think of the weights as the cost related, e.g., to the sensitivity of the measurement devices. Following this interpretation, we should spent approximately 55% of total experimental costs to assure a more accurate sensor for the first trajectory, and approximately 45% for the second trajectory, which requires a less sensitive sensor. In contrast, the second run results in three distinct trajectories with the weights of 0.44464, 0.34726 and 0.2081 (cf. Figs 4.1(a) and 4.2(a)). However, combining the second and third trajectories together with the total weight 0.55536, we can observe that this solution is quite similar to the previous one with only two distinct sensor paths. The differences can be explained in terms of the suboptimality of the solutions for the internal problem in line 5 of the Wynn–Fedorov algorithm (in order to assure the compromise between the computational burden and the quality

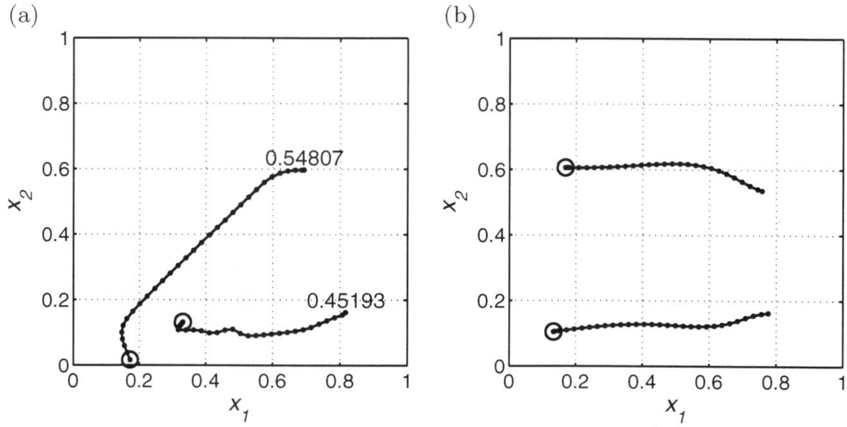

Fig. 4.1 Optimal trajectory of two sensors using weighted (a) and replication-free (b) D-optimal sensor routing

of the solution, in practice we are satisfied with fairly good approximation of the global optimum). Thus, in both simulations we come up only with different suboptimal solutions to our problem, but with acceptable quality, in a sense. The obtained Fisher information matrices are

$$M_{(1)} = \begin{bmatrix} 124.3815 & 68.0614 & 25.7666 \\ 68.0614 & 41.5653 & 13.4240 \\ 25.7666 & 13.4240 & 8.7691 \end{bmatrix}, \tag{4.104}$$

$$M_{(2)} = \begin{bmatrix} 130.0149 & 72.3503 & 26.6154 \\ 72.3503 & 44.2181 & 14.1798 \\ 26.6154 & 14.1798 & 8.6267 \end{bmatrix}, \tag{4.105}$$

with determinants equal to 1787.91 and 1583.19, respectively.

For comparison, we also present results obtained using the optimal control approach without replications of paths for D-optimum trajectories of moving sensors. This is equivalent to the case when the weights are fixed and assumed to be equal for each trajectory. The results are shown in Fig. 4.1(b) (two sensors) and Fig. 4.2(b) (three sensors). The quality of these solutions in terms of the determinant of the FIM are 1272.14 and 1582.91 for two and three sensors, respectively. This means that, using heterogenous nodes with different sensitivity properly chosen prior to the experiment, we are able to increase the quality of measurement data as in our example two sensors in the weighted strategy achieved a better criterion value than three sensors in the replication-free experiment.

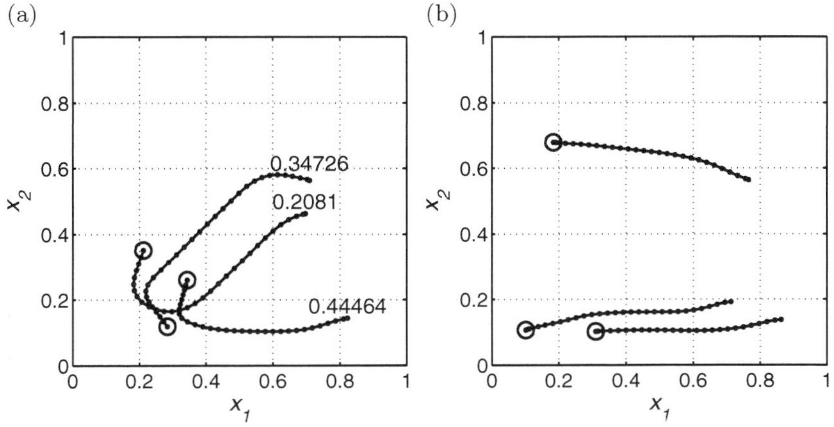

Fig. 4.2 Optimal trajectory of three sensors using weighted (a) and replication-free (b) D-optimal sensor routing

4.4.2 Calibration of Air Pollution Models

As the next example of potential practical application, we consider the mass balance of chemical species and transport of tracers including advection, diffusion and reaction processes. One of the most tangible and interesting practical instances of this phenomenon is monitoring an air pollutants in order to produce accurate forecast of pollutant concentrations [109, 168, 172]. Since proper calibration of the models in the form of the systems of PDEs directly influences the quality of the forecast, the identification accuracy becomes its essential element.

As a suitable example presenting the delineated approach, consider an atmospheric pollutant transport-chemistry process over an urban area being normalized to a unit square. At the point $x_0 = (0.3, 0.6)$, an active source of pollution is located, which leads to changes in the pollutant concentration $y(x,t)$. The entire process over the normalized observation interval $T = [0,1]$ and velocity field $v(x,t)$ varying in space and time according to the following model (cf. Fig. 4.3):

$$v(x,t) = \big(2(y+1/4), 2(t-x)\big) \tag{4.106}$$

can be described by the advection-diffusion-reaction equation:

$$\begin{aligned}\frac{\partial y(x,t)}{\partial t} &+ \nabla \cdot \big(v(x,t)y(x,t)\big) \\ &= \nabla \cdot \big(\kappa(x)\nabla u(x,t)\big) + \zeta(x), \;\; x \in \Omega,\end{aligned} \tag{4.107}$$

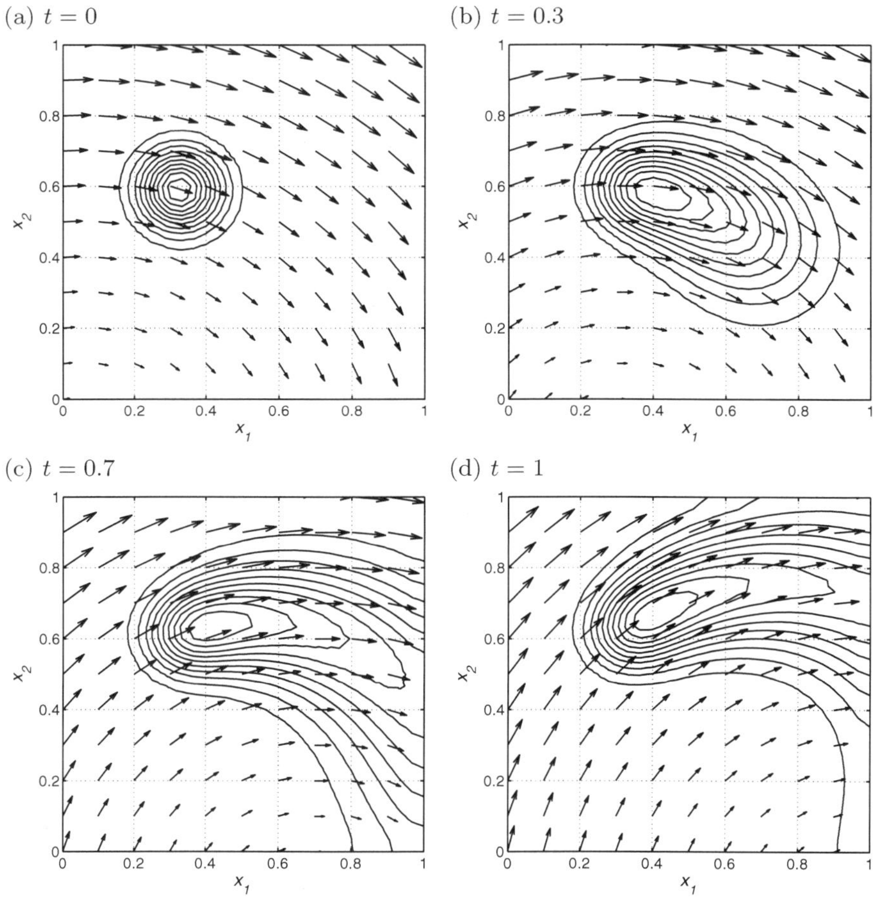

Fig. 4.3 Temporal changes in the wind velocity field and pollutant concentration

subject to the boundary and initial conditions:

$$\begin{cases} \dfrac{\partial y(x,t)}{\partial n} = 0, & \text{in } \Gamma \times T, \\ y(x,0) = 0, & \text{in } \Omega, \end{cases} \tag{4.108}$$

where the term $\zeta(x) = e^{-100\|x-x_0\|^2}$ represents an intensity of the active source of the pollutant, and $\partial y/\partial n$ stands for the partial derivative of y in the direction of the outward normal to the boundary Γ. In simulations, the following form of the distributed diffusion coefficient was applied:

$$\kappa(x) = \theta_1 + \theta_2 x_1^2 + \theta_3 x_2^2, \tag{4.109}$$

so the parameters $\theta = (\theta_1, \theta_2, \theta_3)$ need estimation based on measurement data. The values $\theta_1 = 0.02, \theta_2 = \theta_3 = 0.005$ were taken as the initial estimates of the sought parameters.

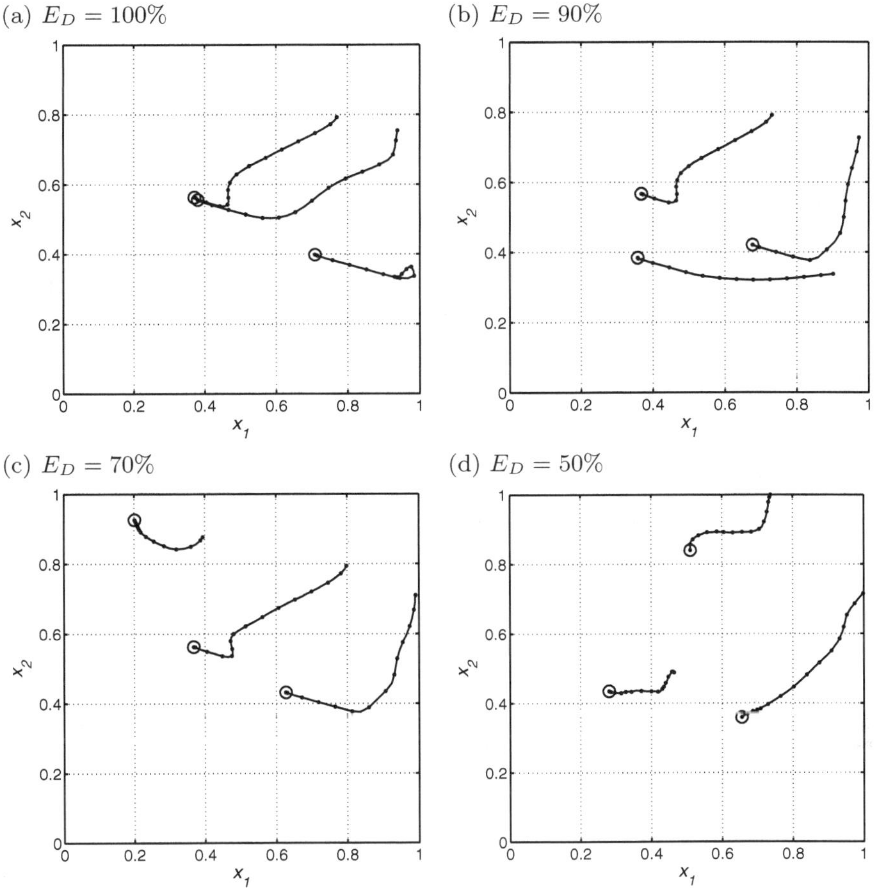

Fig. 4.4 Optimal trajectories of sensors (starting positions marked with open circles) with a guaranteed estimation efficiency level

Distribution of sensor nodes. In the first experiment, the goal was to determine the optimal distribution of $N = 3$ movable sensors subject to the assumed estimation accuracy level. In order to verify the proposed approach, a MATLAB program was written using a PC machine equipped with an Intel Core2Duo T9500 processor (2.5GHz, 3 GB RAM) running Windows Vista and MATLAB 7.5 (R2007b). First, the system of PDEs was solved using some routines of COMSOL 3.4 [43] for a spatial mesh composed of 1800 triangles and 961 nodes and evenly partitioned time interval (30 subintervals). The sensitivity coefficients were then linearly interpolated and stored. Finally, for determining the optimal trajectories, the package RIOTS_95 [247] was applied. From among its three main optimization procedures the routine

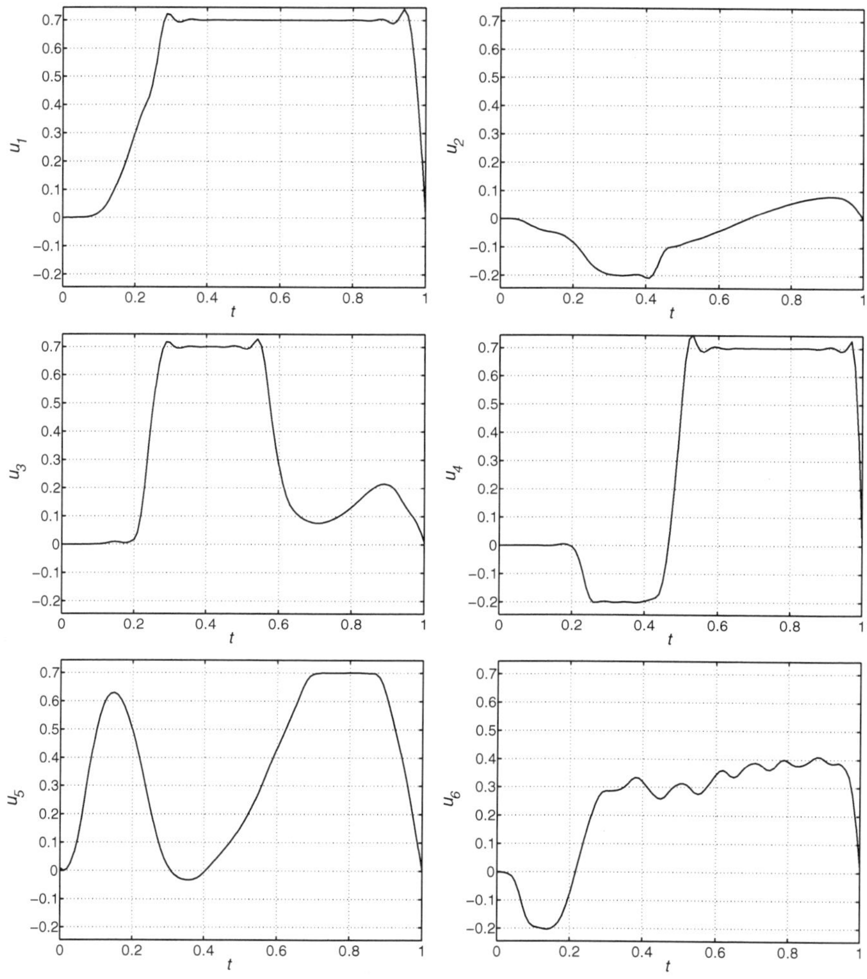

Fig. 4.5 Optimal controls for the efficiency level $E_D = 90\%$

riots was used, which is based on the SQP algorithm. We applied the simple sensors dynamics (4.36) and imposed the following bounds for u:

$$-0.4 \le u_i(t) \le 0.7, \quad \forall t \in T, \quad i = 1, \ldots, 6. \tag{4.110}$$

In order to avoid convergence to the local minima, the simulations were restarted several times from different initial starting points. Each simulation took 5–10 minutes of computation time. The results are presented in Fig. 4.4, where the sensor paths are shown for different accuracy levels, and Fig. 4.5 illustrates corresponding controls. The D-optimal trajectories are shown in Fig. 4.4(a), and they partially coincide trying to follow the complex pollutant concentration changes. This inconvenient effect disappears when the efficiency is slightly decreased (cf. Fig. 4.4(b)). The distribution of sensors improves significantly in Figs. 4.4(c)–(d), but at the cost of the efficiency of parameter estimation.

Limited path lengths. In the second scenario, two approaches represented by Problems 4.5 and 4.6 to constrain the maximal lengths of sensor paths were applied to the DPS considered. For the clarity of interpretation and comparison of solutions, this time only $N = 2$ sensors were applied, following the dynamics given by (4.36) and with bounds for u:

$$\|(u_i(t), u_{i+1}(t))\| \leq 1, \quad \forall t \in T, i = 1, 2 \tag{4.111}$$

(i.e., the maximal velocity in any direction must not exceed 1).

The routine `riots` from the RIOTS_95 package was used to generate the optimal trajectories which are presented in Figs. 4.6(a)–(c). In order to avoid convergence to the local minima, the simulations were restarted several times from different initial starting points. Each simulation took about 4–6 minutes of computation time.

For comparison, Fig. 4.6 (a) shows the D-optimum trajectories obtained for the setting without any constraints imposed on the lengths of sensor paths. We can observe how sensors try to follow the complex pollutant concentration changes, leading to quite sophisticated sensor motions. As for Problem 4.5, after shortening the maximal trajectory length to $s_{\max} = 0.4$, the optimal trajectories are smoother (as intuitively often changes in the movement direction make the paths longer) and still explore the same subregion of the spatial domain. However, the complexity of the process considered induces strong nonlinearity in the dependence between trajectory lengths and the efficiency of the solutions. Hence it is almost impossible to control the quality of observational data gathered by the network. Therefore, aggressive reduction of the maximum length leads in this case to dramatic deterioration of the efficiency in the experiment (which is decreased to $E_D = 41.24\%$). It becomes clear that Problem 4.6 is more flexible to control the quality of the observation process and the lengths of trajectories can be significantly shortened (sparing the valuable resources) with a guaranteed level of efficiency (cf. Fig. 4.6(c)).

Limited kinetic energy of nodes. As the last setting, two approaches, represented by Problems 4.8 and 4.9, to limit the motion energies of robots conveying sensors were tested. Similarly to the previous scenario, to determine the optimal trajectories for $N = 2$ sensors, the dynamics given by (4.36) were used with maximum velocity $u^j(t)$ not exceeding 1 for all sensor nodes. As before, the RIOTS_95 package was used to obtain the solution which is presented in Fig. 4.7.

As for Problem 1, after setting the maximum energy spent on one sensor to $e_{\max} = 0.3$ (the energy was normalized with the maximum possible

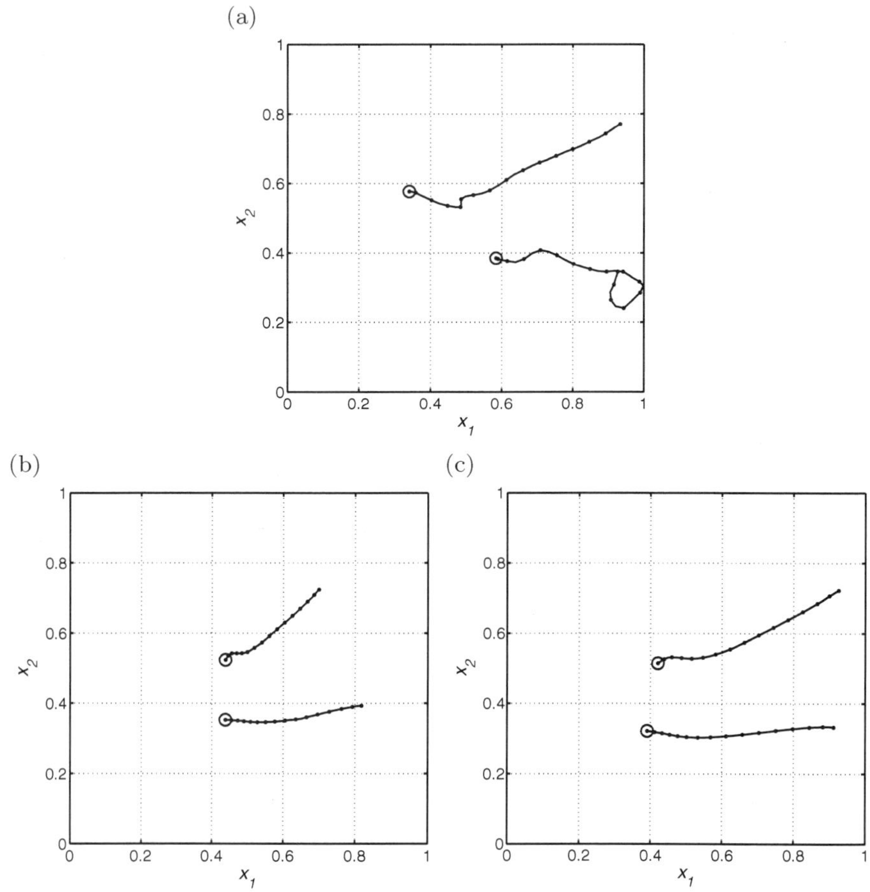

Fig. 4.6 Optimal sensor trajectories for D-optimal solution (the maximal trajectory length is 0.71) (a), Problem 4.5 for maximal trajectory lengths set to $s_{\max} \leq 0.4$ (with the efficiency level $E_D = 41.24\%$) (b), Problem 4.6 for the guaranteed D-efficiency value set to $E_D \geq 0.75$ (the maximal length of trajectory is $\mathcal{J}_2 = 0.55$) (c). The initial sensor positions are marked with open circles, and the sensors positions at the consecutive points of the time grid are denoted by discs

kinetic energy for one sensor related to the maximum velocity), the optimal trajectories are much smoother compared with the D-optimal ones but still explore the same subregion of the spatial domain (cf. Figs. 4.6(a) and 4.7(a)). However, a significant reduction in power consumption limits leads in this case to a serious decrease in the efficiency of the experiment (which was $E_D = 55\%$), and due to the complexity of the process it is extremely difficult to control the level of efficiency. Again, it becomes clear that the guaranteed efficiency approach is more flexible to control the quality of the observation

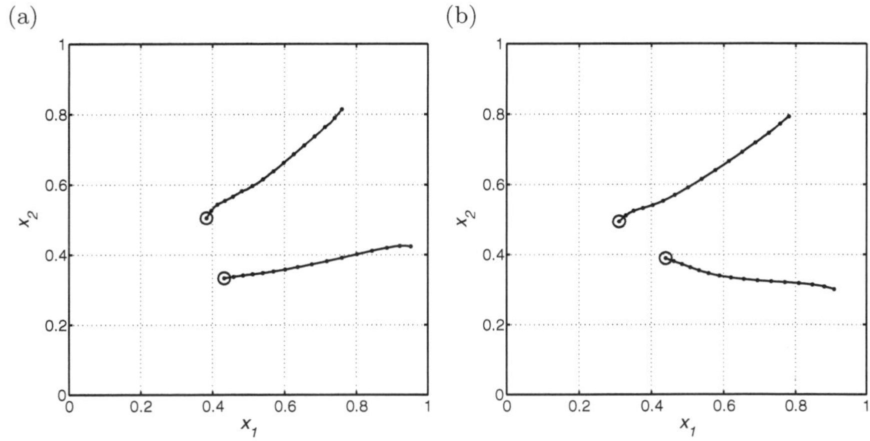

Fig. 4.7 Optimal sensor trajectories for Problem 4.8 with the maximal energy level per sensor set to $e_{\max} \leq 0.3$ (efficiency reached $E_D = 55\%$) (a), and Problem 4.9 for the guaranteed D-efficiency value set to $E_D \geq 70\%$ (the maximal kinetic energy of a single sensor is $\mathcal{J}_2 = 0.35$ compared to the value 0.53 in the case of D-optimal solution) (b). The initial sensor positions are marked with open circles, and the sensors positions at the consecutive points of the time grid are denoted by discs

process while keeping the energy of sensor motions at a reasonably low level (cf. Fig. 4.7(b)). Although the trajectories in Figs. 4.7(a) and (b) are quite similar, their D-efficiency differs by about 15%. This surprising fact explains how important an appropriate trajectory design is, since a slightly different sensor configuration may not measure the important local changes in the state of the DPS, leading to noninformative data. As for the controls, which are interpreted as velocities of nodes, their squares are proportional to the motion energy. Therefore, it can be easily seen that the direct effect of the limited energy is a regularization of the control plots, cf. Fig. 4.8, which are smoother and flat compared to unconstrained D-optimal trajectories.

4.5 Summary

In the first part of this chapter, some fundamental results of modern optimum experimental design theory were adopted to the framework of sensor networks with mobile nodes following the so-called 'direct approach' presented in the seminal paper [232]. The problem reduces to solving at each time instant a separate optimization task to which classical design algorithms can be applied. The drawbacks of this technique are the high demand for computational power due to the complexity of the resulting mathematical programming task, and the fact that only the measurability of the sensor trajectories can be

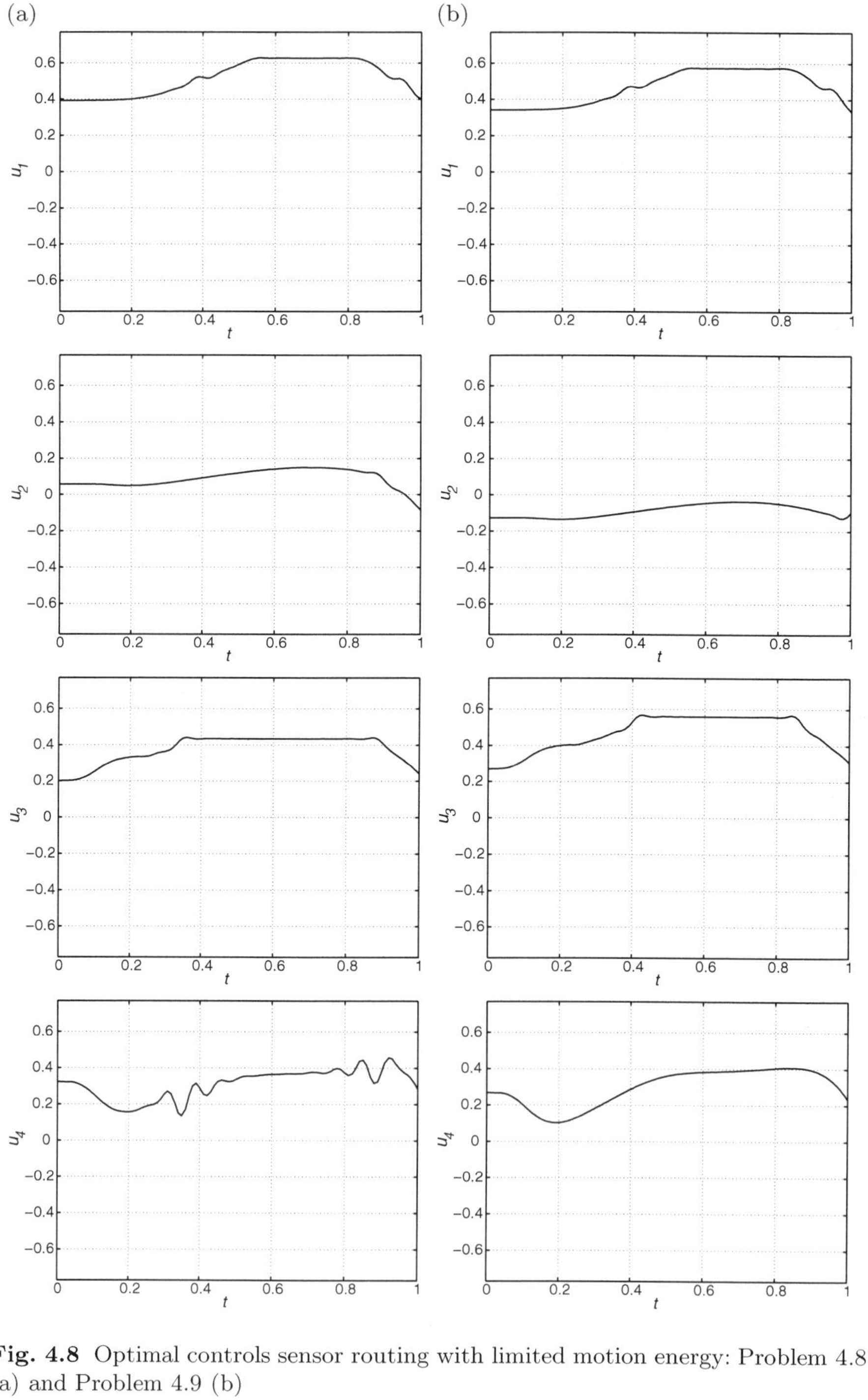

Fig. 4.8 Optimal controls sensor routing with limited motion energy: Problem 4.8 (a) and Problem 4.9 (b)

guaranteed. It was shown that these difficulties can be reduced to some extent by suitable parametrization of the sensor paths. However, imposing additional control constraints (e.g., for collision avoidance or the sensor dynamics) complicates further the computational task and restricts the spectrum of potential applications.

Therefore, a special formulation of the problem of planning the optimal trajectories for a given number of sensor nodes is provided, giving the possibility of allowing for the alternative control objectives. More precisely, a resource-aware approach is proposed taking into account the limited capabilities of sensor nodes, related, e.g., to movement constraints, limited battery power, proper spatial distribution, etc. The main idea is to reformulate the problem in terms of an optimal control task with state-variable inequality constraints representing geometric constraints induced by the admissible measurement regions. The proposed approach is a significant extension of the method presented by Uciński [276], where, taking account of the dynamic models of the vehicles carrying the sensors, the problem was finally reduced to determination of both the control forces of the sensors and initial sensor positions instead of the sensor paths directly. In this work, the problem is further refined toward the objectives related to network resources, which can be considered an attempt to establish an interconnection between the quality of parameter estimation in DPSs and the limited reserves of sensor network nodes. Based on the notion of experimental Ψ-efficiency, a number of formulations of the related problem were proposed, with special attention paid to prospective applications where energy constraints, limited path lengths and coverage of the monitored area are crucial factors. It was also shown that the resulting control problems can be reformulated to an equivalent canonical Mayer or Lagrange form. Although the problem is still hard to solve, this paves the way for direct application of dedicated numerical tools, as in such a form the problem was throughoutly studied in the literature, and a number of effective programming methods exist for its solution [78, 94, 111, 226, 248, 261, 281]. Simulation experiments validate the fact that using nodes with mobility usually results in a significant increase in the values of the adopted quality criteria compared to stationary nodes, and hence to a much greater accuracy in the resulting parameter estimates. A decided advantage of this approach is the independency of a particular form of the partial-differential equation describing the DPS under consideration. The only requirement is the existence of sufficiently regular solutions to the state and sensitivity equations. It is also very flexible with respect to different models of sensor motion dynamics and resources constraints.

As was mentioned above, simultaneous optimization of controls for all sensor nodes is still, computationally, a highly intensive task. Hence, in practice, this technique may be addressed only for networks of relatively small size. Therefore, the last purpose of this chapter was to propose an original approach to decompose the problem and solve it iteratively via a sequence of simpler subproblems focusing on joint optimization for trajectories and measurement accuracies of mobile nodes. As a result, a method is developed which combines the 'direct' and 'control-based' approaches into a very efficient numerical scheme in which we wish to simultaneously optimize the number of sensors and their trajectories, as well as to optimally share the experimental effort. The latter is understood here as allowing for different measurement accuracies of individual sensors, which are quantified by weights steering the corresponding measurement variances. This leads to a much more general setting which most frequently produces an uneven allocation of experimental effort between different sensors and remains in contrast to the existing approaches. The corresponding solutions could then be implemented on a sensor network with heterogeneous mobile nodes. It is demonstrated how to employ the powerful tools of optimum experimental design theory to characterize the optimal solution and adapt the Wynn–Fedorov algorithm to construct its numerical approximation. As a crucial subtask in each iteration, an optimal control problem must be solved, but only for one trajectory at once, which dramatically reduces the total computational burden.

Chapter 5
Decentralized Sensor Scheduling Using Randomized Gossip Algorithms

All the approaches to sensor scheduling discussed so far have relied on centralized techniques, which assume the existence of some superior entity to maintain the whole network and responsible for global optimization of the observation strategy. The distributed nature of the design problem is taken into account very occasionally. But recent advancements in sensor networks necessitate effective, distributed and fault-tolerant algorithms for computation and information exchange. The purpose of the investigations undertaken in this chapter was to establish a practical approach to properly formulate and solve the sensor scheduling problem in a decentralized manner allowing simultaneous parallelization and robustness of the design problem.

Proper estimation of the performance index is achieved by adapting the class of so-called *gossip* algorithms, based on a pairwise communication scheme [31, 311]. This opens possibilities for decentralized strategies of sensor routing which, while being independent of a particular model of the dynamic DPS in question, would be versatile enough to cope with practical monitoring networks consisting of a large number of sensors. The different nodes of the network independently calculate and store the desired quantities, and their final estimates are obtained in a fully decentralized fashion. The advantages of this scheme of information exchange seem to be clear, as no powerful information fusion center is necessary and it is robust with respect to individual sensor faults since the global estimate is recorded at all sensor nodes and can easily be recovered. Moreover, it leads to very simple implementation and great performance of the resulting procedures.

5.1 Conventions and Notation

Consider a connected graph $\widetilde{G} = (\widetilde{V}, \widetilde{E})$, where the vertex set $\widetilde{V}$ contains N nodes and $\widetilde{E}$ is the edge set.

M. Patan: Optimal Sensor Networks Scheduling in Identification of DPSs, LNCIS 425, pp. 135–153.
springerlink.com

Asynchronous time model. Let us assume that each node has a clock which ticks at the times being realizations of a Poisson process with rate 1. This means that the tick times at each node have exponential rate of 1, independent over time and within the nodes. Thus, equivalently, we can consider a single clock ticking according to a Poisson process of rate N. Let $r = 0, 1, 2, \ldots$ be the discrete time index, which partitions the continuous configuration time axis into time slots $Z_r = (z_{r-1}, z_r]$. The lengths of Z_rs are i.i.d. exponentials of rate N. If $i_r \in \{1, \ldots, N\}$ is the number of the node whose clock ticked at time z_r, then i_r are i.i.d. variables distributed uniformly over $1, \ldots, N$. On average, during each configuration time subinterval, the clock ticks N times [31, Lem. 1].

Synchronous time model. In the synchronous time model, time is assumed to be divided commonly across nodes. Within each time subinterval, each node communicates one of its neighbors independently and (not necessarily uniformly) at random. This allows the situation that all nodes communicate simultaneously, in contrast to the asynchronous model, where only one node communicates at a given time. However, still each node contacts pairwise only one other node at a time. Since the analysis for both the asynchronous and the synchronous models is equivalent [31, 117, 311] and leads to the same general conclusions, in the following we assume the asynchronous time model for the configuration process as it is simpler and more convenient for analysis.

Distributed gossip algorithm $\mathcal{A}(P)$. A particular class of time-invariant gossip communication algorithms denoted by $\mathcal{A}$ is considered. This class is characterized by a matrix $P \in \mathbb{S}^N$. The elements of P are nonnegative with $P_{ij} > 0$ only in $(i, j) \in \widetilde{E}$. Further, it is assumed that P is a doubly stochastic matrix, i.e., whose rows and columns sums up to 1. Then, the algorithm associated with P is equivalent to letting the i-th node clock tick during the r-th time slot and communicate i with j with probability P_{ij}. At this time, both nodes exchange the information and update their data according to the specified goal (e.g., averaging or cumulating the data).

Average consensus time. The interest is to determine the time it takes for all nodes to reach the consensus (estimate the desired quantity with satisfactory accuracy).

Definition 5.1. For any $0 \leq \epsilon \leq 1$, the *ϵ-consensus time* of an algorithm $\mathcal{A}(P)$ is defined as

$$T_{\mathrm{c}}(\epsilon, P) = \sup_{\zeta(0)} \inf \left\{\tau : \Pr\left(\mathcal{J}_{\mathrm{c}}(\zeta(\tau)) \geq \epsilon\right) \leq \epsilon\right\}, \tag{5.1}$$

where the $\mathcal{J}_{\mathrm{c}}$ denotes the performance index measuring the distance to the goal of the consensus, and $\zeta(\tau)$ is a network state at a given time (current estimates stored in nodes).

The ϵ-consensus time can be interpreted as the smallest time it takes to achieve a consensus among the nodes with the accuracy ϵ with high probability independently of the initial state of the nodes $\zeta(0)$.

The consensus time is related to the notion of the mixing time for a random walk with transition probabilities derived from the gossip algorithm [31]. In the following, these ideas will be consecutively adopted to provide the distributed versions of scheduling algorithms for sensor networks with scanning and mobile nodes, respectively.

5.2 Distributed Scheduling for Scanning Networks

At this point, assuming that the numbers of candidate and gauged sites are rather large, the efficient clusterization-free designs for scanning sensors discussed in Section 3.1.4 will be adopted to construct a distributed protocol of data fusion leading to the decentralized solution of the scheduling problem. The proposed approach substantially refines the original approach to distributed sensor configuration reported by Patan [187] to the more general setting of scanning sensor networks, where selecting the best subset of sensors to be activated constitutes an inherently discrete large-scale resource allocation problem whose solution may be prohibitively time consuming.

5.2.1 Scheduling Problem Reformulation

The optimal sensor scheduling problem consists in seeking for each time subinterval T_k the best subset of n locations from among N given potential ones. Introducing for each possible location x_k^i ($i = 1,\dots,N$ and $k = 1,\dots,K$) a variable v_k^i, each of them taking the value 1 or 0 depending on whether or not a sensor residing at x_k^i is activated during T_k, the FIM in (3.2) can then be rewritten as

$$M(v_1,\dots,v_K) = \sum_{i=1}^{N}\sum_{k=1}^{K} v_k^i \Upsilon_k(x_k^i), \tag{5.2}$$

where $v^i = (v_1^i,\dots,v_K^i)$. It is straightforward to verify that the matrices $\Upsilon_k(x_k^i)$ are nonnegative definite and, therefore, so is $M(v_1,\dots,v_K)$. Hence, the optimal scheduling problem (see Problem 3.1) takes the following form:

Problem 5.1. Find a sequence $v = (v^1,\dots,v^N)$ to minimize

$$\mathcal{P}(v) = \Psi\big(M(v)\big), \tag{5.3}$$

subject to

$$\sum_{i=1}^{N} v_k^i = n, \quad k = 1, \dots, K, \tag{5.4}$$

$$v_k^i = 0 \text{ or } 1, \quad i = 1, \dots, N, \quad k = 1, \dots, K. \tag{5.5}$$

This constitutes a 0–1 integer programming problem which necessitates an original and efficient solution. A key difficulty in developing successful numerical techniques for such a discrete optimization task is that the number of sensor nodes in the network may be quite large, which considerably increases the complexity of the problem. Another important issue is to properly decompose a computational task in order to provide a possibility for a decentralized procedure of determining an optimal sensor activation schedule.

In Section 3.1.4 a general sequential computational scheme has been proposed to solve this problem approximately based on continuous relaxation operating on sensor densities instead of direct sensor locations. Here, we adopt the idea of (Ψ, ω)-*optimal designs* again, for the purpose of proper decomposition of the optimization task into simpler subproblems and construction of a suitable data exchange protocol. This constitutes the subject of the next section.

5.2.2 Decentralized Exchange Algorithm

Clearly, unless the examined design problem is quite simple, we must employ a numerical algorithm to make the outlined idea useful. Moreover, the key property of the resulting procedure should be an effective distribution of computations between the sensor nodes in a fully decentralized way.

5.2.2.1 Distributed Data Exchange

Owing to Theorem 3.3, $\xi_k^\star(\mathrm{d}x)$ should be nonzero in the areas where $\psi_k(\,\cdot\,,\xi^\star)$ takes on a larger value. Thus the key idea when constructing a computational algorithm for sensor density optimization is to move some measure from areas with smaller values of $\psi_k(\,\cdot\,,\xi^{(r)})$ to those with larger values, as we expect that such a procedure will improve $\xi^{(r)}$ during the next configuration slot. A fundamental question related to this issue is whether the functions $\psi_k(\,\cdot\,,\xi^{(r)})$ can be calculated or estimated in a decentralized manner. From (3.11) we see that the only component of $\psi_k(\,\cdot\,,\xi^{(r)})$ which cannot be calculated independently of other nodes is the global information matrix $M(\xi^{(r)})$. Furthermore, from (5.2) it is clearly seen that this matrix is a weighted average of the local information matrices given by $\Upsilon_k(x_k^i)$. In such a way, the task of FIM estimation is closely related to the problem of distributed averaging on a sensor network, which appears in many applications and has been the subject of extensive studies [23, 31, 33, 35, 117, 311]. In other words, the goal of the consensus across the network in the case considered is to achieve the average of the FIM within the smallest ϵ-averaging time.

One of the simplest and popular techniques of distributed averaging is pairwise communication flooding, also known as a *gossip* scheme, which in its classic version assumes that at the r-th time slot the i-th sensor contacts some neighboring node j with probability P_{ij}, i.e., a pair (i, j) is randomly and independently selected. At this time, both nodes set their values equal to the average of their current values. Denoting by $M^\ell(\xi^{(r)})$ an estimate of a global FIM maintained by the ℓ-th sensor at time slot Z_r can be written as ($\leftarrow$ is the update operator)

$$M^\ell(\xi^{(r)}) \leftarrow \frac{1}{2}\Big(M^i(\xi^{(r)}) + M^j(\xi^{(r)})\Big), \quad \ell \in \{i, j\}. \tag{5.6}$$

Under some assumptions on the communication scheme, such an update leads to the mean of local FIMs of all network nodes, see [311] for detailed conditions for this to happen.

However, in our setting the distributed averaging problem is not typical as not all of the nodes contribute to the global FIM at the r-th configuration slot but only those located at the support of the current design. Therefore, apart from updating local estimates of the FIM, the sensor nodes should be equipped with a mechanism to store and change the global design $\xi^{(r)}$. This can be achieved by the exchange of tokens representing the activation of the sensors at given subintervals T_k. Such tokens are transferred between nodes in the situation where a neighbor node at a particular observation subinterval T_k is more informative in the sense of the function $\psi_k(\,\cdot\,, \xi^{(r)})$ calculated on the current estimate of the FIM (and, obviously, it is not activated yet). A decided advantage of such an approach is that the structure of design $\xi^{(r)}$ is distributed via tokens over the whole network, and hence at the i-th node it is necessary to store only the binary vector $v^i = (v^i_1, \dots, v^i_K)$ indicating whether or not the node owns the token at the k-th time instant.

It is clear that in order to achieve convergence to the mean of the FIMs for selected sensor nodes at a given time subinterval only the nodes possessing tokens should follow the update (5.6). More precisely, we have to apply a weighted average

$$M^\ell(\xi^{(r)}) \leftarrow \sum_{k=1}^{K} M^\ell(\xi^{(r)}_k), \quad \ell \in \{i, j\}, \tag{5.7}$$

where for $v^i_k + v^j_k > 0$

$$M^\ell(\xi^{(r)}_k) = \Big(v^i_k M^i(\xi^{(r)}_k) + v^j_k M^j(\xi^{(r)}_k)\Big)/(v^i_k + v^j_k). \tag{5.8}$$

Algorithm 5.1. Distributed data exchange model. Indexes i and j denote, respectively, data from a local repository and obtained from a neighbor. The quantities $M^i(\xi)$ and v^i are the only ones that have to be stored in the local repository till the next communication event

procedure EXCHANGE_PROTOCOL
  EXCHANGE($M^j(\xi)$,$M^i(\xi)$) ▷ sends and receives FIM ...
  EXCHANGE(v^i,v^j) ▷ ... and activation schedules
  $M^i(\xi) \leftarrow \frac{1}{2}(M^i(\xi) + M^j(\xi))$ ▷ global FIM averaging
  $D(\xi) \leftarrow \overset{\circ}{\Psi}[M^i(\xi)]$ ▷ Calculate the derivative of FIM
  for $k \leftarrow 1, K$ **do**
    if XOR(v_k^i, v_k^j) **then** ▷ the token may be exchanged
      $\psi_k(x_k^i, \xi) \leftarrow \operatorname{trace}\left[D(\xi)\Upsilon_k(x_k^i)\right]$
      EXCHANGE$\big(\psi_k(x_k^i, \xi), \psi_k(x_k^j, \xi)\big)$
      if $\psi_k(x_k^i, \xi) > \psi_k(x_k^j, \xi)$ **then**
        $v_k^i \leftarrow 1$ ▷ activation token goes to the more informative node
      else
        $v_k^i \leftarrow 0$
      end if
    end if
  end for
  $M^i(\xi) \leftarrow \frac{r-1+\frac{1}{K}\sum_{k=1}^{K} v_k^i}{r} M^i(\xi) + \frac{1}{r}\sum_{k=1}^{K} v_k^i \Upsilon_k(x_k^i)$ ▷ Final FIM update
end procedure

The convergence of this averaging scheme with high probability is proven to be of exponential rate, cf. [31, Th. 3], which is crucial for the effectiveness of the optimization task. However, there are a number of drawbacks of such an update of the FIM. First of all, since the weights v_k^i are binary variables, from (5.8) it immediately follows that the information can flow only in one direction, i.e., from the active nodes which own some tokens to the dormant ones without tokens. A direct consequence of the fact that dormant nodes cannot share the information with active ones is that for any time subinterval T_k the nodes owning the tokens have to form a connected graph in order to achieve a consensus in the network. In practice, this may not be the case, especially when the range of each node is very limited. Furthermore, the network is not robust with respect to individual sensor faults, and the network topology becomes of crucial importance.

To overcome those difficulties, instead of (5.8), the following two stage update is proposed as a generalization with simultaneous optimization and data exchange:

1. At the first stage of communication the nodes i and j estimate the global FIM according to the classical averaging scheme (5.6). Then, based on this estimate, the nodes calculate the function $\psi_k(\,\cdot\,,\,\cdot\,)$ for each observation subinterval T_k. Finally, using ψ_k's as criteria for choosing a more informative node at each T_k, the tokens can be exchanged between the nodes.

This problem can be cast in the framework of semidefinite programming and effectively solved in a centralized fashion. For a complete network connectivity graph, it is a simple matter to show that a solution to Problem 5.2 is P with all diagonal entries equal to zero and all off-diagonal entries equal to the value of $\frac{1}{n-1}$. Since a detailed discussion on such important issues is far beyond the scope of this work, the reader can be referred to the seminal papers [31, 311], where also a distributed subgradient scheme is provided to solve this problem in a fully decentralized way.

5.3 Distributed Sensor Routing

For the case of sensor networks equipped with mobile nodes, as was indicated in Section 4.2, once the trajectory design task has been formulated in terms of an optimal control problem, it can be extensively treated using numerous dedicated solvers. On the other hand, it should once again be pointed out that the resulting problem, even if stated in canonical Mayer form, is still nontrivial as it involves dynamic constraints consisting of a large-scale system of ordinary differential equations. In addition, some of those are functionally dependent on the solution of the sensitivity equations associated with the PDE modelling the distributed parameter system in question. Thus, even for a small number of network nodes, we encounter high complexity of the optimization process and potential severe difficulties in determining a solution via a centralized algorithm. Therefore, in order to deal with this issue, a decentralized version of the computation scheme is of critical importance for potential application fields.

5.3.1 Problem Decomposition

A key idea underlaying a distributed policy for optimization of sensor trajectories consists in decomposition of the centralized optimal control problem into a set of simpler subproblems related to a separate design of the trajectory for each particular sensor node. A good starting point for such a procedure is the centralized formulation described by Problem 4.2, which forms a basis for distribution of computations across the network. First, note that, owing to the assumption of uncorrelated observations from distinct network nodes, each sensor contributes to the FIM given by (4.3) independently. Furthermore, the criterion in Problem 4.2 depends only on terminal values of the elements of the FIM. Thus, the FIM entirely determines the sensor trajectories and is the only global quantity required for problem solution which cannot be calculated independently of other nodes. A combination of these two facts leads to the possibility of problem decomposition if the FIM could be calculated or estimated in a decentralized way.

Following the notation in (4.33) and rewriting the FIM given by (4.3) to explicitly express the contributions of individual trajectories, we have

$$M(q) = \frac{1}{N}\sum_{j=1}^{N} M(x^j(t)), \tag{5.13}$$

where

$$M(x^j(t)) = \int_T g(x^j(t),t)C^{-1}(x^j(t),t)g^\mathsf{T}(x^j(t),t)\,\mathrm{d}t. \tag{5.14}$$

It is clear that the average FIM is an arithmetic mean of the local information matrices given by (5.14). Thus, analogously to the scanning sensors case, the estimation of the global FIM is related to the problem of distributed averaging on a network, and this can be achieved in a similar way using randomized gossip algorithms.

Particularly, according to the asynchronous time model, denoting by $M_\ell^{(r)}$ the estimate of the global FIM maintained by the ℓ-th sensor at time slot Z_k, we have the following updates for the communicating pair:

$$M_\ell^{(r)} \leftarrow \frac{1}{2}\Big(M_i^{(r-1)} + M_j^{(r-1)}\Big), \quad \ell \in \{i, j\}. \tag{5.15}$$

For each node, having updated the estimate of the global FIM, the central idea is to determine its trajectory separately treating the contribution of other nodes to the global FIM as fixed variables which are not optimized, as we expect that such a procedure will improve the quality of the current guidance policy. Mathematically, by analogy to Problem 4.2, introducing the state vector

$$w^j(t) = \begin{bmatrix} x^j(t) \\ q^j(t) \end{bmatrix}, \tag{5.16}$$

where $q^j : T \to \mathbb{R}^{m(m+1)/2}$ is the solution to the Cauchy problem

$$\dot{q}^j(t) = \operatorname{svec}\big(g(x^j(t),t)C^{-1}(x^j(t),t)g^\mathsf{T}(x^j(t),t)\big), \quad q^j(0) = \mathbf{0}, \tag{5.17}$$

we have

$$w_0^j = w^j(0) = \begin{bmatrix} x_0^j \\ \mathbf{0} \end{bmatrix}. \tag{5.18}$$

Hence, the control design problem for the j-th node can be stated in the following form.

Problem 5.3. Find a pair $(w_0^j, u^j) \in \overline{\mathcal{Q}}_j$ which minimizes the performance index

$$\mathcal{J}(w_0^j, u^j) = \overline{\Psi}(w^j(t_f)) \tag{5.19}$$

subject to

$$\begin{cases} \dot{w}^j(t) = \overline{h}(w^j(t), u^j(t)), \\ w^j(0) = w_0^j, \\ \overline{\gamma}_l(w^j(t)) \le 0, \quad \forall t \in T, \quad l \in \overline{\nu}, \end{cases} \tag{5.20}$$

where

$$\overline{\mathcal{Q}}_j = \big\{(w_0^j, u^j) \mid x_0^j \in \Omega_{\mathrm{ad}}, u^j : T \to \mathbb{R}^b, \\ u_l \le u^j(t) \le u_u \text{ a.e. on } T\big\}, \tag{5.21}$$

and

$$\overline{h}(w^j(t), u^j(t)) = \begin{bmatrix} h(x^j(t), u^j(t)) \\ \operatorname{svec}\big(g(x^j(t), t)C^{-1}(x^j(t), t)g^{\mathsf{T}}(x^j(t), t)\big) \end{bmatrix}, \tag{5.22}$$

$$\overline{\Psi}(w^j(t)) = \Psi[\tfrac{1}{N}\operatorname{Smat}(q^j(t)) + \tfrac{N-1}{N}M_j], \tag{5.23}$$

$$\overline{\gamma}_l(w^j(t)) = \gamma_l(x^j(t)). \tag{5.24}$$

After solving Problem 5.3 and determining the controls for a given network node, its estimate of the global FIM can be updated according to the weighted mean

$$M_j^{(r)} \leftarrow \frac{N-1}{N} M_j^{(r)} + \frac{1}{N}\operatorname{Smat}(q^j(t_f)). \tag{5.25}$$

The first term is expected to ensure a consensus among the nodes (represents average information from the rest of the network), while the second one accounts for the increase in the total contribution of the node considered.

5.3.2 Adaptive Sensor Routing

Obviously, the solutions of Problem 5.3 are dependent on the current estimates of the global FIM stored in network nodes. Therefore, a very natural idea is to alternate the design and estimation steps. At each configuration interval Z_k, in turn, local sensor trajectories are determined based on the available estimates of the FIM, then the FIM is appropriately updated and propagated through the network via a pairwise communication process. In this general scheme implemented in Algorithm 5.2, it is supposed that each estimation phase adaptively improves the knowledge of the FIM, and this can then be used to improve the quality of the next trajectory design to be performed.

As for convergence, we have following result.

Proposition 5.1. *Assume that* $\{M_j^{(r)}\}$ *is a sequence of iterates constructed by Algorithm 5.2. Then the sequence* $\{\Psi(M_j^{(r)})\}$ *is monotone decreasing and*

Algorithm 5.2. Adaptive distributed sensor routing. Indexes j and i denote, respectively, data from a local repository and obtained from a neighbor. The function EXCHANGE is responsible for both sending and receiving data to/from a connected neighbor (order depending on who initiated the communication), and SOLVEOCP solves Problem 5.3 for a given FIM estimate. The constant r_2 represents an arbitrarily chosen number of pairwise communications which guarantee a satisfactory consensus among the nodes at each iteration of the algorithm

1: **procedure** ADAPTIVE_ROUTING
2: Set $r = 0$. Set arbitrarily r_2, u^j and x_0^j. ▷ Initialization
3: Calculate $x^j(t)$ and $M_j^{(r)} = M(x^j(t))$.
4: **while** TERMINATION CONDITION **do** ▷ Repeat sequentially
5: **for** $s \leftarrow 1, r_2$ **do**
6: Connect to the random neighbor i.
7: EXCHANGE($M_j^{(r)}$,$M_i^{(r)}$) ▷ Sends and receives FIM
8: Update $M_j^{(r)} \leftarrow \frac{1}{2}(M_j^{(r)} + M_i^{(r)})$
9: **end for**
10: $(x_0^j, u^j(t)) \leftarrow$SOLVEOCP($M_j^{(r)}$) ▷ Trajectory design
11: Calculate $x^j(t)$ and $M(x^j(t))$.
12: Update $M_j^{(r)} \leftarrow \frac{N-1}{N} M_j^{(r)} + \frac{1}{N} M(x^j(t))$
13: $r \leftarrow r + 1$
14: **end while**
15: **end procedure**

$$\lim_{r\to\infty} \Psi(M_j^{(r)}) = \min_{M\in\mathbb{M}} \Psi(M).$$

Proof. The convergence of the procedure is a direct consequence of the consensus which is achieved among the network nodes (see [311] for detailed conditions for this to happen). Indeed, assuming that the local estimate of the FIM at each node converges to the average FIM at a given time slot Z_r, Algorithm 5.2 is nothing else than a classical block coordinate descent method. The consensus among the nodes guarantees that each node (block) contributes to every individual local FIM estimate according to the essentially cyclic rule [22, 274] (i.e., each node sufficiently often contributes to the solution). Since the solution to the problem in line 10 of the procedure assures that the resulting Ψ will be nondecreasing, the convergence result can be further proven by proceeding analogously to the main line of reasoning in the convergence analysis of the class of block coordinate descent methods [22, 274]. ■

As regards practical implementation of Algorithm 5.2, it is terminated after exceeding some prescribed finite configuration time and in general it is not simple to propose a suitable decentralized termination condition. It is also clear that the sensor network topology and the choice of a proper communication scheme in terms of probabilities P_{ij} significantly influence the behavior of the delineated procedure similarly to the discussion provided in

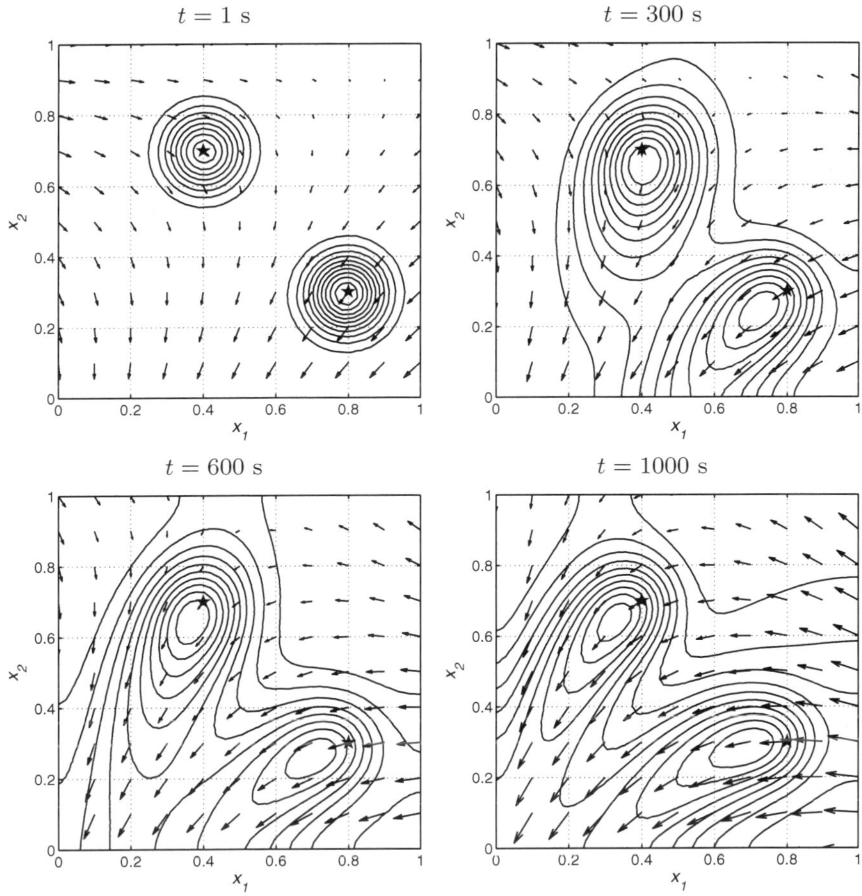

Fig. 5.1 Temporal changes in the wind velocity field and pollutant concentration (stars indicate the locations of pollutant sources)

Section 5.2.2.2. However, in practical settings, in the case of the small and medium number of network nodes, some arbitrary choice of P_{ij} is usually sufficient to provide reasonable results.

5.4 Numerical Examples

Scanning network. As an illustrative example of the presented approach, consider the problem of sensor configuration for parameter estimation in the process of air pollutant transport over a given urban area Ω, being a square with a side of the length of 1 km. In this domain, two active sources of pollution are present, which yields pollutant spatial concentration $y = y(x, t)$.

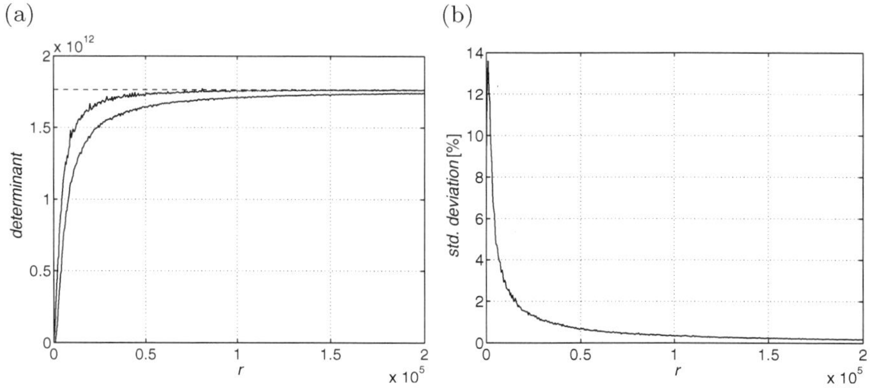

Fig. 5.2 Convergence plots of determinant bounds (max and min value for each iteration) (a) and the coefficient of variation (b)

The evolution of y over the observation interval $T = [0, 1000]$ (in seconds) is described by the following advection-diffusion-reaction equation:

$$\begin{aligned} \frac{\partial y(x,t)}{\partial t} &+ \nabla \cdot \big(v(x,t)y(x,t)\big) + \alpha y(x,t) \\ &= \nabla \cdot \big(\kappa \nabla y(x,t)\big) + f_1(x) + f_2(x), \quad x \in \Omega, \end{aligned} \tag{5.26}$$

subject to the boundary and initial conditions

$$\frac{\partial y(x,t)}{\partial n} = 0, \quad \text{on } \partial\Omega \times T, \tag{5.27}$$

$$y(x,0) = y_0, \quad \text{in } \Omega, \tag{5.28}$$

where the terms $f_\ell(x) = \mu_\ell \exp\big(-100\|x - \chi^\ell\|^2\big)$, $\ell = 1, 2$ represent pollutant sources with emission intensities μ_ℓ located at the points $\chi^\ell = (\chi_1^\ell, \chi_2^\ell)$, $\ell = 1, 2$, and $\partial y/\partial n$ stands for the derivative of y in the direction of the outward normal to the boundary $\partial\Omega$. The average spatio-temporal changes in the wind velocity field over Ω were approximated according to the model (scaled in [km/h])

$$v(x,t) = 7.2 \cdot \big(x_1 + x_2 - t \cdot 10^{-3}, (2x_1 - 1)t \cdot 10^{-3} + x_2 - 1\big).$$

Furthermore, κ denotes an unknown turbulent diffusion coefficient and $\alpha = 0.02\ \text{s}^{-1}$ stands for the absorption rate modelling a slow decay of the pollutant. Figure 5.1 illustrates the resulting complex process dynamics.

Our goal is to identify the sources and the unknown diffusion coefficient, i.e., to estimate the vector $\theta = (\mu_1, \chi_1^1, \chi_2^1, \mu_2, \chi_1^2, \chi_2^2, \kappa)$ using a sensor network with scanning nodes. The observation horizon was split into five evenly partitioned subintervals $T_k = [200(k-1), 200k)$, $k = 1, \dots, 5$.

In order to verify the proposed approach, a MATLAB program was written using a PC equipped with Intel Core i7 processor (1.83GHz, 4 GB RAM) running Windows 7 and MATLAB 2009b. First, the system of PDEs was solved using efficient solvers of the COMSOL environment based on the finite element method [43]. The nominal values of the system parameters were assumed to be $\theta = (12\ \mathrm{kg/s}, 0.4\ \mathrm{km}, 0.7\ \mathrm{km}, 15\ \mathrm{kg/s}, 0.8\ \mathrm{km}, 0.3\ \mathrm{km}, 50\ \mathrm{m}^2/\mathrm{s})$. Calculations were performed for a spatial mesh composed of 978 triangles, 520 nodes and an evenly partitioned time interval (101 subintervals).

The observation grid was assumed to be created at locations selected from among those elements of the above-mentioned 520-point triangulation mesh which do not lie on the outer boundary (there were 460 such nodes, which are indicated with dots in Fig. 5.3). Given $N = 460$ prospective sites in Ω, we aim at selecting at each time subinterval T_k their subset consisting of locations at which measurements made by $n = 100$ sensors would lead to D-optimum least-squares estimates of the parameters θ.

It was assumed that the network was fully connected with a uniform probability distribution for the connection between the selected pair of nodes. The complexity of the system dynamics makes the prediction of the observation locations rather difficult and nonintuitive. The sensors tend to form patterns reflecting the areas of greatest changes in the pollutant concentration. Sensor configurations at different stages of the algorithm are shown in Fig. 5.3. Convergence to the optimal sensor configuration is very fast at the first stage, which is clearly seen in Fig. 5.2(a). Since the algorithm deals with simultaneous optimization and the averaging of the FIM, we can observe that the variance of the distributed estimator increases in this phase (observe the gap between the bounds in Fig. 5.2(a)). Because of the increase in the value of the FIM determinant, a much better measure of the current variability of the averaging estimator is the coefficient of variation plotted in Fig. 5.2(b), where convergence of the second moment is clearly illustrated.

After a relatively small configuration time r, a sensor activation pattern is very similar to the optimal one calculated with the use of the centralized approach (cf. Figs. 5.3(c) and (d)). Then the algorithm considerably slows down, and convergence with a high accuracy requires many pairwise communications.

Mobile network. In order to demonstrate the performance of Algorithm 5.2, consider again calibration of the atmospheric pollution model presented in Section 4.4.2. This time our goal is to determine D-optimal trajectories for the network of $N = 4$ mobile sensor nodes in order to provide the most informative data for estimating the turbulent diffusivity coefficient. In order to verify the proposed decentralized approach, a MATLAB program was written using a PC equipped with Centrino T9300 processor (2.5GHz, 3 GB RAM) running Windows Vista and MATLAB 2007b. In order to adaptively solve optimal control problems, the package RIOTS_95 was applied. In each configuration stage ten pairwise averaging steps were performed before

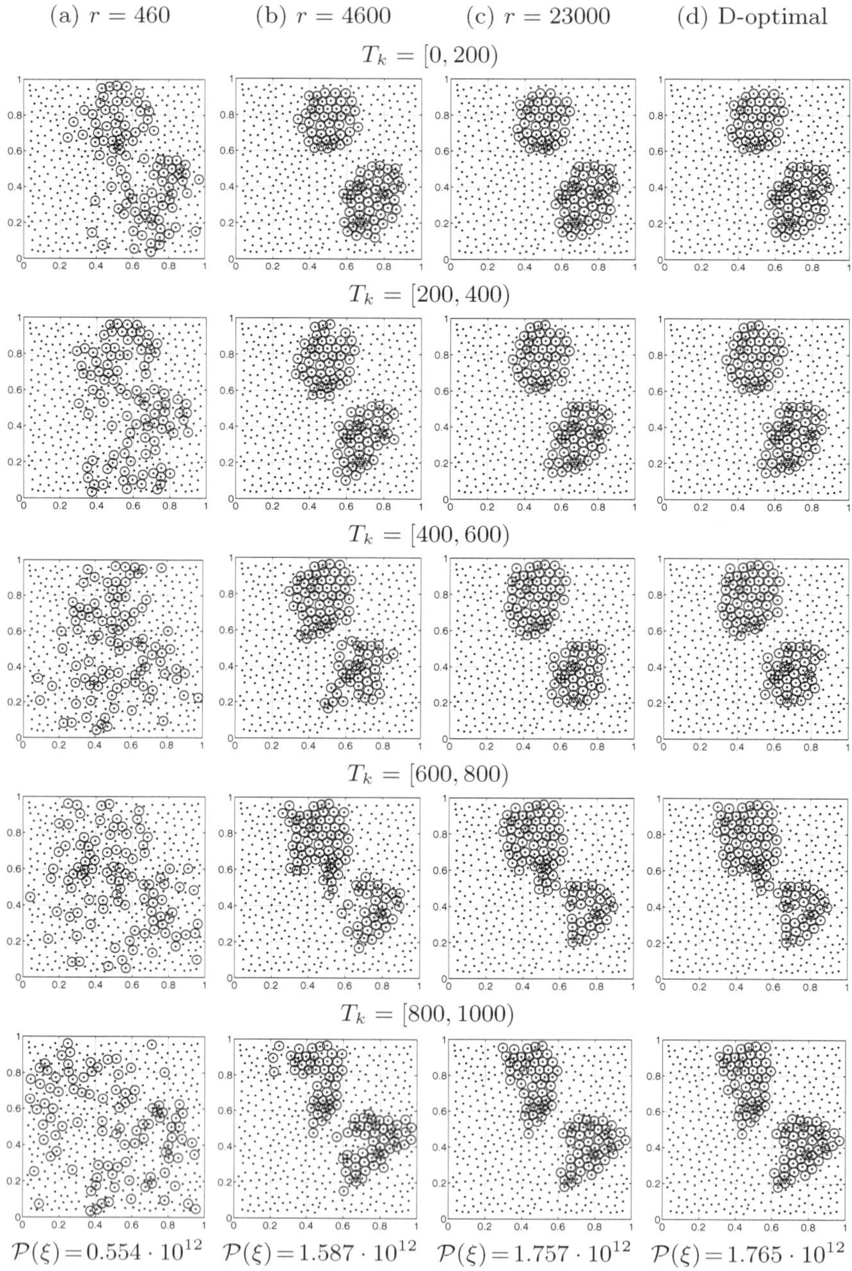

Fig. 5.3 Allocation of active sensors at chosen observation subintervals in consecutive stages of network configuration (a)–(c) and final D-optimal configuration (d)

trajectory design and the uniform probability distribution for communications between the nodes was assumed. As regards sensor dynamics, the following simple model was adopted:

$$\dot{q}(t) = u(t), \quad q(0) = q_0,$$

and the bounds for u were taken as

$$|u_i(t)| \leq 0.3, \quad \forall t \in T. \tag{5.29}$$

The total simulation of $r = 50$ asynchronous configuration steps took about 3 minutes of computation time. Sample results are presented in Fig. 5.4, where

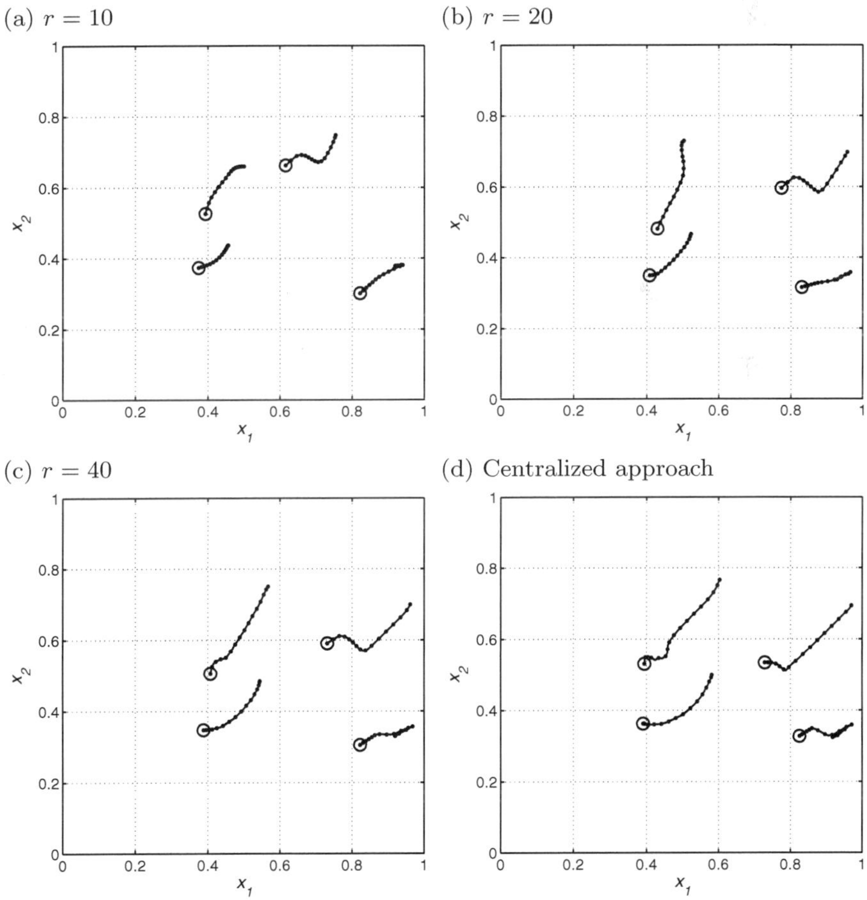

Fig. 5.4 Approximations of D-optimal sensor trajectories obtained in different stages of a distributed configuration scheme (Algorithm 5.2) (a)–(c) versus trajectories determined in a centralized fashion (d). Open circles indicate the initial positions of mobile nodes

we have trajectories calculated at consecutive distributed adaptation stages compared to D-optimal trajectories determined in a fully centralized way. We can observe how sensors try to follow the complex pollutant concentration changes leading to quite sophisticated sensor motions. Also, we can see that Algorithm 5.2 after a relatively small configuration time r provides a solution (cf. Fig. 5.4(b)–(c)) which is very similar to the reference trajectory design shown in Fig. 5.4(d). This is because of a small number of nodes, leading to very effective propagation of FIM estimates through the network.

5.5 Summary

The purpose of this chapter was to establish an important extension of the techniques provided in the preceding two chapters toward fully decentralized numerical schemes for the sensor scheduling task. Making an explicit use of the so-called randomized gossip algorithms, it was possible to develop efficient approaches for networks with scanning and mobile nodes, which distribute the computational burden across the network. In the context of parameter estimation for DPSs, such an idea has not been exploited yet.

As for scanning DSNs, a crucial difficulty is the large scale of the resulting global optimization problem, since the monitoring networks encountered in process industry or environmental engineering may often consist of several hundreds or even thousands of nodes. In particular, the clusterization-free approach has been successfully adopted here, leading to an extremely simple exchange procedure realized in a distributed manner. Although such a class of exchange algorithms is known in optimum experimental design theory, there have been just a few attempts to exploit it in the context of decentralized computation schemes dedicated to sensor networks identifying DPSs [187, 188]. This work contributes to this issue providing a proper mathematical formulation and characterization of the problem in the general setting of scanning observations.

Another problem is a proper decomposition of the relaxed optimization problem in order to offer a possibility of its efficient solution in a decentralized manner. The main contribution here consists in the construction of a specialized protocol of information exchange for simultaneous optimization and estimation of the global optimality criterion at each node. Consequently, the proposed method leads to a fully distributed scheme of calculations which can be implemented with great ease, and our experience provides evidence that, with this tool, even large-scale design problems can be solved using an off-the-shelf PC.

Further in the chapter, the problem of sensor routing was addressed for a monitoring network with mobile nodes. It was demonstrated that the optimal control problem can be fitted to the framework of decentralized trajectory design, and the relevant procedure for its solution is developed, which is dedicated to scalable sensor networks. Proper reformulation of the control problem in the Mayer form makes it possible to directly apply numerous existing solvers for dynamic optimization. A decided advantage of the resulting adaptive procedure for sensor routing is that it not only simplifies the problem solution as each node focuses only on the design of its own moving path and computations are done in parallel, but it also makes the network more robust with respect to node failures since every node preserves global information required for its trajectory design.

Obviously, there still remain some open problems which need close attention. One of such issues is an appropriate refinement of the proposed approach taking into account numerous constraints inherent to network design mentioned in the previous chapters, e.g., energy consumption or time-varying network topology.

Chapter 6
Combinatorial Approach to Sensor Activation

As was elucidated in Chapter 3, although laborious research on the development of strategies for efficient sensor placement has been conducted with numerous contributions and the need for systematic methods was widely recognized, most techniques communicated by various authors usually rely on exhaustive search over a predefined set of candidate solutions and the combinatorial nature of the design problem is taken into account very occasionally [305]. Obviously, such an approach is feasible for a relatively small number of possible sensor locations, and becomes useless as the number of possible candidate locations increases.

The introduction of continuous designs presented in Chapter 3, leading to extremely effective algorithms solving the scheduling task via continuous relaxation, constitutes a possible approach with a great potential for applications. However, bearing in mind that it provides only approximate solutions to the problem, in some practical situations it may be difficult to control the level of suboptimality for the resulting activation schedules. Particularly, within the setting where numerous constraints related to network resources and costs of the experiment have to be imposed, a rounding of experimental design may significantly deteriorate the quality of the solution. Moreover, if the number of activated sensors does not necessarily have to be large, the exchange clusterization-free algorithms discussed in Sections 3.1.4 and 5.2 can fail to provide satisfactory approximation to the global optimum.

The aim of the research reported here was to develop a practical approach to sensor selection which would be versatile enough to cope with practical monitoring networks consisting of many discrete scanning sensors [183, 200, 204, 282, 291]. Specifically, we adopt the setting for which the observation system comprises multiple stationary sensors located at already specified locations, and it is desired to activate only a subset of them during a given time interval while the other sensors remain inactive [50]. A reason for not using all the available sensors could be the reduction of the observation system complexity, the cost of operation and maintenance and/or limited network resources [305]. Since selecting the best subset of sites to locate the

M. Patan: Optimal Sensor Networks Scheduling in Identification of DPSs, LNCIS 425, pp. 155–182.
springerlink.com

sensors constitutes an inherently discrete large-scale resource allocation problem whose solution may be prohibitively time consuming, an efficient guided search algorithm based on the branch-and-bound method is developed, which implicitly enumerates all the feasible sensor configurations, using relaxed optimization problems that involve no integer constraints.

Obviously, this idea is not novel, since the branch-and-bound method constitutes one of the most frequent approaches to solve discrete optimization problems and it has indeed been used in the context of network design, cf., e.g., [25]. The key issue leading to effective application of the branch-and-bound technique is an estimation of accurate bounds for optimal solution at each node of the search tree. Uciński and Patan [291] developed a simple, yet powerful, computational scheme to obtain such bounds for the restricted problems in the case of stationary sensor networks. In their paper, apart of derivation of the optimality conditions, an original algorithm is proposed which can be interpreted as a simplicial decomposition one with the restricted master problem which can be readily solved by the weight optimization algorithms discussed in Section 3.1.2. The proposed combination of the simplicial decomposition algorithm with the known weight optimization algorithm for constructing optimizing probability distributions constitutes a novel approach which is very flexible for different extensions regarding scanning observations [183, 282], parallelization of computations [183], temporal or experimental constraints [204, 205] and a multiobjective formulation [206].

6.1 Scanning Problem Revisited

Here, the fixed switching schedule for scanning sensors introduced in Section 3.1 is adopted, i.e., the time horizon T is arbitrarily partitioned into subintervals $T_k = (t_{k-1}, t_k]$, $k = 1, \dots, K$, where $0 = t_0 < t_1 < \cdots < t_K = t_f$. The state y can be observed (possibly indirectly) by N pointwise sensors, but among them only n_k are activated on T_k.

The optimal sensor scheduling problem considered in what follows consists in seeking for each time subinterval T_k the best subset of n_k locations from among the N given potential ones, so that the problem is reduced to a combinatorial one. In other words, the problem is to divide for each time subinterval the N available sensor nodes into n_k active ones and the remaining $N - n_k$ dormant ones so as to maximize a criterion quantifying the identification accuracy. In order to formulate this mathematically, introduce for each possible location x^i ($i = 1, \dots, N$) a set of variables v_k^i, each of them taking the value 1 or 0 depending on whether or not a sensor residing at x^i is activated during T_k, respectively. The FIM in (3.2) can then be rewritten as

$$M(v_1, \dots, v_K) = \sum_{i=1}^{N} \sum_{k=1}^{K} v_k^i M_k^i, \tag{6.1}$$

where

$$M_k^i = \Upsilon_k(x_k^i) \tag{6.2}$$

and $\Upsilon_k(\,\cdot\,)$ is defined by (3.3).

If we set $v_k = (v_k^1, \dots, v_k^N)$ and apply a suitable design criterion Ψ operating on FIM (cf. Section 2.2.4) then our design problem takes the following form.

Problem 6.1. Find a sequence $v = (v_1, \dots, v_K)$ to minimize $\mathcal{P}(v) = \Psi\big(M(v)\big)$ subject to

$$\sum_{i=1}^{N} v_k^i = n_k, \qquad k = 1, \dots, K, \tag{6.3}$$

$$v_k^i = 0 \text{ or } 1, \qquad i = 1, \dots, N, \qquad k = 1, \dots, K. \tag{6.4}$$

This constitutes a 0–1 integer programming problem which necessitates an ingenious solution. In the following, a general sequential computational scheme is proposed being a generalization of the algorithm developed by Uciński and Patan [200, 291] to solve a similar but simpler problem using the branch-and-bound method, which is a standard technique for such a class of tasks.

6.2 Solution by the Branch-and-Bound Method

6.2.1 General Outline

The Branch-and-Bound (BB) method constitutes a general algorithmic technique for finding optimal solutions of various optimization problems, especially discrete or combinatorial ones [22, 79]. If applied carefully, it can lead to algorithms that run reasonably fast on average.

Principally, the BB method is a tree-search algorithm combined with a rule for pruning subtrees. Suppose we wish to maximize an objective function $\mathcal{P}(v)$ over a finite set V of admissible values of the argument v called the feasible region. BB then progresses by iteratively applying two procedures: branching and bounding. *Branching* starts with smartly covering the feasible region with two or more smaller feasible subregions (ideally, partitioning into disjoint subregions). It is then repeated recursively for each of the subregions until no more division is possible, which leads to a progressively finer partition of V. The consecutively produced subregions naturally generate a tree structure called the BB tree. Its nodes correspond to the constructed subregions, with the feasible set V as the root node and the singleton solutions $\{v\}$, $v \in V$ as terminal nodes. In turn, the core of *bounding* is a fast method of finding upper and lower bounds to the minimum value of the objective function over a feasible subdomain. The idea is to use these bounds to economize computation by eliminating nodes of the BB tree that have no chance of containing an optimal solution. If the lower bound for a subregion V_A from the search tree is greater than the upper bound for any other (previously examined) subregion V_B, then V_A and all its descendant nodes may be safely discarded from the search. This step, termed *pruning*, is usually implemented by maintaining a global variable that records the minimum upper bound encountered among all subregions examined so far. Any node whose lower bound is greater than this value need not be considered further and thereby can be eliminated. It may happen that the upper bound for a node matches its lower bound. That value is then the minimum of the function within the corresponding subregion and the node is said to be solved. The search proceeds until all nodes have been solved or pruned, or until some specified threshold is met between the best solution found and the lower bounds on all unsolved problems.

In what follows, we will use the symbol E to denote the set of index pairs $\{(i,k)|\ i=1,\dots,N \text{ and } k=1,\dots,K\}$ identifying sensor locations and time subintervals. Further, $E^{(\ell_1)}$ will be understood as a subset of E containing all the pairs with location fixed at index ℓ_1, and $E_{(\ell_2)}$ will denote a subset of E consisting of pairs with time index fixed at ℓ_2.

Consider a relaxation of Problem 6.1, which starts by partitioning the feasible set V consisting of all the sequences $(v_1,\dots,v_K)$ satisfying constraints (6.3)-(6.4) into subsets of the form

$$V(E_0,E_1)=\{v\in V \mid v_k^i=0,\ \forall(i,k)\in E_0,\ v_k^i=1,\ \forall(i,k)\in E_1\}, \quad (6.5)$$

where E_0 and E_1 are disjoint subsets of E. Consequently, $V(E_0,E_1)$ is the subset of V such that a sensor is activated at the locations and time subintervals specified with indices in E_1, no sensor is activated at the time and locations with indices in E_0, and a sensor may be activated or stay dormant at the remaining time moments and locations. Each subset $V(E_0,E_1)$ can be directly identified with a node in the BB tree.

The key assumption in the BB method is that for every nonterminal node $V(E_0,E_1)$, i.e., the node for which $E_0\cup E_1\neq E$, there is an algorithm that determines a lower bound $\underline{\mathcal{P}}(E_0,E_1)$ to the minimum design criterion over $V(E_0,E_1)$, i.e.,

$$\underline{\mathcal{P}}(E_0,E_1)\le \min_{v\in V(E_0,E_1)} \mathcal{P}(v), \quad (6.6)$$

and a feasible solution $\bar{v}\in V$ for which $\mathcal{P}(\bar{v})$ can serve as an upper bound to the minimum design criterion over V. We may compute $\underline{\mathcal{P}}(E_0,E_1)$ by solving the following relaxed problem.

Problem 6.2 ($\mathbf{R(E_0, E_1)}$). Find a sequence $\underline{v}$ to minimize $\mathcal{P}(v)$ subject to the constraints

$$\sum_{i=1}^{N} v_k^i = n_k, \quad k = 1, \dots, K, \tag{6.7}$$

$$v_k^i = 0, \quad \forall (i,k) \in E_0, \tag{6.8}$$

$$v_k^i = 1, \quad \forall (i,k) \in E_1, \tag{6.9}$$

$$0 \le v_k^i \le 1, \quad \forall (i,k) \in E \setminus (E_0 \cup E_1). \tag{6.10}$$

In Problem $\mathrm{R}(E_0, E_1)$, all 0–1 constraints on the variables v_k^i are relaxed by allowing them to take any value in the interval $[0, 1]$, except for the variables v_k^i, $(i,k) \in E_0 \cup E_1$ which are fixed at either 0 or 1. A simple and efficient method for its solution is given in Section 6.3. As a result of its application, we set $\underline{\mathcal{P}}(E_0, E_1) = \mathcal{P}(\underline{v})$.

As for the upper bound $\bar{v}$, we can specify it as the best feasible solution (i.e., an element of V) found so far. If no solution has been found yet, we can either set the upper bound to $+\infty$, or use any initial guess about the optimal solution (experience provides evidence that the latter choice leads to much faster convergence).

6.2.2 Branching Rule and the Ultimate Algorithm

The result of solving Problem $\mathrm{R}(E_0, E_1)$ can be used to construct a branching rule for the binary BB tree. We adopt here the approach in which the node/subset $V(E_0, E_1)$ is expanded (i.e., partitioned) by first picking out all fractional values from among the values of the relaxed variables, and then rounding to 0 and 1 a value which is the most distant from both 0 and 1. Specifically, we apply the following steps:

1. Determine
$$(i_\star, k_\star) = \arg \min_{(i,k) \in E \setminus (E_0 \cup E_1)} |v_k^i - 0.5|. \tag{6.11}$$
2. Partition $V(E_0, E_1)$ into $V(E_0 \cup \{(i_\star, k_\star)\}, E_1)$ and $V(E_0, E_1 \cup \{(i_\star, k_\star)\})$ whereby two descendants of the node in question are defined.

Combination of this branching rule with a search strategy to incrementally explore all the nodes of the BB tree constitutes a sequential procedure which starts from the root corresponding to $V(\emptyset, \emptyset) = V$ and the fully relaxed problem. Here we use a common depth-first technique [238, 244] which always expands the deepest node in the current fringe of the search tree. The reason behind this decision is that the search proceeds immediately to the deepest level of the search tree, where the nodes have no successors [85]. In this way, upper bounds on the optimal solution can be found or improved as fast as possible.

A recursive version of the resulting depth-first branch-and-bound is embodied in Algorithm 6.1. The operators involved in this implementation are as follows:

- CONSTRAINTS-TEST(E_0, E_1) returns true only if the constraints (6.3) are violated, i.e., $|E_{1(k)}| > n_k$ or $|E_{0(k)}| > N - n_k$.
- SINGULAR-FIM(E_0, E_1) returns true only if expansion of the current node will result in a singular FIM (cf. [291, Prop. 2] for a very simple singularity test for FIM in the stationary case which can be generalized to the setting considered).
- RELAXED-SOLUTION(E_0, E_1) returns a solution to Problem R(E_0, E_1).
- Ψ-FIM(v) returns the Ψ-optimality criterion value for the FIM corresponding to v.
- INTEGRAL-TEST(v) returns true only if the current solution v is integral.
- INDEX-BRANCH(v) returns the index defined by (6.11).

Algorithm 6.1. A sequential recursive version of the depth-first BB method. *UPPER* and *v_best* are the global variables which are respectively the minimal value of the $\mathcal{P}(v)$ over feasible solutions found so far and the solution at which it is attained

```
procedure RECURSIVE-DFBB(E0, E1)
    if CONSTRAINTS-TEST(E0, E1) or SINGULAR-FIM(E0, E1) then
        return      ▷ Violated constraints on sensor allocation or/and singular FIM within the whole branch
    end if
    v_relaxed ← RELAXED-SOLUTION(E0, E1)
    Ds_relaxed ← D-FIM(v_relaxed)                      ▷ Bounding
    if Ds_relaxed ≤ UPPER then
        return                                          ▷ Pruning
    else if INTEGRAL-TEST(v_relaxed) then
        v_best ← v_relaxed
        UPPER ← Ds_relaxed
        return                    ▷ Relaxed solution is integral
    else
        (i*, k*) ← INDEX-BRANCH(v_relaxed)               ▷ Branching
        RECURSIVE-DFBB(E0 ∪ {(i*, k*)}, E1)
        RECURSIVE-DFBB(E0, E1 ∪ {(i*, k*)})
    end if
end procedure
```

6.3 Simplicial Decomposition Algorithm for Solving the Relaxed Problem

6.3.1 Separability Form of Optimality Conditions

One of the crucial elements of efficient processing of non-leaf nodes of the BB tree is an effective algorithm for estimation of an upper bound to the maximum criterion value over $V(E_0, E_1)$, leading directly to Problem $\mathrm{R}(E_0, E_1)$. This section provides a detailed exposition of a simplicial decomposition method which is particularly suited for its solution.

For notational convenience, we replace the variables v_k^i, $(i,k) \in E \setminus (E_0 \cup E_1)$ with w_j, $j = 1, \dots, q$, where $q = |E \setminus (E_0 \cup E_1)|$, since there exist a bijection π from $\{1, \dots, q\}$ to $E \setminus (E_0 \cup E_1)$ such that $w_j = v_{\ell_2}^{\ell_1}$, $(\ell_1, \ell_2) = \pi(j)$, $j = 1, \dots, q$. Additionally, setting

$$r_k = n_k - |E_{1(k)}|, \tag{6.12}$$

$$q_k = N - |E_{1(k)}| - |E_{0(k)}|, \tag{6.13}$$

we obtain the following formulation.

Problem 6.3 ($\mathbf{R}'(\mathbf{E_0}, \mathbf{E_1})$). Find $w = (w_1, \dots, w_q)$ to minimize

$$\mathcal{Q}(w) = \Psi\Big(G(w)\Big) \tag{6.14}$$

subject to

$$\sum_{(i,k)\in E_{2(k)}} w_{\pi^{-1}(i,k)} = r_k, \quad k = 1, \dots, K, \tag{6.15}$$

$$0 \le w_j \le 1, \quad j - 1, \dots, q, \tag{6.16}$$

where $E_2 = E \setminus (E_0 \cup E_1)$ and

$$G(w) = A + \sum_{j=1}^{q} w_j S_j, \quad A = \sum_{(i,k)\in E_1} M_k^i, \quad S_j = M_{\ell_2}^{\ell_1}, \quad (\ell_1, \ell_2) = \pi(j).$$

Let W stand for the set of all vectors $w \in \mathbb{R}^q$ satisfying (6.15)–(6.16). Note that it forms a polyhedron in $\mathbb{R}^q$. Recall that the criterion Ψ is assumed to be convex and strictly convex over the cones $\mathbb{S}_+^m$ and $\mathbb{S}_{++}^m$, respectively, cf. [32, 222]. Thus the objective function $\mathcal{Q}(w)$ is convex as the composition of Ψ with an affine mapping, see [32, p. 79]. We wish to minimize it over the polyhedral set W. If the FIM corresponding to an optimal solution $w^\star$ is nonsingular, then a special form of the optimality conditions can be derived, being the generalization of the result of Uciński and Patan [291].

Proposition 6.1. *Suppose that the matrix $G(w^\star)$ is nonsingular for some $w^\star \in W$. The vector $w^\star$ constitutes a global solution to Problem $R'(E_0, E_1)$ if, and only if, there exist numbers $\lambda_k^\star$, $k = 1, \dots, K$ such that*

$$\varphi(i,k,w^\star) \begin{cases} \geq \lambda_k^\star & \text{if } w_j^\star = 1, \\ = \lambda_k^\star & \text{if } 0 < w_j^\star < 1, \\ \leq \lambda_k^\star & \text{if } w_j^\star = 0, \end{cases} \tag{6.17}$$

where

$$\varphi(i,k,w) = \operatorname{trace}\big[G^{-1}(w)S_j\big], \quad j = \pi^{-1}(i,k), \quad (i,k) \in E_2. \tag{6.18}$$

Proof. Problem $\mathrm{R}'(E_0, E_1)$ can be rewritten as follows: Find $w^\star \in \mathbb{R}^q$ to minimize

$$\mathcal{Q}(w) = \Psi\big(G(w)\big) \tag{6.19}$$

subject to the constraints

$$\sum_{(i,k)\in E_{2(k)}} w_{\pi^{-1}(i,k)} - r_k = 0, \quad k = 1, \dots, K, \tag{6.20}$$

$$-w_j \leq 0, \quad j = 1, \dots, q, \tag{6.21}$$

$$w_j - 1 \leq 0, \quad j = 1, \dots, q. \tag{6.22}$$

Associating the dual variables

$$\lambda_k \in \mathbb{R}, \quad \mu_j \in \mathbb{R}_+, \quad \nu_j \in \mathbb{R}_+, \quad j = 1, \dots, q \tag{6.23}$$

with the constraints (6.20), (6.21) and (6.22), respectively, we define the Lagrangian of (6.19)–(6.22) as

$$\begin{aligned} \mathcal{L}(w,\lambda,\mu,\nu) = \Psi\big(G(w)\big) &+ \sum_{k=1}^{K} \lambda_k \Big(\sum_{(i,k)\in E_{2(k)}} w_{\pi^{-1}(i,k)} - r_k \Big) \\ &- \sum_{j=1}^{q} \mu_j w_j + \sum_{j=1}^{q} \nu_j (w_j - 1). \end{aligned} \tag{6.24}$$

Simple computation shows that

$$\frac{\partial \mathcal{L}}{\partial w_j} = -\varphi(i,k,w) + \lambda_k - \mu_j + \nu_j, \tag{6.25}$$

where $(i,k) = \pi(j)$.

Let us examine the first-order Karush–Kuhn–Tucker (KKT) conditions for our problem [22]:

$$-\varphi(i,k,w) + \lambda_k - \mu_j + \nu_j = 0, \qquad j = 1,\dots,q, \tag{6.26}$$

$$\mu_j w_j = 0, \qquad j = 1,\dots,q, \tag{6.27}$$

$$\nu_j(w_j - 1) = 0, \qquad j = 1,\dots,q, \tag{6.28}$$

$$\mu_j \geq 0, \qquad j = 1,\dots,q, \tag{6.29}$$

$$\nu_j \geq 0, \qquad j = 1,\dots,q, \tag{6.30}$$

$$0 \leq w_j \leq 1, \qquad j = 1,\dots,q, \tag{6.31}$$

$$\sum_{(i,k)\in E_{2(k)}} w_{\pi^{-1}(i,k)} = n_k - |E_{1(k)}|, \qquad k = 1,\dots,K. \tag{6.32}$$

For problems with constraints which are both linear and linearly independent, as is the case here, the KKT conditions are necessary for optimality. Additionally, the convexity of the objective function (6.19) implies that they also become sufficient. Consequently, the optimality of $w^\star$ amounts to the existence of some values of λ_k, μ_j and ν_j, $k = 1,\dots,K$, $j = 1,\dots,q$, denoted by $\lambda_k^\star$, $\mu_j^\star$ and $\nu_j^\star$, respectively, such that (6.26)–(6.32) are satisfied.

Suppose that $w_j^\star = 1$ for some index j. Then from (6.27) it follows that $\mu_j^\star = 0$ and, therefore, (6.26) reduces to

$$\varphi(i,k,w^\star) = \lambda_k^\star + \nu_j^\star \geq \lambda_k^\star, \tag{6.33}$$

the last inequality owing to (6.30). In turn, on account of (6.28), the assumption $w_j^\star = 0$ yields $\nu_j^\star = 0$, and then (6.26) simplifies to

$$\varphi(i,k,w^\star) = \lambda_k^\star - \mu_j^\star \leq \lambda_k^\star, \tag{6.34}$$

which is due to (6.29). Finally, by (6.27) and (6.28), the assumption $0 < w_j^\star < 1$ clearly forces $\mu_j^\star = \nu_j^\star = 0$, for which (6.26) gives

$$\varphi(i,k,w^\star) = \lambda_k^\star. \tag{6.35}$$

Conversely, having found $w^\star \in \mathbb{R}^q$ and $\lambda_k^\star \in \mathbb{R}$ for which (6.17) is fulfilled, we can define

$$\begin{aligned} \mu_j^\star &= \max(\lambda_k^\star - \varphi(i,k,w^\star), 0), \\ \nu_j^\star &= \max(\varphi(i,k,w^\star) - \lambda_k^\star, 0), \quad j = 1,\dots,q, \quad (i,k) = \pi(j), \end{aligned} \tag{6.36}$$

which guarantees satisfaction of (6.26)–(6.32). This means that $w^\star$ is a KKT point and this is equivalent to its global optimality. ∎

Remark 6.1. The above result can be basically inferred by fitting Problem $\mathrm{R}'(E_0, E_1)$ into the abstract setting of Theorem 1 of [219] dealing with a very general case of directly constrained probability measures. Here, however, we have obtained it in a quite elementary manner, taking account of the specific form of our problem.

Proposition 6.1 reveals one characteristic feature of the optimal solutions, namely, that when identifying them the function φ turns out to be crucial and optimality means separability of the components of $w^\star$ in terms of the values of this function. Specifically, for each $k = 1, \dots, K$, the values of $\varphi(\,\cdot\,, k, w^\star)$ for the indices corresponding to the fractional components of $w^\star$ must be equal to some constant $\lambda_k^\star$, whereas for the components taking the value 0 or the value 1, the values of $\varphi(\,\cdot\,, k, w^\star)$ must be no larger and no smaller than $\lambda_k^\star$, respectively.

6.3.2 Discarding Singular Information Matrices

Note that the assumption that $G(w)$ is nonsingular can be dropped, since there is a very simple method to check whether or not the current relaxed problem will lead to an FIM which is nonsingular.

Proposition 6.2. *The FIM corresponding to the solution to Problem $R'(E_0, E_1)$ is singular if and only if so is $G(\bar{w})$, where*

$$\bar{w}_j = \frac{r_k}{q_k}, \quad (i,k) = \pi(j), \quad j = 1, \dots, q. \tag{6.37}$$

Proof. Observe that the following Loewner ordering holds true for any $w = (w_1, \dots, w_q)$:

$$\begin{aligned} 0 \preceq G(w) &= A + \sum_{j=1}^{q} w_j S_j = A + \sum_{(i,k)\in E_2} w_{\pi^{-1}(i,k)} S_{\pi^{-1}(i,k)} \\ &\preceq \tilde{w}\Big(A + \sum_{(i,k)\in E_2} \frac{r_k}{q_k} S_{\pi^{-1}(i,k)}\Big) = \tilde{w}\, G(\bar{w}), \end{aligned} \tag{6.38}$$

where $\tilde{w} = \max_k (q_k/r_k)$. A fundamental property of the determinant is that it preserves this monotonicity [103, Corr. 7.7.4], which gives

$$0 \le \det\big(G(w)\big) \le (\tilde{w})^m \det\big(G(\bar{w})\big). \tag{6.39}$$

This makes the claim obvious. ■

Consequently, a test of the singularity of the matrix $G(\bar{w})$ can be built into the BB procedure in order to drop the corresponding node from further analysis and forego the examination of its descendants. Otherwise, the vector defined by (6.37) may serve as a good starting point for the simplicial decomposition algorithm outlined in what follows.

Remark 6.2. A solution to Problem $\mathrm{R}'(E_0, E_1)$ is not necessarily unique. Note, however, that for nonsingular cases (after all, pruning discards such cases from further consideration), the resulting FIM is unique. Indeed, Problem $\mathrm{R}'(E_0, E_1)$ can equivalently be viewed as minimization of the criterion (6.14) over the convex and compact set of matrices $\mathbb{M} = \{G(w) \mid \sum_{(i,k)\in E_{2(k)}} w_\pi^{-1}(i,k) = r_k, k = 1,\dots,K,\ 0 \le w_j \le 1,\ j = 1,\dots,q\}$. But if only Ψ is strictly convex over the cone of positive-definite matrices, $\mathbb{S}^m_{++}$, which constitutes the interior of $\mathbb{S}^m_+$ relative to $\mathbb{S}^m$, then this fact implies the uniqueness of the optimal FIM.

6.3.3 Simplicial Decomposition Scheme

Simplicial Decomposition (SD) constitutes an important class of methods for solving large-scale continuous problems in mathematical programming with convex feasible sets [22, 100, 208]. In the original framework, where a concave objective function is to be maximized over a bounded polyhedron, it iterates by alternately solving a linear programming subproblem (the so-called *column generation problem*) which generates an extreme point of the polyhedron, and a nonlinear *restricted master problem* (RMP) which finds the maximum of the objective function over the convex hull (a simplex) of previously defined extreme points. This basic strategy of simplicial decomposition has appeared in numerous references [96, 97, 300], where possible improvements and extensions have also been discussed. A principal characteristic of an SD method is that the sequence of successive solutions to the master problem tends to a solution to the original problem in such a way that the objective function strictly monotonically approaches its optimal value.

Problem $\mathrm{R}'(E_0, E_1)$ is perfectly suited for the application of the SD scheme. In this case, it boils down to Algorithm 6.2. Here $\nabla\mathcal{Q}(w)$ signifies the gradient of $\mathcal{Q}$ at w, and it is easy to check that

$$\nabla\mathcal{Q}(w) = \left[\mathrm{trace}\big(G^{-1}(w)S_1\big), \dots, \mathrm{trace}\big(G^{-1}(w)S_q\big)\right]^{\mathsf{T}}. \tag{6.40}$$

Crucial operators related to its implementation are as follows:

- Initial-Solution() generates an interior point of W. Under the assumption that such a feasible solution exists, it is not difficult to provide a simple scheme for its derivation. For example, a weight $\bar{w}$ resulting from Proposition 6.2 can be applied.
- SolveCGS($w^{(c_{\mathrm{sd}})}$) solves the column generation subproblem:

$$z = \arg\min_{w\in W} \nabla\mathcal{Q}(w^{(c_{\mathrm{sd}})})^{\mathsf{T}}(w - w^{(c_{\mathrm{sd}})}). \tag{6.41}$$

- SolveRMP($\mathrm{Z}^{(c_{\mathrm{sd}})}$) returns the solution of the restricted master problem:

$$w^{(c_{\mathrm{sd}})} = \arg\min_{w\in \mathrm{co}(\mathrm{Z}^{(c_{\mathrm{sd}})})} \mathcal{Q}(w), \tag{6.42}$$

where $\mathrm{co}(U)$ denotes the convex hull of set U. As for effective solution of this convex optimization problem, various alternatives exist. In particular, the algorithms for optimization of experimental effort studied in Section 3.1.2 are perfectly suited for this task.

- $\textsc{Purge}(\mathrm{Z}^{(c_{\mathrm{sd}})})$ is an implementation of the *column dropping rule* [208], according to which any extreme point with zero weight in the expression of $w^{(c_{\mathrm{sd}})}$ as a convex combination of elements in $\mathrm{Z}^{(c_{\mathrm{sd}})}$ is removed. This rule makes the number of elements in successive $\mathrm{Z}^{(c_{\mathrm{sd}})}$ sets reasonably low.

Algorithm 6.2. Algorithm model for simplicial decomposition. The input parameter $\epsilon > 0$ is used to stop the procedure. The function returns the optimal solution $w^{(c_{\mathrm{sd}})}$

1: **function** SIMPLICIAL DECOMPOSITION(ϵ)
2: $\quad c_{\mathrm{sd}} \leftarrow 1$ ▷ Initialization
3: $\quad w^{(0)} \leftarrow$ INITIAL-SOLUTION()
4: $\quad \mathrm{Z}^{(0)} \leftarrow \{w^{(0)}\}$
5: $\quad$ **while** 1 **do**
6: $\qquad z \leftarrow$ SOLVECGS($w^{(c_{\mathrm{sd}})}$) ▷ Col. generation subproblem
7: $\qquad$ **if** $\nabla \mathcal{Q}(w^{(c_{\mathrm{sd}})})^{\mathsf{T}}(w^{(c_{\mathrm{sd}})} - z) \le \epsilon$ **then**
8: $\qquad\quad$ **return**($w^{(c_{\mathrm{sd}})}$) ▷ Termination check
9: $\qquad$ **end if**
10: $\qquad \mathrm{Z}^{(c_{\mathrm{sd}})} \leftarrow \mathrm{Z}^{(c_{\mathrm{sd}}-1)} \cup \{z\}$
11: $\qquad z \leftarrow$ SOLVERMP($\mathrm{Z}^{(c_{\mathrm{sd}})}$) ▷ Restricted master problem
12: $\qquad \mathrm{Z}^{(c_{\mathrm{sd}})} \leftarrow$ PURGE($\mathrm{Z}^{(c_{\mathrm{sd}})}$)
13: $\qquad c_{\mathrm{sd}} \leftarrow c_{\mathrm{sd}} + 1$
14: $\quad$ **end while**
15: **end function**

Since we deal with minimization of a convex function $\mathcal{Q}$ over a bounded polyhedral set W, the convergence of Algorithm 6.2 in a finite number of RMP steps is automatically guaranteed [22, 100, p. 221]. Observe that Step 12 implements the *column dropping rule* [208], according to which any extreme point with a zero weight in the expression for $w^{(c_{\mathrm{sd}})}$ as a convex combination of elements in $Z^{(c_{\mathrm{sd}})}$ is removed. This rule makes the number of elements in successive $Z^{(c_{\mathrm{sd}})}$ sets reasonably low.

The SD algorithm may be viewed as a form of modular nonlinear programming, provided that one has an effective computer code for solving the restricted master problem, as well as access to a code which can take advantage of the linearity of the column generation subproblem [97]. The latter issue can be easily settled, as in the linear programming problem the feasible region W is defined by K equality constraints (6.15) and q bound constraints (6.16). The positive definiteness of $G^{-1}(w^{(c_{\mathrm{sd}})})$ and the nonnegative definiteness of S_j taken in conjunction with (6.40) imply that $\partial \mathcal{Q}(w^{(c_{\mathrm{sd}})})/\partial w_j \ge 0$. Hence it is easily seen that the column generation subproblem has an explicit solution where for each $k = 1, \dots, K$ the relaxed weights $w_{\pi^{-1}(i,k)}$ corresponding to r_k largest values selected from among $\partial \mathcal{Q}(w^{(c_{\rm sd})})/\partial w_j$ are set to unity, the others being zero. Consequently, no specialized linear programming code is needed. This constitutes a clear advantage of the presented approach.

However, if some additional constraints on the network resources have to be added (such a situation will be considered further in the chapter), the simple form of the set W can still be exploited directly, since numerous techniques have been proposed to achieve considerable speedup, ranging from improvements on the simplex method (cf. its upper-bounding version described by Pierre [213, p. 224]) to large-scale interior-point methods which are accessible in popular numerical packages (cf. the primal-dual interior-point variant of Mehrotra's predictor-corrector algorithm implemented in MATLAB's Optimization Toolbox [164], cf. [173].

6.3.4 Restricted Master Problem

Suppose that in the $c_{\rm sd}$-th iteration of Algorithm 6.2, we have

$$Z^{(c_{\rm sd})} = \{z_1, \dots, z_s\}, \tag{6.43}$$

possibly with $s \le c_{\rm sd}$ owing to the built-in deletion mechanism of points in $Z^{(c)}$, $1 \le c \le c_{\rm sd}$, which did not contribute to the convex combinations yielding the corresponding iterates $w^{(\ell)}$. Step 11 of Algorithm 6.2 involves minimization of the design criterion (6.14) over $\mathrm{co}(Z^{(c_{\rm sd})})$.

From the representation of any $w \in \mathrm{co}(Z^{(c_{\rm sd})})$ as

$$w = \sum_{\ell=1}^{s} \alpha_\ell z_\ell, \tag{6.44}$$

or, in component-wise form,

$$w_j = \sum_{\ell=1}^{s} \alpha_\ell z_{\ell,j}, \quad j = 1, \dots, q, \tag{6.45}$$

$z_{\ell,j}$ being the j-th component of z_ℓ, it follows that

$$G(w) = A + \sum_{j=1}^{q} w_j S_j = \sum_{\ell=1}^{s} \alpha_\ell \Big(A + \sum_{j=1}^{q} z_{\ell j} S_j \Big) - \sum_{\ell=1}^{s} \alpha_\ell G(z_\ell). \tag{6.46}$$

From this we see that the RMP can equivalently be formulated as the following problem.

Problem 6.4 (RMP). Find the sequence of weights $\alpha = (\alpha_1, \dots, \alpha_s)$ to maximize

$$\mathcal{T}(\alpha) = \Psi\big(H(\alpha)\big) \tag{6.47}$$

subject to $\alpha \in \mathbb{P}_s$, where

$$H(\alpha) = \sum_{\ell=1}^{s} \alpha_\ell G(z_\ell). \tag{6.48}$$

In order to fully exploit the specific form of the objective function (6.47), Problem RMP can be provided with an equivalent probabilistic formulation. Specifically, the nonnegativeness of the weights $z_{\ell,j}$, $j = 1, \dots, q$ and the nonnegative definiteness of the matrices A and S_j, $j = 1, \dots, q$ imply that $G(z_\ell) \succeq 0$, $\ell = 1, \dots, q$. Defining $\mathcal{X}$ as a discrete random variable which may take values in the set $\{1, \dots, s\}$ and treating the weights α_ℓ, $\ell = 1, \dots, s$ as probabilities attached to its possible numerical values, i.e.,

$$p_{\mathcal{X}}(\ell) = \mathrm{P}(\mathcal{X} = \ell) = \alpha_\ell, \quad \ell = 1, \dots, s, \tag{6.49}$$

we can interpret $p_{\mathcal{X}}$ as the probability mass function of $\mathcal{X}$, and $H(\alpha)$ as the weighted mean of the function $\mathcal{G} : \ell \mapsto G(z_\ell)$. Therefore, Problem RMP can be thought of as that of finding a probability mass function maximizing the objective operating on the mean of G. Consequently, the restricted master problem can be efficiently solved using directly any of the algorithms dedicated to experimental effort optimization presented in detail in Section 3.1.2 as those are especially well suited for such a class of optimization tasks.

6.4 Parallel Realization of the Branch and Bound

Clearly, in order to overcome the excessively growing computational costs for large-scale network scheduling problems, the parallelization of Algorithm 6.1 constitutes an additional problem of great practical relevance. Fortunately, the branch-and-bound approach has become a popular method for solving combinatorial problems and until now numerous strategies have been employed for its effective parallelization (for a general review of parallel BB methodology, see [83]).

Two classical approaches are known to accelerate the BB search: *node-based* and *tree-based*. Node-based strategies aim to accelerate the search by executing in parallel a particular operation mainly associated to the subproblem evaluation, e.g., essential numerical tasks as matrix inversions, parallel mathematical programming or meta-heuristic methods used to compute upper bounds and to derive feasible solutions. This class of strategies has also been identified as low-level [83, 84] techniques of parallelization, because they do not aim to modify the search trajectory, neither the dimension of the BB tree nor its exploration. Speeding up the calculations is the only objective.

On the contrary, tree-based strategies are far more interesting as they can change the search path and dramatically reduce the dimension of the resulting

Algorithm 6.3. MASTER process. It uses `IDLE_Q` (idle processes queue), `BUSY_Q` (busy processes queue) and `TASK_Q` (task queue) for the maintenance of task redistribution. Main vital messages from the *Master* to *Worker* processes are: `new_task` (sends E_0, E_1 and *UPPER* to a worker node), `update_upper_bound` (sends *UPPER* to a worker node) and `end_calc` (finishing the calculations when all worker nodes are done)

```
procedure MASTER
    SOLVE PDEs                              ▷ Solve the underlying PDE system
    INITIALIZATION    ▷ Spawn np workers and broadcast the solution of PDEs
    repeat
        if (LENGTH(IDLE_Q)> 0) and (LENGTH(TASK_Q)> 0) then
            SEND new_task TO IDLE NODE          ▷ Pop subproblem from TASK_Q
        end if
        WAIT FOR A MESSAGE FROM BUSY NODES
        EXECUTE THE MESSAGE REQUEST ▷ Action depending on message type
    until (LENGTH(IDLE_Q)=np) and (LENGTH(TASK_Q)=0)
    FINALIZATION              ▷ Broadcast end_calc and gather the results
end procedure
```

BB tree. Furthermore, they have been the subject of a broad and very comprehensive research effort [84, 141], therefore here we focus on this strategy and in the following it will be demonstrated how to effectively parallelize the exploration of the BB tree.

The main idea of parallel implementation of the computational scheme considered stems from the dichotomy of branching process suitably splitting the main problem into subproblems which can be solved by different processors [84, 141]. A fully decentralized version of the branch-and-bound leads to numerous difficulties from the point of view of implementation, such as avoidance of overheads due to load balancing, communications, dynamical task assignments, redundancy, etc. Since detailed algorithmic solutions to cope with such impediments are far beyond the main topics of this book, here we focus on the hierarchical *Master-Worker* classical configuration scheme applicable using the Message Passing Interface (MPI). Currently, the MPI is a one of the most common ways of programming parallel systems with distributed memory (i.e., computing clusters or grid networks) as it unifies the way of writing programs (regardless of the programming language). A great advantage of the MPI is its simplicity. Although it provides over 100 routines, a very small subset of only few of them is enough to build quite complex parallel functionality of programs.

In the centralized scheme proposed here, the *Master* process takes responsibility for distribution of jobs to *Worker* processes, coordination of information interchange among the processes, and maintains the progress of computations. The maintenance of job redistribution is based on the busy process queue, the idle process queue and the task queue. Each *Worker* process

Algorithm 6.4. WORKER process. Uses `local_stack` to explore the right branch of subproblem. Main vital messages from the *Worker* to the *Master* process are `task_results` (sends E_0, E_1 and *v_best* to the master node), `new_task` (sends E_0, E_1 to the master node) and `im_idle` (if a worker has finished processing a subproblem).

procedure WORKER
  Read the PDEs solution
  while (RECEIVEMESSAGE()=`end_calc`) **do**
    RECEIVE THE DATA ▷ Update local bound and/or push the new task to the `local_stack`
    while LENGTH(`local_stack`)> 0 **do**
      RUN SEQUENTIAL BB ▷ lines 2–12 of Algorithm 6.1
      $(i_\star, k_\star) \leftarrow$ INDEX-BRANCH(*v_relaxed*) ▷ Branching
      SEND $(I_0 \cup \{(i_\star, k_\star)\}, I_1)$ TO MASTER ▷ Left branch to another node
      PUSH $(I_0, I_1 \cup \{(i_\star, k_\star)\})$ TO `local_stack` ▷ Right branch fathomed
    end while
    REQUEST THE MASTER FOR NEW DATA
  end while
  FINALIZATION ▷ Send the results to the master node
end procedure

performs the actual problem solution and fathoming the nodes in the BB tree, independently working on their own part of the subproblem. The generation of subproblems by the *Worker* process is done dynamically. A depth-and-breadth-first combined strategy is used to explore the BB tree. The *Worker* process always explores the right branch of the subproblem (depth-first approach). The left branch is assigned to another *Worker* process (breadth-first approach) via *Master*'s task queue. An extension here is the possibility of updating not only the local current best solution and upper bound, but also *Master*'s respective global values, which enhance the prunning process.

Since the complexity of a given subproblem cannot be determined in advance, proper load balancing becomes a key issue of high performance of parallel BB. Load balancing is a well studied concept; here it is achieved via a periodic query and acknowledgments from *Worker* processes. A general scheme of implementation is given in Algorithms 6.3 and 6.4.

6.5 Resource-Constrained Sensor Activity Scheduling

At this juncture, an important practical extension of the proposed approach to the situation when apart from the identification accuracy some additional resource related constraints have to be simultaneously satisfied should be addressed. In particular, an observation policy is proposed to provide a proper distribution of sensors over the spatial domain considered using a minimal number of sensors and reducing the accumulated activation time. Clearly, taking these alternative objectives under consideration implies an additional layer of complexity. For example, if communication connectivity is to be maintained, we must ensure that each node remains within the range of some other nodes. We must also take into account that sensor activity consumes a considerable amount of energy, which amplifies the need for various forms of power control. As a result, the existing activity scheduling algorithms for wireless sensor networks abstract away from the mathematical models of the observed physical phenomena, and the classical techniques invented for optimal sensor location in DPSs are not adapted to take account of various practical constraints associated with the operation of sensor networks. It turns out that the developed BB approach to sensor selection reported in the preceding sections can be properly extended to cover such a more general setting of multi-criteria scheduling for scanning sensor networks. This constitutes the subject of this section.

6.5.1 Minimum Cover

Starting with the optimal sensor scheduling task defined by Problem 6.1 which consists in seeking, for each time subinterval T_k, the best subset of n_k locations from among the N given potential ones so as to minimize the criterion $\mathcal{P}_1(v) = \Psi(M(v))$ associated with the expected accuracy of the parameters to be estimated, we reduce the problem to a combinatorial one. However, as one of crucial issues for sensor networks is to assure that communication connectivity will be maintained during the monitoring process, we must ensure that each node remains within the range of some other nodes. Moreover, it is often required to achieve this possibly with the lowest cost of operation and equipment.

To formulate this mathematically, given a sensor set $S = \{s_1, \dots, s_N\}$ and a partitioning $\Omega_1, \dots, \Omega_L$ of the domain Ω, such that $\Omega = \cup_{\ell=1}^{L} \Omega_\ell$ and Ω_ℓs are disjoint sets, we can introduce the area coverage matrix $C_{L \times N}$, where each element c_ℓ^i is 1 if Ω_ℓ lies within the range of sensor s_i and 0 otherwise. The idea is illustrated in Fig. 6.1, where, e.g., the area Ω_2 is covered with sensors 1 and 2 only, so that the second row of C is $[1, 1, 0, 0, 0, 0]$.

Attaching to each sensor s_i its associated cost b_i of operation, a minimum cover can be understood as sensor configuration minimizing the maximum cost over the given observational subintervals

$$\mathcal{P}_2(v) = \max_{k=1,\dots,K} \sum_{i=1}^{N} v_k^i b_i \tag{6.50}$$

subject to

$$\sum_{i=1}^{N} c_\ell^i v_k^i \geq 1, \quad \ell = 1, \dots, L, \quad k = 1, \dots, K. \tag{6.51}$$

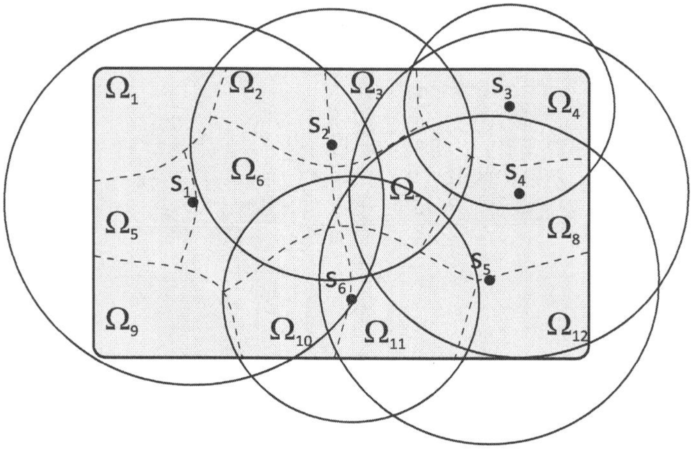

Fig. 6.1 Example of monitoring regions and sensor coverage

6.5.2 Balanced Operation Scheduling

Another critical resource constraint which cannot be neglected results from the fact that the available battery power for each node is limited. Therefore, the activation of sensor nodes along the time slots T_k has to be properly balanced. This means that, if possible, sensor nodes should equally share the effort of providing measurements, i.e., any of the sensors should not be activated much more frequently than others. Since the total activation time for the i-th node is $\tau_i = \sum_{k=1}^{K} v_k^i \Delta t_k$, where $\Delta t_k = t_k - t_{k-1}$, in order to provide a balanced sensor schedule, the criterion

$$\mathcal{P}_3(v) = \max_{i=1,\dots,N} \tau_i \tag{6.52}$$

has to be minimized. In effect, the observations are distributed evenly among the network nodes enforcing them to share the experimental effort. The idea is illustrated in Fig. 6.2.

6.5.3 Goal Attainment Formulation

Multiobjective sensor scheduling. Incorporating the objectives (6.50) and (6.52) into the analysis, our design problem may be cast as a multicriteria optimization one.

Problem 6.5. Find a sequence $v = (v_1, \dots, v_K)$ to minimize

$$\mathcal{P}_1(v), \mathcal{P}_2(v) \text{ and } \mathcal{P}_3(v) \tag{6.53}$$

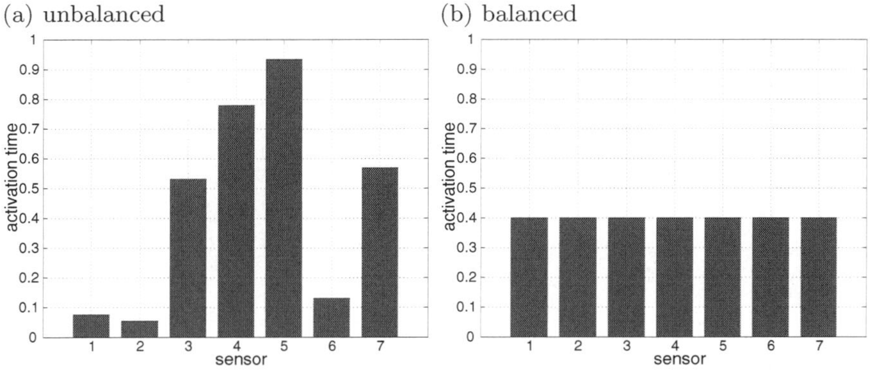

Fig. 6.2 Illustration of time unbalanced (a) and balanced (b) sensor scheduling

subject to

$$\sum_{i=1}^{N} c_\ell^i v_k^i \geq 1, \qquad \ell = 1,\dots,L, \ \ k = 1,\dots,K, \tag{6.54}$$

$$v_k^i = 0 \text{ or } 1, \quad i = 1,\dots,N, \ \ k = 1,\dots,K. \tag{6.55}$$

This constitutes a binary programming problem whose solution is highly nontrivial due to the nonlinearity of all criteria. Moreover, the minimax character of $\mathcal{P}_2$ and $\mathcal{P}_3$ makes a direct application of nonlinear integer programming techniques even more difficult. Therefore, in order to make the BB approach useful for solving the multicriteria scheduling problem above, its further conversion is required.

Application of goal attainment method. The majority of methods developed for solving multiobjective optimization problems are based on the concept of the *Pareto optimal* solution, which is understood as a solution that cannot be improved in the sense of any objective function without simultaneous decrease of at least one other objective. The main idea of such a group of techniques is to generate a set of Pareto optimal solutions and use some additional rule to select from this set one particular solution to become the ultimate one [236]. Within this class of methods, the *goal attainment method* is one of the most popular ones due to its flexibility in applications and the intuitive formulation leading to a single aggregate objective function [236].

For each criterion $\mathcal{P}_j$, a goal π_j is set on its value and a nonnegative weight β_j expressing the importance of the j-th objective function relative to the other criteria in meeting the goal π_j. The goals can be set a priori or can be determined as optimal solutions to the problem when only one particular objective function is selected. Applying these ideas to the sensor network design formulated in Problem 6.5, we can restate the task in terms of single objective optimization.

Problem 6.6. Find a sequence v to minimize

$$\mathcal{P}(v, b_{\max}, \tau_{\max}, \gamma) := \gamma \tag{6.56}$$

subject to

$$\mathcal{P}_1(v) - \gamma\beta_1 \le \pi_1, \tag{6.57}$$

$$b_{\max} - \gamma\beta_2 \le \pi_2, \tag{6.58}$$

$$\tau_{\max} - \gamma\beta_3 \le \pi_3, \tag{6.59}$$

$$\sum_{j=1}^{N} v_k^i b_i \le b_{\max}, \qquad k = 1, \dots, K, \tag{6.60}$$

$$\sum_{k=1}^{K} v_k^i \Delta t_k \le \tau_{\max}, \qquad i = 1, \dots, N, \tag{6.61}$$

$$\sum_{i=1}^{N} c_\ell^i v_k^i \ge 1, \qquad \ell = 1, \dots, L, \quad k = 1, \dots, K, \tag{6.62}$$

$$v_k^i = 0 \text{ or } 1, \qquad i = 1, \dots, N, \quad k = 1, \dots, K, \tag{6.63}$$

$$\beta_1 + \beta_2 + \beta_3 = 1. \tag{6.64}$$

The formulation of the scheduling task in terms of the goal attainment technique embodied in Problem 6.6 forms a basis for the application of guided search based on the BB algorithm. On the analogy to Problem 6.2, the relaxation of Problem 6.6 can be derived.

Problem 6.7 (relaxed goal attainment problem). Find a sequence $\underline{v}$ to minimize (6.56) subject to (6.57)–(6.62) and

$$0 \le v_k^i \le 1, \ \ \forall (i,k) \in E \setminus (E_0 \cup E_1), \tag{6.65}$$

$$v_k^i = 0, \ \ \forall (i,k) \in E_0, \quad v_k^i = 1, \ \ \forall (i,k) \in E_1. \tag{6.66}$$

In particular, the minimax criteria $\mathcal{P}_1$ and $\mathcal{P}_2$ are eliminated and we come up with the optimization of the linear objective function over a convex set of sensor configurations. Since the constraint (6.57) is nonlinear (but still convex), the simplicial decomposition scheme presented in Section 6.3 cannot be applied here. However, the problem can be readily solved using any suitable method dedicated to solve convex optimization problems (e.g., sequential quadratic programming).

6.6 Application Examples: Air Pollution Transport

Parallel sensor scheduling. The following example serves as a vehicle to display some salient features of the proposed approach. Consider simultaneous advection and diffusion of an air pollutant over an urban area normalized to the unit square $\Omega = (0,1)^2$. Inside this domain, an active source of pollution is present, which influences the pollutant spatial concentration $y = y(x,t)$. The evolution of y over the normalized observation interval $T = (0,1]$ is described by the following advection-diffusion equation:

$$\frac{\partial y(x,t)}{\partial t} + \nabla \cdot \big(v(x,t)y(x,t)\big) = \nabla \cdot \big(\kappa(x)\nabla y(x,t)\big) + f(x),\ x \in \Omega \tag{6.67}$$

subject to the boundary and initial conditions:

$$\frac{\partial y(x,t)}{\partial n} = 0, \quad \text{on } \Gamma \times T, \qquad y(x,0) = y_0, \quad \text{in } \Omega, \tag{6.68}$$

where the term $f(x) = 50\exp\big(-50\|x-c\|^2\big)$ represents a source of pollutant located at point $c = (0.3, 0.3)$, and $\partial y/\partial n$ stands for the derivative of y in the direction of the outward normal to the boundary Γ. The mean spatio-temporal changes of the wind velocity field over the area were approximated by $v = \big(2(x_1 + x_2 - t),\ x_2 - x_1 + t\big)$. The assumed functional form of the spatial-varying diffusion coefficient $\kappa(x)$ is

$$\kappa(x) = \theta_1 + \theta_2 x_1 x_2 + \theta_3 x_1^2 + \theta_4 x_2^2. \tag{6.69}$$

The subject of interest here is identification of the parameters $\theta = (\theta_1, \dots, \theta_4)$. Therefore, these parameters need estimation based on measurement data from monitoring stations.

In our simulation studies, the described branch-and-bound technique was applied to determine the D-optimal activation schedule for different numbers of sensors to maximize the reliability of the detection. The nominal values of the parameters were fixed at $\theta_1^0 = 0.02$, $\theta_2^0 = 0.01$ and $\theta_3^0 = \theta_4^0 = 0.005$. We solved the resulting system of PDEs using routines of the MATLAB PDE toolbox for a spatial mesh composed of 682 triangles and 378 nodes. As for the numerical integration required to evaluate information matrices for admissible observation sites, the trapezoidal rule was applied with the time step equal to 0.04, based on the sensitivity vector interpolated at the nodes representing admissible locations. In the scenario considered, the observation grid was assumed to be created at locations selected from among those elements of the aforementioned 378-point triangulation mesh which do not lie on the outer boundary (there were 312 such nodes, which are indicated with dots in Fig. 6.3). The time interval was divided into 13 equal subintervals (which gives the total number of 4056 variables).

A MATLAB program was written to implement the components of parallel version of the BB procedure embodied by Algorithms 6.3 and 6.4, and the `mpich-1.2.6` implementation of the MPI for the message passing library. Computations were performed on the homogenous local network of workstations equipped with Pentium IV 3.0GHz processors and 3GB RAM each, running under the control of Windows XP and connection between nodes realized via Gigabit Ethernet. For the implementation of its principal part, which is the solution of the relaxed problem with the use of the SD scheme given in Algorithm 6.2, the following setting was assumed:

- The accuracy of the solution of the relaxed problem (the termination check in line 7) was set on the level of $\epsilon = 10^{-5}$.
- The tolerance and the maximum number of iterations for the multiplicative algorithm (Algorithm 3.4 was used) for solving the restricted master problem in line 11 were set as $\eta = 10^{-5}$ and $c_{\max} = 1000$, respectively.

Finally, to take full advantage of the efficient gradient projection scheme performed at each node of the BB tree, Algorithm 6.4 was extended to incorporate a rounding procedure after solving the relaxed problem. Thus, this solution provides not only a lower bound to the currently processed branch, but it may also lead to great improvement in the upper bound *UPPER*. Since this choice is complicated by the presence of sites with identical weights, the target sites are chosen randomly. In effect, it is not necessary to descend to the bottom level of the BB tree in order to update *UPPER*. This fact is of crucial importance for large-scale problems, where such an action costs many recursive calls of Algorithm 6.1. Thus the rounding of relaxed solutions significantly speeds up the algorithm, as evidenced by numerous simulation experiments.

D-optimal sensor configurations for chosen time moments and different numbers of allocated sensors are shown in Fig. 6.3. It is clear that the complexity of the system dynamics makes the choice of the observation locations rather difficult and nonintuitive. The sensors tend to form a pattern reflecting the areas of greatest changes in the pollutant concentration, but the observations are averaged over time and it is not trivial to follow the dynamics of the observation strategy. Surprisingly, measurements in the closest vicinity of the pollution source turned out not to be very attractive for parameter estimation. The intuition fails in this case and it is very difficult to predict the solution when armed only with the experimenter's experience.

The results concerning algorithm performance are summarized in Tab. 6.1 and the speedups for sample problem size are presented in Fig. 6.4. Although the presented examples are rather medium-scale, we have to remember that in the worst case (i.e., when the number of sensors is closest to half the number of available sites) the cardinalities of the search space for the spatio-temporal grid considered can reach even $3.06 \cdot 10^{1203}$. Examination of the data from Tab. 6.1 leads to interesting conclusions. Unexpectedly, with an increased number of sensors (and the size of the corresponding search space), the

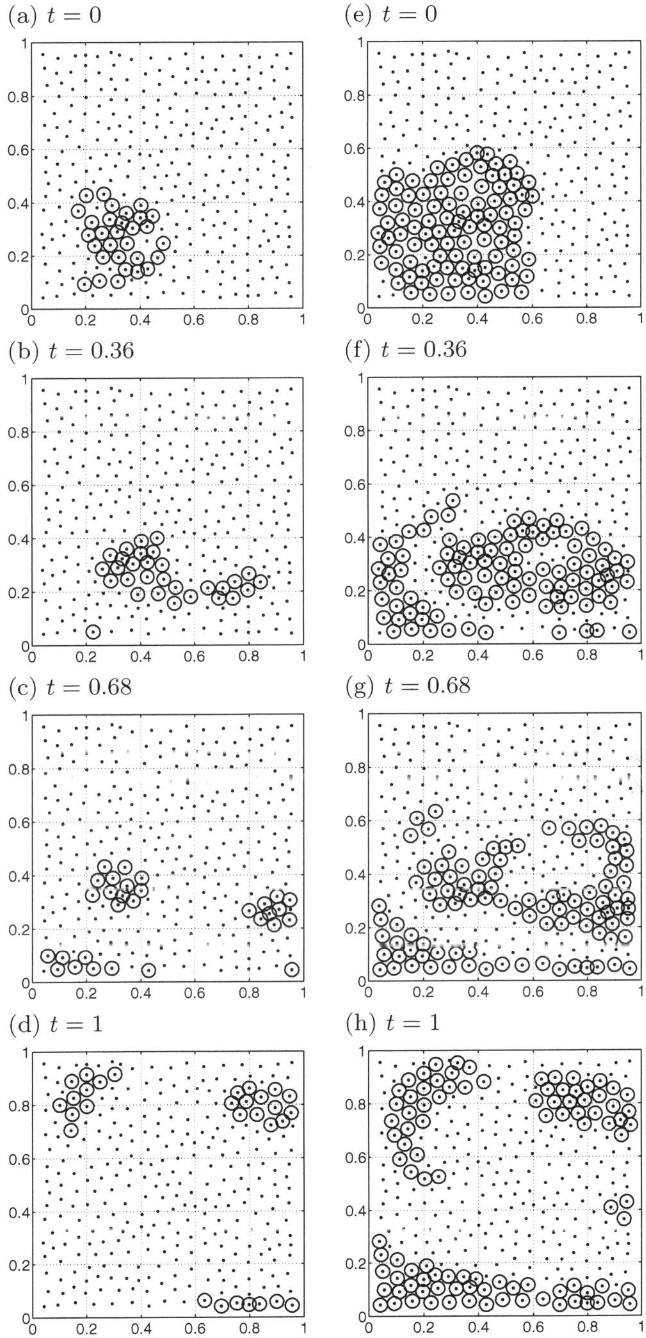

Fig. 6.3 *D*-optimal activation configurations for 30 (a)–(d) and 100 (e)–(h) sensors.

Table 6.1 Comparison of algorithm performance for different numbers of sensors

No. of sensors	Number of processors 1	2	4	6	10	20
	CPU time [h:min:sec]					
10	17:28:08.47	8:54:45.95	4:31:32.35	3:10:13.51	2:04:37.82	1:16:54.14
30	11:33:46.29	6:07:04.49	3:03:03.19	2:10:24.49	1:23:35.22	0:50:58.53
60	1:48:54.91	0:59:50.61	0:29:50.39	0:20:54.30	0:13:17.91	0:08:09.94
80	1:27:02.54	0:48:37.62	0:24:02.69	0:17:10.09	0:11:16.50	0:06:41.70
100	0:18:36.43	0:11:24.93	0:05:13.60	0:03:44.63	0:02:28.66	0:01:26.96
150	0:08:19.87	0:05:14.38	0:02:23.64	0:01:45.46	0:01:07.82	0:00:41.10

pruning process becomes more efficient (only in the case of a finer grid without rounding did we observe the influence of the 'curse of dimensionality'). This effect can be explained by observing that a higher density of sensors leads to a better estimate of the upper bound to the optimal value of the design criterion, which results in an increased efficiency of pruning and whereupon the search is speeded up. Moreover, the increase in the sensor density seems to make the rounded solution of the relaxed problem closer to the optimal one in terms of the criterion value. Therefore, the influence of rounding cannot be overestimated since in our example it significantly decreases the number of recursive function calls and, surprisingly, a low number of sensors leads to a more difficult situation from the computational point of view.

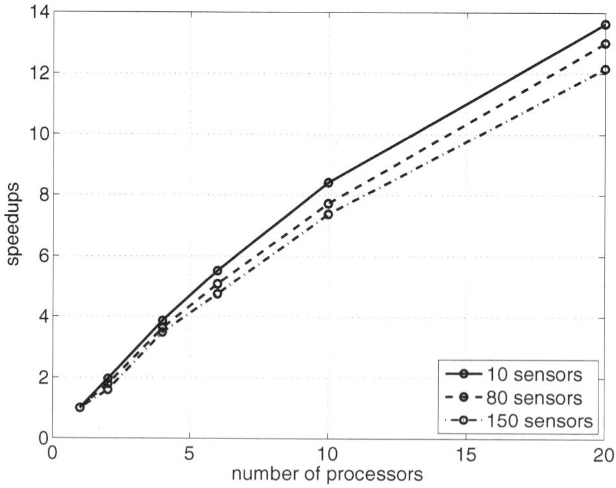

Fig. 6.4 Speedups for different problem sizes

Resource-aware sensor activation. As an illustration of the multiobjective sensor scheduling, reconsider the experimental setting of the air pollutant transport process from two stationary sources described in the first paragraph of Section 5.4. Figure 5.1 illustrates the resulting complex process dynamics.

This time our goal is to identify the intensities of the pollution sources, i.e., the vector of $\theta = (\mu_1, \mu_2)$, using a homogenous sensor network with scanning nodes with the range of 300 m. Simultaneously, we are interested in minimizing the number of sensors used per time slot (i.e., cost per each sensor is assumed to be $b_i = 1$) and the number of time slots each sensor is activated. The weights for particular objectives were set to $\beta_1 = 0.5$, $\beta_2 = 0.3$ and $\beta_3 = 0.2$, ordering their importance. The observation horizon was split into five evenly partitioned subintervals $T_k = [200(k-1), 200k)$, $k = 1, \dots, 5$. In order to verify the proposed approach, a MATLAB program was written using a PC equipped with Intel Core i7 processor (1.6GHz, 4 GB RAM) running Windows 7 and MATLAB 2009b. The nominal values of the system parameters were assumed to be exactly the same as in Section 5.4.

The observation grid was assumed to be located on the internal points of a spatial triangulation mesh (there was 52 such points, which are indicated with dots in Fig. 6.5(a)). A separate triangulation mesh composed of 68 triangles was used to generate the partitioning of the domain. The size of the mesh should be a compromise between reasonably small number of constraints and good decomposition of the domain, giving sensors the possibility to cover many different subdomains. The potential sensor locations and area partitioning are shown in Fig. 6.5(a). As an effective solver for the relaxed problem within the BB search, the `fgoalattain` function from the MATLAB Optimization Toolbox was applied. The algorithm started from a randomly generated initial solution and converged within 98 recursive calls to the procedure solving the relaxed optimization problem. Optimal sensor configurations at the consecutive time slots are shown in Fig. 6.5. From the point of view of estimation accuracy, the sensors should follow the areas of greatest changes in the pollutant concentration, but due to the limited activation time and the limited number of nodes used, the sensors switch to other areas and it is hard to interpret the changes in the observation strategy. Evidently, the coverage of the monitored area is guaranteed with no more than 12 sensors used (the goal for this objective was 9) and no more than 3 time slots allocated for each sensor (the goal was 1). The value of the determinant of the information matrix dropped to 52% of that achieved when the D-optimality criterion was the only performance index.

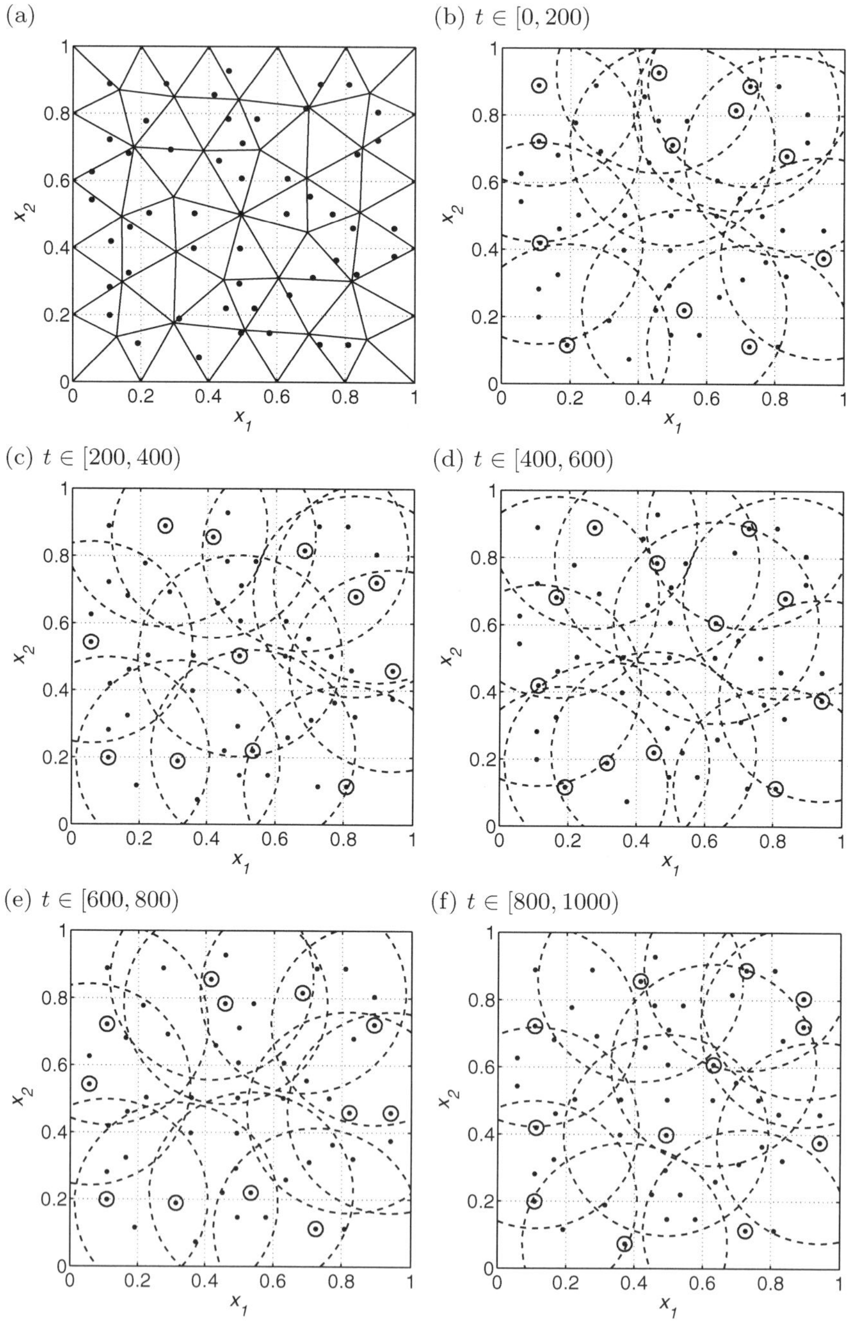

Fig. 6.5 Spatial partitioning of the monitored area with admissible sensor positions (a) and allocation of active sensors with their range in the consecutive stages of the experiment (b)–(f)

6.7 Summary

This chapter addresses the problem of selecting optimal observation points for scanning networks in view of a combinatorial optimization approach. In the existing formulations, an optimal sensor activation schedule is computed as that which globally minimizes a criterion directly connected with the expected quality of the parameter estimates. But then the key difficulty becomes the large scale of the resulting global optimization problem, since the monitoring networks encountered in process industry or environmental engineering may often consist of several hundreds of stations. Obviously, this makes the exhaustive search on a candidate-by-candidate basis practically intractable and creates a need for techniques which would implement guided search and have acceptable performance.

A starting point here is the most common formulation, in which the measurement system has a finite number of sensor candidate positions and the aim is to select the best subset of points of desired cardinality. Choosing the best subset translates to minimizing a criterion defined on the Fisher information matrix and fits into the framework of nonlinear 0–1 integer programming. The solution of this combinatorial design problem using the branch-and-bound method constitutes a quite natural option, but the main problem when trying to implement it has been the lack of a low-cost procedure to obtain lower bounds to the optimal values of the optimality criterion. The main contribution of this chapter consists in the development of an original and efficient computational scheme to produce such bounds. This was possible by adapting specialized algorithms for experimental effort optimization delineated in Chapter 3, which are in common use by statisticians concerned with optimum experimental design. The link to plug this algorithms into the proposed scheme was the simplicial decomposition perfectly suited for large-scale problems which can be encountered here. The idea of its application in the context of optimal sensor scheduling for parameter estimation is new. Consequently, the proposed method can be implemented with great ease and our experience provides evidence that, with this tool, even large-scale design problems can be solved using an off-the-shelf PC. A further significant speedup can be achieved by employing a simple rounding heuristic, as described in Section 6.6.

As a by-product of the process of constructing the ultimate algorithm, characterization of optimal solutions to the specific subproblem discussed in Proposition 6.1 was obtained. In principle, it can be derived as particular case of some general results in optimum experimental design, but this would require from the reader a fluent knowledge of measure theory. Because of this, the corresponding proof, being a generalization of result from [291], proceed here independently on a quite elementary level, based on the Karush–Kuhn–Tucker conditions.

Let us remark that an alternative approach to select a best n-element subset from among a given N-element set of candidate sites could be to employ an exchange class of algorithms presented in Section 3.1.4 within a sensor-network setting, based on the concept of replication-free designs. Even more efficient extensions of this idea could be to adapt the fast algorithm based on multiple simultaneous exchanges [142, 197, 295]. A step in this direction was made by Liu et al. [154], who refined it and applied the resulting 'sort-and-cut' technique to solve an E-optimum sensor selection problem. It is beyond doubt that this approach outperforms the BB technique proposed here as far as the running time is concerned. On should note, however, that exchange algorithms are heuristics and thus are only capable of finding globally competitive solutions (i.e., nearly optimal ones), with an explicit trade of global optimality for speed. The approach presented here is superior in the sense that it always produces global maxima and, what is more, does it within tolerable time.

Certainly, there is room for refinements and further theoretical developments. Since computing on clusters of PCs has become as common as computing on scientific workstations had been a decade ago, a parallel version of the branch-and-bound application aimed at solving large-scale problems is also developed within this chapter. This task is facilitated by the specific structure of the proposed branch-and-bound algorithm whose components are particularly suited for parallelization, which is achieved without complex extensions. Consequently, a powerful tree-based approach is developed, capable of solving scheduling tasks of serious complexity on a grid of computers.

Another practical extension addressed in this chapter is a resource-aware scheduling approach based on the formulation of the sensor activation problem as a multiobjective optimization one [206], in which we are interested in simultaneous maximization of identification accuracy and minimization of the number of time slices each node is active, employing a minimal number of sensors which guarantee the coverage of whole monitored area. The problem is solved using the goal attainment method, which is suitably incorporated into the BB technique.

Chapter 7
Sensor Location under Parametric and Location Uncertainty

In Section 2.2.5.4 it was indicated that, in general, for nonlinear parametrization of the system responses, optimum experimental conditions strongly depend on the unknown parameter values which only have to be estimated. This causes one of the main complications related to the determination of optimal experimental conditions. A common approach is then to design the experiment for some reasonable nominal parameter values whose knowledge is a prerequisite for applying the locally optimal sensor location methods described in the previous chapters. Since the uncertainty of those nominal values is not taken into account, practical application of such procedures is limited to situations when system responses change slowly in the set of admissible parameters.

Often, the parameter estimates at hand are far from their true values and, in addition to this, properties of locally optimal sensor locations can be very sensitive to changes in these parameters [80]. Neglecting the problem of parametric uncertainty may make the solution far from optimal. Consequently, this may lead to a question about practical usefulness of the nonlinear experimental design [306]. To overcome those difficulties, several attempts have been made in the literature regarding general optimum experimental design theory either by application of the sequential approach [70, 80, 306, 307], modification of the obtained locally optimal designs [143], introduction of designs in the average sense or the so-called minimax designs [180, 220, 221, 276, 277, 281, 307]. However, none of the listed techniques is without drawbacks and the problem still remains open.

Another difficulty appearing within the sensor location problem is that the existing approaches rely on the common assumption that the potential sensor positions are known with an arbitrary accuracy. In practical experimental settings, such a situation is often impossible to achieve because of the technological constraints of the network, structural dislocations, positioning errors or limitations of the method used for deployment of sensors over a spatial area (e.g., nodes are deployed using an airplane). Specifically, an

M. Patan: Optimal Sensor Networks Scheduling in Identification of DPSs, LNCIS 425, pp. 183–206.
springerlink.com

experimental design for estimating the parameters of a DPS is considered when the actual sensor locations used during the measurement stage may randomly change around the values specified at the network design stage. In the context of DPSs, such a kind of observation model uncertainty has received very little attention so far [185, 202].

The main purpose of this chapter is to briefly present the methods of sensor scheduling for monitoring networks whose inherent characteristic is some level of uncertainty regarding the model parameters and/or sensor positions. Apart from the appropriate problem formulation, suitable design criteria are provided, and efficient configuration algorithms originating from the experimental design and statistical learning theory are presented as well.

7.1 Sequential Design Techniques

Since it is clear that for any design ξ the elements of the FIM depend on the true vector of parameters θ_{true} which is obviously unknown, determination of the optimal design will require finding some compact domain Θ_{ad} containing θ_{true} and where $M(\xi,\theta)$ varies insignificantly for $\theta \in \Theta_{\mathrm{ad}}$. In the case when initial information about the parameters is missing, the only solution is to conduct an additional analysis for predetermination of the region Θ_{ad}. Nevertheless, it is impossible to construct an optimal design for such an experiment without a sufficient knowledge about the parameters, because most often optimal sensor locations depend on θ.

A very natural idea is to alternate experimentation and estimation steps. Accordingly, the total time horizon is divided into several contiguous parts and each of them is related to the corresponding stage of the experiment. At each stage, in turn, locally optimal sensor configurations are determined based on the available parameter estimates (nominal parameter values can be assumed as initial guesses for the first stage), measurements are taken at the newly calculated sensor positions, and the data obtained are then analyzed and used to update the parameter estimates (see Fig. 7.1(a)). In this general scheme, it is intuitively supposed that each estimation phase improves our knowledge about the parameters and this knowledge can then be used to improve the quality of the next experiment to be performed. Owing to its simplicity, sequential design is commonly considered as a universal panacea for the shortcomings of local designs. Moreover, the idea is relatively easy to implement and could be used in conjunction with a wide class of design optimization algorithms without substantial improvements. However, some important and nontrivial questions are to be faced, cf. [80], e.g., regarding the influence of initial estimates on the design or asymptotic properties of sequential procedures. Some developments regarding a theoretical justification for the sequential approach and its convergence properties can be found, e.g., in [80, 306, 307].

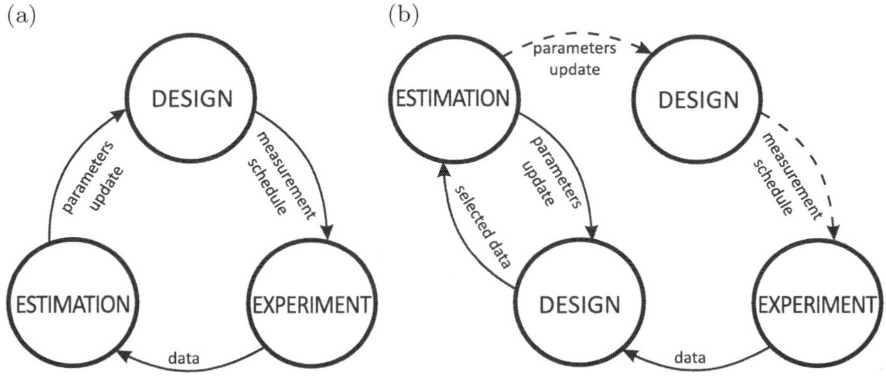

Fig. 7.1 General scheme of adaptive sequential design (a) and sequential design with an additional iterative design loop (dashed arrows denote optional steps) (b)

At this point, it should be mentioned that some of the approaches considered within the framework of this book are especially suited for adaptation when implementing sequential design strategies. Particularly, a scanning strategy provides by definition the division of the experimental resources (time interval) into separate parts. This fact can often be exploited to naturally extend the approaches from Sections 3.1 and 6.2 using sequential techniques in such a way as to obtain a higher level of robustness to parametric uncertainty [180, 197]. Sequential designs have been employed here to incorporate the hitherto collected measurements into the process of constructing an optimized sensor schedule. Starting from the sensor scheduling task defined by Problem 6.1, the critical issue is to build the sensor activation schedule based on a dynamic-data driven scheme of computations, i.e., the currently selected measurements from the sensors must serve as a basis for the update of parameter estimates and these, in turn, can be used to correct the sensor measurement schedule.

7.1.1 Measurement Selection via Iterative Design

One of the main drawbacks of classical adaptive sequential design is the fact that it sometimes becomes impractical, because it is always related to an additional cost of the experiment (e.g., the experimental time may be too long or resources spent on the experiment may be too excessive). In addition to this, sequential designs necessitate the experimentation stage to be renewable. Since for some classes of systems this is rarely possible (e.g., industrial processes), it becomes clear that such an approach cannot be treated as a universal remedy for the discussed shortcomings of local designs. Finally, it is often desired to choose only a subset of measurements gathered by sensors during a given time interval. A reason for not using all the available sensor data could be the rejection of noisy uninformative observations which deteriorate the quality of the estimation process as well as reduction of problem complexity and/or the cost of observation system operation. To overcome this difficulty, the classical sequential scheme can be enhanced by inclusion of an internal design loop, cf. Fig. 7.1(b). The purpose of this additional design phase is the preprocessing of the experimental data so as to select the most informative observations for the estimation task. In such a way we may improve the quality of both the design and parameter estimates without the necessity of numerous replications of the experiment, which becomes optional (Fig. 7.1(b)).

A heuristic scheme of such a procedure is embodied in Algorithm 7.1. The operators involved in this implementation are as follows:

- CollectMeas(ξ_r) returns the system response measured by sensor nodes according to schedule ξ_r.
- EstimateParams(Z_{sel}) determines the estimate of system parameters for the current design interval.
- MeasDesProb($\widehat{\theta}$) returns the locally optimal sensor activation schedule determined for the current parameter estimate $\widehat{\theta}$.
- SelDesProb($\hat{y}, Z_{\text{meas}}$) returns the solution to the data selection design problem formulated by Problem 6.1 for the set of admissible locations defined by support of design $\xi^{(r)}$. As a result, we obtain the schedule ζ and the corresponding selected subset of observations $Z_{\text{sel}} \subset Z_{\text{meas}}$.
- SolvePDEs($\widehat{\theta}$) generates the solution to the system response and its sensitivities based on the estimated system parameters.
- StopEstim() returns the least-square problem termination condition generated by the solver in EstimateParams.

7.2 Optimal Designs in the Minimax Sense

When trying to develop some alternative approaches to sequential designs to tackle the parametric uncertainty of the model, one of the possible solutions is to optimize the performance of the experiment for the worst possible case in the set of admissible parameters Θ_{ad}. The most attractive property of this idea is that no further hypothesis about the distribution of the parameters is needed.

Algorithm 7.1. Sequential-iterative design procedure
Input variables: R is a prescribed upper limit of experimental phases, $\widehat{\theta}_0$ is an initial estimate of system parameters, ξ_0 is an initial schedule of measurements

```
procedure SequentialDesign(R, θ̂₀, ξ^(0))
    r ← 0                                       ▷ Initialization of sequential design counter
    Z_meas ← ∅                                  ▷ Initialize set of observations
    repeat                                      ▷ External adaptive loop of sequential design
        Z_meas ← Z_meas ∪ CollectMeas(ξ^(r))    ▷ EXPERIMENT PHASE
        repeat                                  ▷ Internal iterative loop of design
            ŷ ← SolvePDEs(θ̂)                    ▷ Approximation of PDE system evolution
            [ζ, Z_sel] ← SelDesProb(ŷ, Z_meas)  ▷ INTERNAL DESIGN PHASE
            θ̂ ← EstimateParams(Z_sel)           ▷ ESTIMATION PHASE
        until StopEstim()                       ▷ Internal iteration stop condition
        ξ^(r+1) ← MeasDesProb(θ̂)                ▷ EXTERNAL DESIGN PHASE
        r ← r + 1
    until r ≤ R
end procedure
```

7.2.1 Problem Statement

Focusing all attention and effort on providing possibly maximum information to a parameter vector θ which is most difficult to identify in O_{ad}, the problem can be brought down to the minimization of the criterion

$$\mathcal{J}_{\mathrm{MM}}(\xi) = \max_{\theta \in \Theta_{\mathrm{ad}}} \Psi[M(\xi, \theta)]. \tag{7.1}$$

The most popular choice in the literature is the minimization of the MMD-optimality criterion [180, 278, 281, 307],

$$\mathcal{J}_{\mathrm{MMD}}(\xi) = \max_{\theta \in \Theta_{\mathrm{ad}}} \big[\ln\det M^{-1}(\xi, \theta)\big]. \tag{7.2}$$

The minimax criterion (7.1) defines the optimal design as

$$\xi^\star_{\mathrm{MM}} = \arg \min_{\xi \in \Xi(X)} \big\{ \max_{\theta \in \Theta_{\mathrm{ad}}} \Psi[M(\xi), \theta] \big\}, \tag{7.3}$$

where Θ_{ad} is assumed to be compact. To characterize the minimax designs, assume that both $\Psi(\,\cdot\,,\,\cdot\,)$ and $\partial\Psi(\,\cdot\,,\,\cdot\,)/\partial M$ are continuous. Then, it can be shown (cf. [73, Th. 2.6.1, p.42] or [281, Th. 3.5, p.69]) that a necessary and sufficient condition for $\xi^\star_{MM}$ to be optimal is the existence of a probability measure $\omega^\star$ on $\Theta_{\mathrm{ad}}(\xi) = \big\{\hat{\theta} : \Psi[M(\xi, \hat{\theta})] = \max_{\theta \in \Theta_{\mathrm{ad}}} \Psi[M(\xi, \theta)]\big\}$ such that

$$\min_{x \in X} \int_{\Theta_{\mathrm{ad}}(\xi)} \psi(x, \xi^\star_M, \theta)\, \omega^\star(\mathrm{d}\theta) \geq 0.$$

Additionally, the Carathéodory theorem ensures the existence of $\omega^\star$ with no more than $m+1$ support points.

Clearly, practical application of these results is by much more difficult than the corresponding conditions for locally optimal designs. The optimization of the minimax criteria provides reasonably good results for the worst possible values of parameters (which may potentially not be unique), but for the other parameter vectors the quality of the optimal design might be questionable. As a final comment, let us indicate one more drawback of minimax designs, namely, that invariant criteria with respect to transformations of θ for calculation of locally optimal designs may not maintain this valuable property with respect to the minimax approach [80].

7.2.2 Minimax Optimization via Relaxation

There exist very few efficient numerical algorithms devoted to solving general minimax problems and, in addition to this, most of them are limited to the situations where optimization is performed over finite sets. Since both ξ and θ belong to infinite sets, the solution of the problem (7.1) becomes a rather hard task except for rare situations where for some regression models the minimax Ψ-optimal design can transformed into a conventional Ψ-optimal one [221]. However, in [281], the discussed optimization problem (7.1) is proposed to be exchanged with minimization of a scalar α, with respect to

$$\max_{\theta\in\Theta_{\mathrm{ad}}} \Psi[M(\xi,\theta)] \le \alpha. \tag{7.4}$$

This is equivalent to the set of constraints

$$\{\Psi[M(\xi,\theta)] \le \alpha, \quad \theta\in\Theta_{\mathrm{ad}}\}, \tag{7.5}$$

which is obviously infinite.

The so redefined problem can be solved with the use of some numerical procedures for inequality constrained Semi-Infinite Programming (SIP) [214, 237]. In this vein, an intuitive approach can be proposed for construction of a finite set containing representative values of θ, which is equivalent to relaxing the problem by taking into account only a finite number of constraints (7.5). This is embodied in the simple relaxation algorithm given by Shimizu and Aiyoshi [251], and its well performance was proven in nonlinear experimental design problems for LPSs [221, 307] and DPSs [180, 281]. The general form of this procedure can be represented by the scheme below.

Shimizu and Aiyoshi [251] showed that the above routine terminates in a finite number of iterations for any ϵ. Practical usefulness of the algorithm was confirmed for the case of stationary sensors [180, 276], but its applicability to mobile sensors is rather limited due to problems with solving the minimax

Algorithm 7.2. Minimax design via relaxation
Input variables: R is a maximal iteration number, the prescribed upper limit of experimental phases, θ^0 is an initial estimate of system parameters, ϵ is a predefined solution tolerance

1: **procedure** MINIMAXDESIGN(R, θ^0, ϵ)
2: $\quad r \leftarrow 0$
3: $\quad \Theta^r_{\mathrm{rep}} \leftarrow \{\theta^0\}$ ▷ Initialize the set of representative parameter values
4: $\quad$**repeat**
5: $\quad\quad \xi^{(k)} \leftarrow \arg\min_{\xi \in \Xi(X)} \left\{ \max_{\theta \in \Theta^r_{\mathrm{rep}}} \Psi[M(\xi, \theta)] \right\}$ ▷ Solve the current relaxed problem
6: $\quad\quad \theta^{r+1} \leftarrow \arg\max_{\theta \in \Theta_{\mathrm{ad}}} \Psi[M(\xi^{(r+1)}, \theta)]$ ▷ Solve the maximization problem
7: $\quad\quad$**if** $\Psi[M(\xi^{(r)}, \theta^{r+1})] \leq \max_{\theta \in \Theta^r_{\mathrm{rep}}} \Psi[M(\xi^{(r)}, \theta)] + \epsilon$ **then return**
8: $\quad\quad$**end if**
9: $\quad\quad r \leftarrow r + 1$
10: $\quad$**until** $r \leq R$
11: **end procedure**

optimization problem in line 5. Some indications of possible alternatives can be found in the works of Uciński [277, 281]; however, to the best of the author's knowledge, the procedure has not been verified yet in the case of scanning sensors.

7.3 Optimal Designs in the Average Sense

A common opinion is that methods based on the minimax interpretation of the parametric robustness of design lead to overly conservative solutions [303]. This flaw stems from the worst-case nature of the associated performance index. Indeed, it is possible that a design $\xi_1(\theta)$ will be preferred to a design $\xi_2(\theta)$, even though the former seems to be worse than the latter for 'most' realizations of θ. Therefore, if a minimax performance index is used, then all the efforts will be spent on determining the worst-case situation, which might not be a representative case. Sometimes, it is more reasonable to waive such strong requirements and consider designs which are satisfactory for 'most' realizations of $\theta \in \Theta_{\mathrm{ad}}$. This leads directly to designs in the *average* sense, which will be discussed in the following section.

7.3.1 Problem Statement

When the set Θ_{ad} of possible values of θ is compact, then a very logical way to capture this idea is to employ an approach based on a probabilistic description of the prior uncertainty of the unknown system parameters. This uncertainty is assumed to be characterized by the distribution μ which is deduced from the a priori knowledge about the system considered (e.g., it results from previously made observations collected on similar systems). This distribution represents experimenters' additional knowledge which is accessible a priori and regards the confidence level about representative values of the unknown parameters. For such an interpretation of the statistical relations between the parameters and the system behavior, it is possible to define the criterion to be minimized as the expectation of the corresponding 'local' performance index, i.e.,

$$\mathcal{J}(\xi) = \mathrm{E}_{\theta}\left\{\Psi[M(\xi,\theta)]\right\} = \int_{\Theta_{\mathrm{ad}}} \Psi[M(\xi,\theta)]\,\mu(\mathrm{d}\theta). \tag{7.6}$$

A common assumption is to choose the measure μ as

$$\mu(\mathrm{d}\theta) = p(\theta)\,\mathrm{d}\theta, \tag{7.7}$$

where p denotes the prior probability density function for θ. In real-world problems the direct transformation of the a priori knowledge about parameters into a statistical distribution in closed form is rather difficult, or such knowledge is far from being complementary. Then, commonly encountered examples of p can be applied [264, 276]:

- In the case when the parameter vector θ_{true} is known with a probability very close to unity, the prior distribution

$$p(\theta) = \delta(\theta - \theta^0)$$

  can be used.
- In the case when θ is limited to the region Θ_{ad}, but no further information is provided and the experimenter cannot distinguish any particular realization of θ, the uniform distribution on Θ_{ad} may be assumed, i.e.,

$$p(\theta) = \begin{cases} 1/V(\Theta_{\mathrm{ad}}) & \text{for } \theta \in \Theta_{\mathrm{ad}}, \\ 0 & \text{otherwise,} \end{cases}$$

  where $V(\Theta_{\mathrm{ad}})$ denotes the volume of Θ_{ad} in the Lebesgue sense.
- In the case when the θ distribution can be estimated with the accuracy up to second order characteristics (e.g., the expected value $\mathrm{E}(\theta)$ and the covariance matrix $\operatorname{cov}(\theta)$ are available), but there is no other information, the following multidimensional normal distribution can be assumed:

$$p(\theta) = (2\pi)^{-m/2}\big(\det(\operatorname{cov}\theta)\big)^{-1/2}\exp\Big(-\frac{1}{2}\big(\theta - \mathrm{E}(\theta)\big)^{\mathsf{T}}(\operatorname{cov}\theta)^{-1}\big(\theta - \mathrm{E}(\theta)\big)\Big).$$

  The appropriate regularization and normalization should be imposed if values of $p(\theta)$ in the vicinity of the Θ_{ad} limits are not negligible.

According to (7.6), a design that is optimal in the average sense can be defined as

$$\xi_E^\star = \arg \min_{\xi \in \Xi(X)} \int_{\Theta_{\mathrm{ad}}} \Psi[M(\xi, \theta)] \, \mu(\mathrm{d}\theta), \tag{7.8}$$

and is often called a Bayesian design. As the integral in (7.8) plays the role of a linear operator, Theorems 2.3 (p. 23) and 2.4 (p. 25) can be relatively easily reformulated by introducing the following functions:

$$\psi(x, \xi) = \int_{\Theta_{\mathrm{ad}}} \hat{\psi}(x, \xi, \theta) \, \mu(\mathrm{d}\theta), \tag{7.9}$$

$$\phi(x, \xi) = \int_{\Theta_{\mathrm{ad}}} \hat{\phi}(x, \xi, \theta) \, \mu(\mathrm{d}\theta), \tag{7.10}$$

$$\varsigma(\xi) = \int_{\Theta_{\mathrm{ad}}} \hat{\varsigma}(\xi, \theta) \, \mu(\mathrm{d}\theta), \tag{7.11}$$

where $\hat{\psi}(x, \xi, \theta)$, $\hat{\phi}(x, \xi, \theta)$ and $\hat{\varsigma}(\xi, \theta)$ are the equivalents of (2.37)–(2.39), respectively, calculated for a particular realization of the parameter vector θ. In such a way, the optimality conditions and equivalence results can be generalized to the case of Bayesian designs. As regards the existence of optimal solutions, Theorem 2.2 can also be adopted, but with the essential difference that the maximal number of support points cannot be guaranteed to be less than or equal to the value of $m(m+1)/2$. This is a consequence of the fact that the Carathéodory theorem cannot be directly applied since $\mathcal{J}_E$ depends on different matrices $M(\xi, \theta)$ for different vectors θ.

At this point, it should be emphasized that the class of Bayesian criteria is much wider than the corresponding class of their local counterparts. For example, only for D-optimality various choices of suitable functions [307] exist, e.g.,

- ED-optimality criterion

$$\mathcal{J}_{\mathrm{ED}}(\xi) = \mathrm{E}_\theta \big\{ - \det M(\xi, \theta) \big\}, \tag{7.12}$$

- EID-optimality criterion

$$\mathcal{J}_{\mathrm{EID}}(\xi) = \mathrm{E}_\theta \big\{ [\det M(\xi, \theta)]^{-1} \big\}, \tag{7.13}$$

- ELD-optimality criterion

$$\mathcal{J}_{\mathrm{ELD}}(\xi) = \mathrm{E}_\theta \big\{ - \ln \det M(\xi, \theta) \big\}. \tag{7.14}$$

The above criteria usually lead to different optimal solutions, and therefore their practical usefulness depends on some prior experience of the experimenter (cf. [307]).

Since analytical solutions can be obtained only for very simple cases, some efficient numerical techniques are required. Unfortunately, Theorem 2.4

rewritten in this new framework yields efficient solutions only in the case when the distribution μ is discrete with a reasonably moderated number of support points. The main complication in applying any numerical approach from the previous chapters arises from the fact that in order to solve the problem (7.8), respective expectations have to be evaluated, which leads to time-consuming calculations of the multi-dimensional integrals (cf. (7.9)–(7.11)). However, to overcome those difficulties, some stochastic algorithms can be applied to avoid direct numerical evaluation of the statistical expectations, which will be briefly presented in the next section.

7.3.2 Bayesian Optimization via Statistical Learning

As has been indicated earlier, a direct application of numerical algorithms developed in the previous chapters to solving (7.8) is highly complicated by the necessity of tremendous calculations related to the evaluation of the expectations of a local optimality criterion. However, the situation when a criterion is not given explicitly or its evaluation requires great computational efforts is standard for stochastic-approximation techniques. For that reason, those methods seem attractive as they iteratively yield an approximated solution without determination of the functional to be optimized. Such an approach was introduced and successfully applied to the finite-dimensional case by Walter and Pronzato [307] and extended to the sensor location problem for DPSs by Uciński [276], who used a simple classical Robbins–Monro algorithm [212, 260], also known as the stochastic-gradient algorithm.

However, within the framework of this work, another alternative approach will be proposed, which is based on statistical learning theory. It was originally developed in the context of robust controller design by Vidyasagar [301, 302, 303] and adopted to the sensor location problem by Patan [180] and Uciński [281]. Further extensions toward the scheduling of a sensor network constitute the subject of the following sections.

7.3.2.1 Relaxed Notions of Minima

First, assume that the criterion $\Psi[M(\xi,\theta)]$ can take only nonpositive values (this assumption is not too restrictive, as, e.g., the D-optimality criterion satisfies it). Then, consider a measure of the system performance for both the vector θ and design ξ, which takes values from the unit interval. Such a rescaling is essential for the randomized approach presented in what follows. This can be achieved, e.g., by the following transformation of Ψ:

$$\mathcal{P}(\xi,\theta) = \frac{1}{1-\Psi[M(\xi,\theta)]}. \tag{7.15}$$

Once a probability distribution μ over a set Θ_{ad} is chosen, the objective function to be minimized can be defined as

$$\mathcal{J}(\xi) = \mathrm{E}_\theta[\mathcal{P}(\xi, \theta)], \tag{7.16}$$

which implies that a design ξ is occasionally permitted to be inadequate for conditions which are not likely to occur, i.e., which are associated with low probability values. In such a way the problem of determining a Bayesian design is reduced to finding the minimum of a function $\mathcal{J}(\xi) : \Xi(X) \to [0, 1]$.

It is clear that, apart from the NP-hard character of the optimization problem, another crucial barrier is the very high computational cost for calculation of the criterion (7.16) at given ξ. In order to overcome these impediments, we have to weaken the definition of the minimum for our problem. After Vidyasagar [303], who considers several variants of so-called 'near' minima, let us start with the simplest definition of such a quantity [303].

Definition 7.1. Let $\mathcal{J} : \Xi(X) \to \mathbb{R}$ and suppose that $\epsilon > 0$ is a given number. A number $J_0 \in \mathbb{R}$ is said to be a *Type 1 near minimum of* $\mathcal{J}(\,\cdot\,)$ *to accuracy* ϵ, or an approximate near minimum of $\mathcal{J}(\,\cdot\,)$ to accuracy ϵ if

$$\inf_{\xi \in \Xi(X)} \mathcal{J}(\xi) - \epsilon \leq J_0 \leq \inf_{\xi \in \Xi(X)} \mathcal{J}(\xi) + \epsilon. \tag{7.17}$$

An approximate near minimum is probably most commonly identified with the notion of a 'near' minimum. Nevertheless, for the robust approach considered in the average sense, the determination of even an approximation to $J^\star$ is related to a high computational cost, which makes its practical application almost impossible. Therefore, it is necessary to look for other notions of near minima, which will be more useful. One of possibilities is provided with the following definition [303].

Definition 7.2. Let $\mathcal{J} : \Xi(X) \to \mathbb{R}$, and assume that ν is a given probability measure on $\Xi(X)$, and that $\alpha > 0$ is a given number. A number $J_0 \in \mathbb{R}$ is said to be a *Type 2 near minimum of* $\mathcal{J}(\,\cdot\,)$ *to level* α, or a probable near minimum of $\mathcal{J}(\,\cdot\,)$ to level α, if $J_0 \geq J^\star$ and, in addition,

$$\nu\{\xi \in \Xi(X) : \mathcal{J}(\xi) < J_0\} \leq \alpha.$$

In other words, J_0 is a probable near minimum of $\mathcal{J}(\,\cdot\,)$ to level α if there exists a set S with probability $\nu(S) \leq \alpha$, such that

$$\inf_{\xi \in \Xi(X)} \mathcal{J}(\xi) \leq J_0 \leq \inf_{\xi \in \Xi(X) \setminus S} \mathcal{J}(\xi). \tag{7.18}$$

Examples of some algorithms for finding Type 2 near minima can be found in the literature for the situation where the probability distribution is continuous [268] and where this assumption is removed [301]. However, the main impediment is the fact that those algorithms require the performance index $\mathcal{J}(\xi)$ to be computable for any given $\xi \in \Xi(X)$. In the case of (7.16), this requirement is too strong (the precise values of an expectation involve computation of multivariate probability integrals). To further relax this requirement, the following ultimate notion of a near minimum can be proposed.

Definition 7.3. Let $\mathcal{J} : \Xi(X) \to \mathbb{R}$, and suppose that ν is a given probability measure on $\Xi(X)$, and that $\epsilon, \alpha > 0$ are given numbers. A number $J_0 \in \mathbb{R}$ is said to be a *Type 3 near minimum of $\mathcal{J}(\,\cdot\,)$ to accuracy ϵ and level α*, or a probably approximate near minimum of $\mathcal{J}(\,\cdot\,)$ to accuracy ϵ and level α, if $J_0 \geq J^\star - \epsilon$ and, in addition,

$$\nu\{\xi \in \Xi(X) : \mathcal{J}(\xi) < J_0 - \epsilon\} \leq \alpha.$$

This can be interpreted as the existence of an 'exceptional' set $Z \subseteq \Xi(X)$ with probability $\nu(S) \leq \alpha$ such that

$$\inf_{\xi \in \Xi(X)} \mathcal{J}(\xi) - \epsilon \leq J_0 \leq \inf_{\xi \in \Xi(X) \setminus Z} \mathcal{J}(\xi) + \epsilon. \tag{7.19}$$

It becomes clear that a probably approximate near minimum (Type 3) is a combination of Type 1 and 2 near minima according to the relationships between them established by the formulae (7.17)–(7.19).

7.3.2.2 Randomized Approach to Finding Approximate Near Minima

Although evaluation of the expectations (7.16) is an excessively costly task, it is possible to approximate them to an arbitrarily chosen accuracy. This is achieved by approximating the expectations by sample means based on a series of randomly generated parameters. More precisely, assume that $\theta_{\mathrm{ms}} = [\theta^1, \dots, \theta^s]^\mathsf{T} \in \Theta_{\mathrm{ad}}^s$ is a collection of independent identically distributed samples from Θ_{ad}, generated according to the probability measure μ. With such a multisample θ_{ms}, for any design ξ it is possible to define the empirical mean of the criterion $\mathcal{J}(\,\cdot\,)$ as

$$\widehat{\mathcal{J}}(\xi) = \widehat{\mathrm{E}}\{\mathcal{P}(\xi,\theta)|\theta_{\mathrm{ms}}\} = \frac{1}{s}\sum_{j=1}^{s} \mathcal{P}(\xi,\theta^j), \quad \xi \in \Xi(X). \tag{7.20}$$

Defining the quantity

$$q(s,\epsilon) = \mu^s\Big\{\theta_{\mathrm{ms}} \in \Theta_{\mathrm{ad}}^s : \sup_{\xi \in \Xi(X)} |\widehat{\mathcal{J}}(\xi) - \mathcal{J}(\xi)| > \epsilon\Big\}, \tag{7.21}$$

it can be said with the confidence $1 - q(s,\epsilon)$ that every single empirical mean approximates the corresponding true value $\mathcal{J}(\xi)$ with accuracy ϵ. Choosing s large enough so that $q(s,\epsilon) < \delta$, where the parameter δ is given a priori, it can be said with confidence $1 - \delta$ that the function $\widehat{\mathcal{J}}(\xi)$ is a sufficiently close approximation to the original criterion $\mathcal{J}(\xi)$. This implies that the exact minimizer of the former function is an approximate near minimizer of the latter to accuracy ϵ, cf. [303] for a detailed discussion.

Note that there always exists a nonzero probability, i.e., $q(\,\cdot\,)$, that the algorithm may fail to produce an approximate near minimum of $\mathcal{J}(\,\cdot\,)$. The law of large numbers implies that these Monte Carlo estimates converge to $\mathcal{J}(\xi)$ as s tends to ∞. As for a measure of the closeness of $\widehat{\mathcal{J}}(\xi)$ to $\mathcal{J}(\xi)$ for a finite s, the bound given by Hoeffding's inequality [301, p. 24] can be used:

$$q(s,\epsilon) \le 2\exp(-2s\epsilon^2). \tag{7.22}$$

This means that once an s-element i.i.d. sample has been drawn from ξ, we may assert that $\widehat{\mathcal{J}}(\xi)$ is at a maximum distance of ϵ to $\mathcal{J}(\xi)$ with confidence coefficient $1-2\exp(-2s\epsilon^2)$. Consequently, if we wish to estimate (7.16) with a predefined accuracy ϵ and a confidence coefficient $1-\delta$, it is sufficient to draw an s-element i.i.d. sample $(\theta^1,\dots,\theta^s)$ from μ, where

$$s \ge \frac{1}{2\epsilon^2}\ln\left(\frac{2}{\delta}\right), \tag{7.23}$$

and to determine the corresponding arithmetic mean $\widehat{\mathcal{J}}(\xi)$. Having clarified how to steer the accuracy of approximating expectations, we can return to the key idea of the algorithm for finding near minima of $\mathcal{J}$ and make the necessary modifications so as to replace the exact values of $\mathcal{J}$ by the corresponding arithmetic means. This leads to the extremely simple numerical procedure, which can be embodied in the following scheme.

Algorithm 7.3. Randomized algorithm for calculation of Type 3 near minima. Input constants ϵ, α, $1-\delta$ denote respectively the accuracy of the solution, the significance level for approximation, and the confidence that a criterion is approximated with the accuracy ϵ; ν is some known *a priori* probability distribution for designs ξ

---

procedure AVERAGEDESIGN(ϵ,α,δ)
    Choose integers $r \ge -\frac{\ln(2/\delta)}{\ln(1-\alpha)}$ and $s \ge \frac{1}{2\epsilon^2}\ln\left(\frac{4r}{\delta}\right)$
    for $i \leftarrow 1, r$ **do**
        Sample ξ^i from distribution ν
        for $j \leftarrow 1, s$ **do**
            Sample θ^j from distribution μ
        end for
        $\widehat{J}_i \leftarrow \frac{1}{s}\sum_{j=1}^{s}\mathcal{P}(\xi,\theta^j)$ ▷ Monte-Carlo estimate of the criterion
    end for
    $i^\star \leftarrow \arg\min_{1\le i\le r}\widehat{J}_i$
    $\xi^\star \leftarrow \xi^{i^\star}$ ▷ probably approximate near minimizer of $\mathcal{J}$ to accuracy ϵ and level α with confidence $1-\delta$
end procedure

The proof of the correctness of this algorithm can be provided analogously to the results in [281, p.193]. Algorithm 7.3 is extremely easy to implement and, in addition to that, it does not require any assumptions about the character of the criterion $\mathcal{J}$ apart from the requirement that its range is a subset of the interval $[0, 1]$. It is only the necessity of the normalization of (7.15), which may result in a very flat surface of the criterion in the vicinity of the minimum that requires some attention. A practical realization of the routine for very small values of parameters ϵ, δ, α may lead to excessively high numbers r and s. Consequently, if the requirements for the approximation accuracy are too strong, the resulting efficiency may be comparable to standard methods.

Example 7.1. In order to clarify differences between the various robust approaches with respect to parametric uncertainty, consider as an example a one-dimensional homogeneous and inductance-free transmission line of length π. The propagation of the voltage signal y over the time horizon $T = [0, 1]$ is described by the parabolic equation

$$\frac{\partial y(x,t)}{\partial t} = \theta_1 \frac{\partial^2 y(x,t)}{\partial x^2} + \theta_2 y(x,t), \quad (x,t) \in \Omega \times T = [0,\pi] \times [0,1]. \quad (7.24)$$

The ends of the line are short-circuited and the initial signal distribution along the line is known, which is reflected by the following boundary and initial conditions:

$$\begin{cases} y(0,t) = y(\pi,t) = 0, & 0 \le t \le 1, \\ y(x,0) = \sin(x) + \sin(2x), & 0 \le x \le \pi. \end{cases} \quad (7.25)$$

The solution of the problem considered can be found in closed form as

$$y(x,t) = e^{(\theta_2 - \theta_1)t} \sin(x) + e^{(\theta_2 - 4\theta_1)t} \sin(2x). \quad (7.26)$$

In our example the unknown parameter vector $\theta = (\theta_1, \theta_2)$ was assumed to belong to the compact set $\Theta_{\mathrm{ad}} = [0, 1]^2$ with the uniform distribution of θ. To simplify our problem, the set of admissible support points was restricted to the uniform grid resulting from 99 divisions of Ω (i.e., evenly distributed 100 nodes). In such a way, the potential measurement sites were the elements of the set $X = \{x^j : x^j = j\pi/99, \ j = 0, \dots, 99\}$. The observation horizon T was split into two equal subintervals $T_1 = (0, \frac{1}{2}]$ and $T_2 = (\frac{1}{2}, 1]$.

The main objective was to determine the activation schedule $\xi = (\xi_1, \xi_2)$ to estimate θ as accurately as possible, based on measurements taken at points from X. To this end, ED- and MMD-optimum design procedures were implemented in MATLAB and run on Intel Centrino Core2 Duo PC (2.5GHz, 3 MB RAM). In particular, Algorithm 3.1 for optimizing experimental effort was adopted with direct numerical evaluation of the expectations (integration of the criterion was performed with use of the procedure DBLQUAD with the adaptive integration based on the Gauss–Kronrod rule implemented by the function QUADGK). Such an approach was compared with Algorithm 7.3 based on the randomized estimation of mathematical expectations. Furthermore, to provide a more comprehensive context, the minimax relaxation (Algorithm 7.2) was also used to obtain an approximation to the robust design in the minimax sense. For all procedures, the initial design was randomly generated, and in the final solutions the points with weights below 0.01 were removed and their experimental effort uniformly distributed among the other supports. In the case of the minimax procedure, the starting set of representative values of θ was chosen as $\Theta^1_{\mathrm{rep}} = \{(0.5, 0.5)\}$, and the final design was obtained for the worst case corresponding to $\theta = (0.0, 1.0)$. All of the results are presented in Table 7.1.

Table 7.1 Comparison of robust approaches

Algorithm	Iterations	Time [h:m]	Approx. solution
Weight optimization with numerical integration ($\epsilon = 10^{-2}$)	1531	01 : 34.2	$\left\{\begin{Bmatrix} 0.92 & 0.952 & 0.984 \\ 0.03 & 0.94 & 0.03 \end{Bmatrix} \begin{Bmatrix} 2.19 & 2.22 \\ 0.98 & 0.02 \end{Bmatrix}\right\}$
Statistical learning ($\delta = \alpha = \epsilon = 10^{-2}$)	$r = 528$ $s = 61303$	00 : 10.1	$\left\{\begin{Bmatrix} 0.92 & 0.952 \\ 0.02 & 0.98 \end{Bmatrix} \begin{Bmatrix} 2.19 \\ 1 \end{Bmatrix}\right\}$
Minimax relaxation ($\epsilon = 10^{-2}$)	6	00 : 07.3	$\left\{\begin{Bmatrix} 0.952 \\ 1 \end{Bmatrix} \begin{Bmatrix} 2.19 \\ 1 \end{Bmatrix}\right\}$

To fully understand the obtained results, note that for both average and minimax designs it is possible to obtain closed-form solutions which have the same form, i.e.,

$$\xi^\star_E = \xi^\star_{MM} = \left\{\begin{Bmatrix} \arctan(\sqrt{2}) \\ 1 \end{Bmatrix} \begin{Bmatrix} \pi - \arctan(\sqrt{2}) \\ 1 \end{Bmatrix}\right\}. \tag{7.27}$$

This means that at each observational subinterval we focus all observational effort on a single point and switch between symmetrically located sites with respect to the center point of the line. Unfortunately, due to the limited space, it is not possible to present detailed calculations (e.g., the closed form of the ED-optimality criterion possesses several hundreds of components). However, this fact proves the quality of the obtained approximations as the support of any design represents the discrete locations from X which are closest to the optimal values. It becomes clear that, from a practical point of view, integration is a critical operation which extends the time of computations. The approach via statistical learning might be a kind of remedy, but it requires some experience regarding selection of suitable values for the algorithm parameters. But its decided advantage is that it offers an extremely easy parallelization of the computations, therefore outperforming standard integration-based approaches. ★

7.4 Configuration of a Sensor Network under Uncertain Location of Nodes

A great majority of contributed works related to the robustness of sensor scheduling address the parametric or structural model uncertainty. However, in practical realizations of sensor networks, another tangible problem, namely, the uncertainty related to sensor locations, should also be taken into account. In various experimental settings this is a result of imprecise deployment methods for sensor nodes (e.g., air drop) or of the fact that the nodes are moving within the environment and their localization methods (e.g., using GPS) are not accurate or the locations can change according to some structural dislocations of the observed system (e.g., transport of medium or displacements in object geometry). Surprisingly, this factor is most often neglected and in the context of DPSs this problem has received very little attention so far. Therefore, the aim of this section is to report a technique for sensor selection able to cope with practical monitoring networks whose inherent characteristic is some level of uncertainty for the sensor positions.

7.4.1 Observations under Uncertain Locations

For the sake of simplicity, in what follows, we consider the observations provided by N stationary pointwise sensors, namely,

$$z^{\ell}(t) = y(x^{\ell}, t; \theta) + \varepsilon(x^{\ell}, t), \quad t \in T, \tag{7.28}$$

where $z^{\ell}(t)$ is the scalar output and $x^{\ell} \in X$ stands for the location of the ℓ-th sensor ($\ell = 1, \dots, N$), X signifies the part of the spatial domain Ω where the measurements can be made, and $\varepsilon(x^{\ell}, t)$ denotes the zero-mean, Gaussian measurement noise, assumed to be spatially uncorrelated and white. There is no loss of generality as extensions to scanning and mobile sensor networks are not difficult.

A very natural idea of taking sensor location uncertainty into account is to assume at the design stage that the x^{ℓ}s are independent random variables, distributed with probability measures ξ_{ℓ} depending on some hyperparameters λ_{ℓ} characterizing the typical values and variability of locations. In the following, let $\lambda = [\lambda_1^{\mathsf{T}}, \dots, \lambda_N^{\mathsf{T}}]^{\mathsf{T}}$. Then, at the estimation stage, given the model $y(x, t; \theta)$ of a DPS and the outcomes of the measurements $z^{\ell}(\,\cdot\,)$, $\ell = 1, \dots, N$, a global minimizer of the output least-squares error criterion can be applied to estimate θ with $\widehat{\theta}$:

$$\mathcal{J}(\theta) = \sum_{\ell=1}^{N} \int_T \{z^\ell(t) - y(x^\ell, t; \theta)\}^2 \, dt. \tag{7.29}$$

Inevitably, the covariance matrix $\operatorname{cov}(\widehat{\theta})$ of the above least-squares estimator depends on the particular realizations of sensor locations x^ℓ. This fact suggests that we may attempt to select them so as to yield the best estimates of the system parameters.

Introducing the multidimensional random variable $\chi = (x^1, \dots, x^N)$, the average FIM in our setting is given by [185, 202]

$$M(\chi, \lambda) = \frac{1}{N} \sum_{\ell=1}^{N} \int_X \Upsilon(x^\ell) \, \xi_\ell(\mathrm{d}x; \lambda), \tag{7.30}$$

where

$$\Upsilon(x) = \frac{1}{t_f} \int_T G^\mathsf{T}(x,t) C^{-1}(x,t) G(x,t) \, \mathrm{d}t, \tag{7.31}$$

with $G(x,t)$ given by (2.86).

Note that we can estimate θ by the least squares method and characterize the precision of the estimation by the FIM because x^ℓ is known when the observation z^ℓ is made. The situation would be quite different if the observations were taken with uncertain locations at the experiment stage.

7.4.2 Conversion to an Average Design Problem

As for a specific form of the criterion which takes into account uncertainty with respect to locations x^ℓ, similarly to the parametric uncertainty, a consistent and logical approach is to make use of the probabilistic description of the prior knowledge about the system considered. Such knowledge reflects the confidence level about representative values of the unknown hyperparameters λ. For such an interpretation of the statistical relations between the hyperparameters and the system behavior, by analogy to (7.6), it is possible to define the criterion to be minimized as the expectation of the 'local' performance index operating on the particular realization of χ,

$$\mathcal{J}(\lambda) = \mathrm{E}_\lambda \{\Psi[M(\chi, \lambda)]\} = \int_{\Lambda_{\mathrm{ad}}} \Psi[M(\chi, \lambda)] \, \rho(\mathrm{d}\lambda), \tag{7.32}$$

where Λ_{ad} is the set of admissible hyperparameters and ρ denotes their probability distribution.

Minimization of (7.32) with respect to λ is equivalent to minimization of the average volume (with respect to realizations of χ) of the uncertainty ellipsoid for system parameter estimates. It can be considered a direct analogue to the 'averaging' criteria for parametric uncertainty discussed in the previous section. A great advantage of such a formulation of the problem is the possibility of almost straightforward application of the randomized approach to finding approximate near minima presented in Section 7.3.2.2. Indeed, Algorithm 7.3 can be easily adopted to the case of uncertain locations of sensor nodes, the only differences in its implementation being as follows.

- instead of sampling designs ξ from distribution ν in line 4 we have to sample hyperparameters λ from given distribution ρ;
- instead of sampling parameters θ from distribution μ in line 6 we have to sample locations x^ℓ from given distributions $\xi^\ell(\lambda)$.

As a result of Algorithm 7.3 tailored to the case considered, the probably approximate minimizer $\lambda^\star$ is determined characterizing the random behavior of positions of sensor nodes.

7.5 Application Examples

7.5.1 Identification of the Heating Source

As the first application briefly illustrating the potential of iterative sequential design in solving inverse problems, we consider the heat transfer process in a thin copper plate under the influence of the moving heat source operating on the upper face of the plate. This represents a typical experimental setting for the welding process or a solar furnace. Mathematically, the spatial distribution of temperature $y = y(x,t)$ over the time interval $T = [0, 300\text{ s}]$ in the three-dimensional domain $\Omega = [0, 0.01\text{ m}] \times [0, 0.05\text{ m}] \times [0, 0.002\text{ m}]$ can be described by the following PDE of the diffusion type:

$$C(y)\frac{\partial y}{\partial t} - \nabla \cdot \big(\lambda(y)\nabla y\big) = 0, \quad x \in \Omega \tag{7.33}$$

subject to the initial

$$y(x,0) = y_0 = 291\text{ K}, \quad \text{in } \Omega, \tag{7.34}$$

and the boundary conditions

$$\lambda(y)\frac{\partial y(x,t)}{\partial n} = 0, \quad \text{on lateral faces of the plate,} \tag{7.35}$$

$$\lambda(y)\frac{\partial y(x,t)}{\partial n} = h(y)(y - y_0), \quad \text{on lower face,} \tag{7.36}$$

$$\lambda(y)\frac{\partial y(x,t)}{\partial n} = h(y)(y - y_0) + f(x,t), \text{ on upper face,} \tag{7.37}$$

where $\lambda(y) = 0.0129y + 10.03$ $[\text{Wm}^{-1}\text{K}^{-1}]$ is the model of thermal conductivity, $C(y) = (81213 - 0.444y)(0.22y + 432.7)$ $[\text{Jm}^{-3}\text{K}^{-1}]$ is the volumetric heat capacity, $h(y) = 10 + 0.9\sigma(y_0 - y)(y_0 + y)(y^2 + y_0^2)$ $[\mathrm{Wm^{-2}K^{-1}}]$ is the convective heat transfer coefficient, $\sigma = 5.67 \cdot 10^{-8}$ $[\mathrm{Wm^{-2}K^{-4}}]$ being a Stefan–Boltzmann constant, $\partial y/\partial n$ stands for the derivative of y in the direction of the outward normal to the boundary Γ and where the term

$$f(x,t) = \mu(t)\big(-\arctan(10^{-6}(\|x-\xi\| - 0.005)) + \pi/2\big)/\pi$$

represents a continuous approximation of the moving heat source operating on the area of a circle of radius 5 mm along the trajectory $\xi = (0.05 + 0.035\cos(2\pi t/300); 0.025 + 0.015\sin(2\pi t/300))$.

The functional form of the time-varying intensity of the heat source $\mu(t)$ was assumed to be modeled with linear piecewise splines, i.e.,

$$\mu(t) = \theta_j + t(\theta_{j+1} - \theta_j)/50, \quad j = 0, 1, \dots, 6 \tag{7.38}$$

and $\mu(0) = \theta_0 = 0$. The subject of interest here is reconstruction of the source intensity (i.e., identification of the parameters $\theta = (\theta_1, \dots, \theta_6))$ based on the observations of the system. Due to the technical limitations, temperature measurements are possible only on the lower face (e.g., a typical situation can be a welding process where the heated face is unavailable) of the plate and are provided by $N = 9$ thermocouples located on the uniform spatial grid $[0.02, 0.05, 0.08] \times [0.01, 0.025, 0.04]$ forming a 3×3 sensor array.

In our simulation studies, the described sequential iterative design technique was applied to determine the D-optimal measurement schedule. Because the heat source operates on the material locally, in some areas the changes in the temperature are extremely small and observational noise can significantly affect the measurements there. Therefore, in order to provide the most informative data and the continuity of observations along a time interval, we aim at choosing only one sensor per every time instant and use such selected measurements for identification of θ.

In order to verify the proposed approach, a MATLAB program was written using a PC equipped with Intel Core i7 processor (1.6GHz, 4 GB RAM) running Windows 7 and MATLAB 2009b. To determine the elements of the sensitivity vector required to calculate the FIM and update the parameter estimates, the direct-differentiation method was implemented. We solved the resulting system of PDEs using routines of the COMSOL environment based on the finite element method [43]. The experiment phase was assumed to be run once and the provided measurements were further processed in the iterative internal design loop. The Levenberg–Marquardt iteration for the estimation phase was used with the Jacobian matrix based on the calculated parameter sensitivities. As for D-optimal design, the above-mentioned exchange type algorithm [180, 198] was applied.

The optimal switching schedule of sensors used at given time subintervals is presented in Fig. 7.2. The sensors tend to reflect the areas of the greatest changes in the temperature but due to the inertia of the heating and cooling processes and dynamic changes in the source intensity they do not always follow the source trajectory (e.g., Fig. 7.2 (d) and (i)). Furthermore, although during the whole experiment the sensor located in the middle provides the most informative data from among all sensors, surprisingly its measurements are not used since for any given time moment there is always another sensor which provides better data.

To verify the quality of the resulting optimum designs in the sense of better statistical properties of parameter estimates, a number of identification experiments were simulated. The true values of system parameters were assumed to be $\theta_{\mathrm{true}} = (20, 100, 100, 80, 40, 100) \cdot 10^3$ [W]. The prior estimates of the parameter vector θ^0 were generated each time randomly with uniform distribution on the region $\theta_{\mathrm{true}} \pm 20\%$. The system output was corrupted with simulated Gaussian noise with standard deviation of 1 K. The accuracy of the estimation process was taken at the level of $\delta \leq 10^{-3}$. The trials not converging to the assumed accuracy were rejected along with trials leading to gross errors. Three scenarios were compared, namely, D-optimal robust iterative design, D-optimal local design for a nominal value $\theta_0 = (50, 50, 50, 50, 50, 50) \cdot 10^3$ (fixed source intensity) and identification using data from all nine sensors. The statistical properties of the estimates are

Table 7.2 Characteristics of estimation for different measurement data schedules

param.	sample mean $\bar{\theta}_j$	95% confidence interval	
D-optimal robust design			
θ_1	19829.48	18855.11	20803.85
θ_2	99194.96	97742.88	100647.05
θ_3	99927.62	99159.86	100695.39
θ_4	80945.53	76596.24	85294.83
θ_5	39556.62	35715.32	43397.92
θ_6	101750.11	98658.41	104841.81
D-optimal local design			
	sample mean $\bar{\theta}_j$	95% confidence interval	
θ_1	21760.68	18973.18	24548.17
θ_2	98053.72	94411.98	101695.46
θ_3	101592.07	98940.86	104243.28
θ_4	81063.75	75477.20	86650.29
θ_5	38430.42	31383.60	45477.23
θ_6	97878.92	86758.16	99999.67
all measurements			
	sample mean $\bar{\theta}_j$	95% confidence interval	
θ_1	18704.60	14396.90	23012.29
θ_2	106928.24	100506.87	113349.61
θ_3	97133.05	93738.32	100527.79
θ_4	96104.46	76873.11	115335.80
θ_5	22674.88	1263.48	44086.28
θ_6	116157.29	102487.09	129827.49

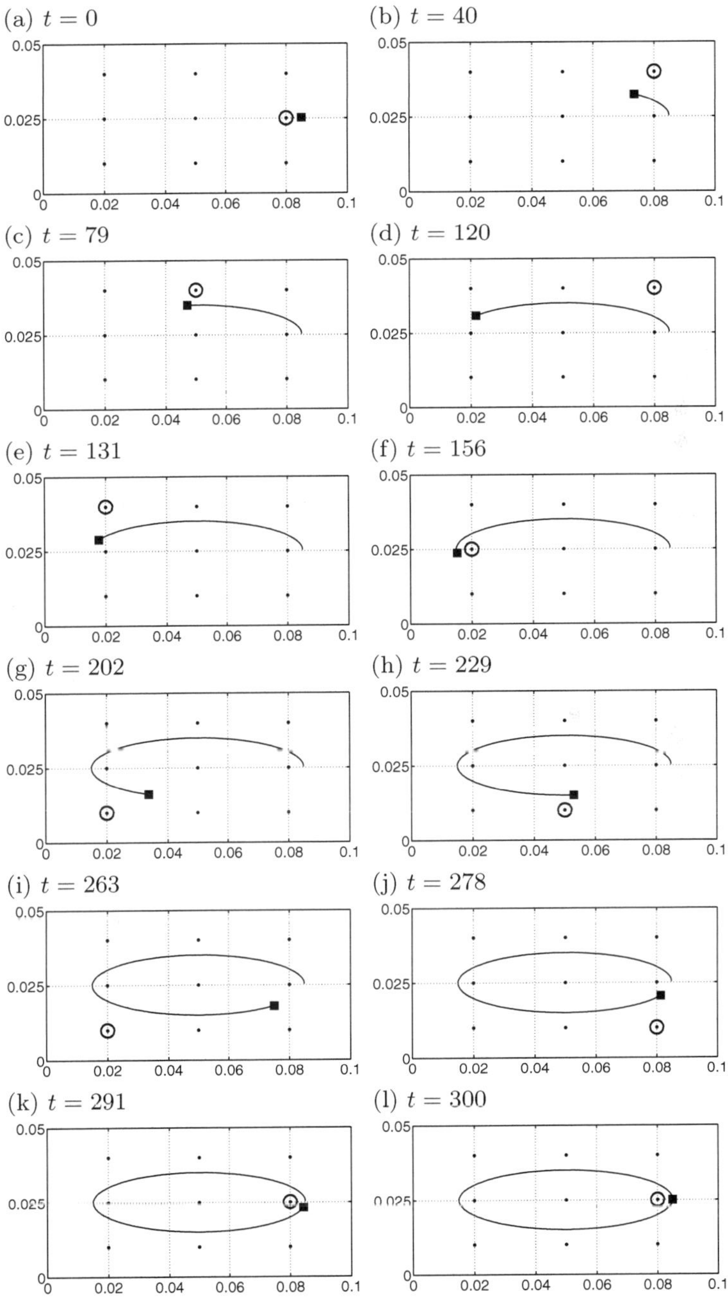

Fig. 7.2 *D*-optimal robust switching schedule of sensing generated by Algorithm 7.1 (selected sensors are indicated with an open circle, a solid line with a square illustrates the trajectory of the moving source and its location at the end of the current time interval)

summarized in Table 7.2. It becomes clear that using all available observations increases the amount of noisy or irrelevant information in the data set, leading not only to biased average values of parameter estimates, but several times larger deviation.

7.5.2 Air Pollution Transport with Uncertain Sensor Locations

As for the second example, an illustration of the randomized approach to sensor network design under the presence of location uncertainty for the process of air pollution monitoring is considered. Within the spatial urban area, which is normalized to the unit square $\Omega = (0,1)^2$, an active source of pollution is present, which influences the pollutant spatial concentration $y = y(x,t)$. The evolution of y over the normalized observation interval $T = (0,1]$ is described by the following advection-diffusion equation:

$$\frac{\partial y(x,t)}{\partial t} + \nabla \cdot \big(v(x)y(x,t)\big) = \nabla \cdot \big(\kappa(x)\nabla y(x,t)\big) + f(x), \quad x \in \Omega \tag{7.39}$$

subject to the boundary and initial conditions:

$$\frac{\partial y(x,t)}{\partial n} = 0 \quad \text{on } \Gamma \times T, \tag{7.40}$$

$$y(x,0) = y_0 \quad \text{in } \Omega, \tag{7.41}$$

where the term $f(x) = 50\exp\big(-50\|x - x^s\|^2\big)$ represents the source of a pollutant located at the point $x^s = (0.3, 0.3)$. The mean spatio-temporal changes of the wind velocity field over the area were approximated by $v = (v_1, v_2)$, where

$$v_1 = 2(x_1 + x_2 - t), \quad v_2 = x_2 - x_1 + t. \tag{7.42}$$

The assumed functional form of the spatial-varying diffusion coefficient $\kappa(x)$ is

$$\kappa(x) = \theta_1 + \theta_2 x_1 x_2 + \theta_3 x_1^2 + \theta_4 x_2^2. \tag{7.43}$$

The subject of interest here is to determine potential areas for the deployment of $N = 5$ monitoring stations which would guarantee possibly the best quality for estimates of system parameters θ in order to produce good predictions of system behavior. However, since the sensors are deployed from the aircraft, their final locations are uncertain to some extent. In order to take these uncertainties into account, the coordinates of sensor spatial positions are assumed to be modeled as normal random variables $x^j \sim \mathcal{N}(\mu^j, \sigma^2 I_N)$. Furthermore, the standard deviation σ in our simulation setting is assumed

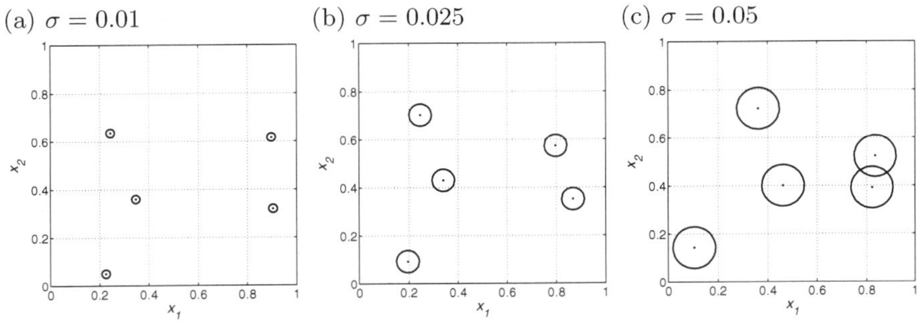

Fig. 7.3 Optimal 90% confidence areas of sensor deployment for different levels of uncertainty

to be known a priori, but the locations μ^j are unknown and, since there is no prior knowledge about their values, μ^js are assumed to be sampled from the uniform distribution ρ over the domain Ω.

In our simulation studies, the described statistical learning technique was applied to determine the locations of stationary sensors. A MATLAB program was written to implement the procedure embodied by Algorithm 7.3. In order to determine the elements of the sensitivity vector required to calculate the FIM, the nominal values of the parameters $\theta_1^0 = 0.02$, $\theta_2^0 = 0.01$ and $\theta_3^0 = \theta_4^0 = 0.005$ were used. The resulting system of PDEs was solved using a MATLAB compatible COMSOL 3.5 environment for the mesh containing 1528 triangles. As for the numerical integration required to evaluate information matrices for admissible observation sites, the trapezoidal rule was applied with the time step equal to 0.04, based on the sensitivity vector interpolated at the nodes representing admissible locations x^ℓ. The pollutant spreads out over the entire domain reflecting the complex combination of diffusion and advection and follows the temporary direction of the wind being the dominant transport factor.

Using the parameters $\delta = \epsilon = 0.01$ and $\alpha = 0.0001$, the proposed algorithm run under Windows Vista on a PC equipped with Intel Centrino T9300 2.53GHz processor and 3GB of RAM produced solutions within 25 min of computation time. Optimal sensor configurations for different values of standard deviation are shown in Fig. 7.3. Again, the complexity of the system dynamics makes an intuitive choice of the observation locations rather difficult and the measurements in the closest vicinity of the pollution source turned out not to be very attractive for parameter estimation. It is clear that with an increase in the standard deviation σ the pattern changes significantly. This effect can be explained by observing that at higher uncertainty represented by high values of σ the spatial average of all observations for a particular sensor dominates the location of maximum performance, leading to a meaningful change in the designed deployment position (cf. Fig. 7.3 (a) and (c)).

7.6 Concluding Remarks

Within the scope of this chapter it was evidenced that the parametric and localization uncertainties are major impediments encountered in the problem of optimal sensor location for DPSs, which makes it highly nontrivial. However, in the former case, it was shown that difficulties arising from the dependence of the optimal solutions on the parameters to be identified can be circumvented to some extent by the introduction of the so-called sequential and robust designs.

In the first part of the chapter, an approach was proposed based on enhancement of sequential experimental design techniques to provide a numerical scheme for computation of experimental designs. Although sequential designs constitute a well-known technique, the resulting time and cost of the experiment may be too long or too large. Introduction of the additional internal design loop makes the sequential approach more flexible, leading to a seamless design and estimation phase. As a result, we obtain both the robust sensing schedule and reliable estimates of the system parameters. From a practical point of view, the approach is very simple for implementation and very easy to combine with other approaches.

Alternatively, as more systematic approaches, designs in minimax and average senses can be introduced on the analogy of the standard procedure in optimum experimental design for nonlinear regression models. Suitable characterizations of the optimality conditions for continuous designs as well as some known numerical algorithms for obtaining approximate solutions were discussed. Specifically, the original statistical learning approach is developed extending the issue of determination of robust designs in the average sense, being an alternative to the standard methods aimed at the local case, since it allows avoiding multi-dimensional integration. Another benefits of this technique are simple parallelization and great flexibility for adapting for different types of system uncertainty, as was shown based on the analysis of uncertain location of network nodes. In particular, the problem was suitably tailored to the framework of randomized design to determine fine approximations to optimal solutions.

Nevertheless, despite a critical character of the problem, no technique exists which is free from drawbacks and there is no doubt that the subject of robust experimental design is still open for research.

Chapter 8
Sensor Network Design for Fault Diagnosis in DPSs

Recently, one can observe an extremely fast development of methods of Fault Detection and Isolation (FDI) for dynamical systems. A wide variety of techniques with many potential applications is described in the rich literature. For surveys, the interested reader is referred to [17, 39, 41, 60, 124, 125, 209]. Nevertheless, a great majority of contributions focus on the methodology dedicated to lumped systems, and there are no effective methods tailored to spatio-temporal systems. Proper recognition of an abnormal behavior of the examined process leads to the necessity of very precise fitting of a nominal model corresponding to the conditions of normal work of real physical phenomena associated with it, as well as the need for appropriate models of abnormal work. In the context of analytical models, the diagnosis is a practical realization of selection of an appropriate model structure in accordance with the current system performance mode. On the other hand, in any possible working conditions, the calibration of a model is strongly dependent on the strategy of taking measurements. In such a way, the problem of sensor location becomes one of the most important issues in fault detection in DPSs. But within the framework of FDI systems, the optimization of the data acquisition process which increases the reliability of the diagnosis is usually neglected and there is a substantial lack of techniques dedicated to this area. In fact, despite its importance, this subject has received little attention so far. In order to fill this gap, the main aim of this chapter is to extend the existing methods of sensor location for parameter estimation and develop the missing interconnection between the properties of the measurement process and optimization of fault detection in DPSs.

In the background of diagnostic problems, the main difficulty is definition of suitable relations between the quality of system diagnosis and the observational strategy. For this purpose, classical hypothesis testing is used for model-based diagnosis for DPSs. This allows us to construct a qualitative criterion of sensor allocation using the notion of Ψ_s-optimum designs originated in optimum experiment design theory. Then, proper fault detection and localization schemes are developed based on the results reported by Patan

M. Patan: Optimal Sensor Networks Scheduling in Identification of DPSs, LNCIS 425, pp. 207–230.
springerlink.com

[194] for stationary sensors and further extended for scanning and mobile sensor networks [183, 198, 199, 200, 201, 203, 207, 290, 292, 293, 294].

Although motivations stem directly from various engineering domains (e.g., computer-assisted tomography, monitoring of intelligent buildings, inspection of civil constructions, etc.), the problem of contaminating source identification is one of the most representative and challenging examples of real-world applications, which is especially interesting and important in the context of sensor networks. This problem has been increasingly receiving significant research interest due to its applications in the fields of security, environmental and industrial monitoring, pollution control, etc. Examples include the detection of a potential biochemical attack in an urban environment, e.g., from a crop-duster spreading toxins in aerosol, sensing explosives mounted on a vehicle, detecting leakage of dangerous biochemical materials from a tank carried by a vehicle, or detecting a life-threatening contaminant source dropped intentionally or unintentionally into a water reservoir. In a typical scenario, after some type of biological or chemical contamination has occurred, there is a developing plume of dangerous or toxic material. Clearly, the evolution of the cloud will be affected by the weather conditions and the surrounding geography. It goes without saying that a crucial factor in responding to a chemical or biological attack is speed. Emergency services wish to quickly know what the toxin is, where its sources are located and what their strengths are. This knowledge would allow them to take appropriate measures to counteract or at least to reduce the impact of the release.

There are a number of difficulties inherent in the above diagnosis problem. First, we note that this is an inverse problem in which, given a model, it is necessary to identify the system parameters from the available information about the process. As such, we demand certain guarantees that the parameters are *identifiable* taking account of the typical case when sensors permit finite-dimensional measurements. Second, as a rule, this inverse problem is ill-posed and has to be regularized to obtain a reasonable approximation to the solution. This question can hardly be overestimated since, using observations provided by sensors which are usually coarsely spaced and without resorting to regularization, we cannot hope to recover diagnostic components that are more oscillatory than dictated by the spacing of the sensors. These issues are often neglected in the engineering literature, which may lead to erroneous results, cf. [256].

The situation is even more complicated when there are multiple faults which can influence the behavior of the process locally in time and space. Then, additionally, from the online implementation viewpoint, we would prefer a recursive procedure for estimating time-dependent parameters characterizing the system faults so that the estimate tracks the measurement data and the new measurements can be effectively incorporated. Unfortunately, no universal solutions to the above problems have been proposed yet.

The purpose of the research presented in this chapter is thus to develop a computationally efficient approach to detection and localization of faults evolving both in time and space using a sensor network with multiple cooperative mobile nodes (since it is the most general setting covering cases of networks with scanning and stationary nodes). In contrast to conventional approaches commonly used in mobile sensor networks, the knowledge of the mathematical model of the distributed parameter system in question is incorporated to form a basis for optimization. Consequently, the important information about the model governing the physical phenomenon in question will not be lost and will be to the profit of early detection of potential system faults.

8.1 Parameter Estimation Approach to Fault Detection in DPSs

Parameter estimation is one of the fundamental methods from among all analytical techniques of fault detection [124, 125]. Its role is of great significance in situations when the abnormal system state appears not only in the form of output changes but also as fluctuations of model parameters. This is a very common situation in practice if only parameters have a physical interpretation built upon proper analysis of various quantities, which are crucial for the process considered. Unfortunately, they are usually nonmeasurable directly and the application of effective parameter estimation algorithms is required in order to obtain their estimates. An elementary idea of fault diagnosis is to compare the resulting parameter estimates with the corresponding known nominal values, treating possible differences as residuals which contain information about potential faults. Based on some thresholding techniques, the appropriate decision making system could be constructed to detect abnormal situations in system functioning, see [194, 198, 200].

Consider a spatio-temporal process whose nominal mode of work $y(x,t;\theta)$ is described by the mathematical model (2.66)–(2.68) with $\theta = \theta^0$, where θ^0 is a vector of constant parameters, which are characteristic for this state. Assuming that the process states are observed over some finite time horizon T using N mobile sensors, moving along the trajectories x^j, $j = 1\dots,N$, the observations $z = [z^1(t),\dots,z^N(t)]$ can be represented by (2.77).

A preliminary version of our fault identification problem is formulated as follows: We seek the parameters that yield a simulated concentration field which best matches the data from the sensors, i.e., those that minimize the mismatch functional $\mathcal{J}(\theta)$ given by (2.81) over the set of admissible parameters Θ. The vector $\hat{\theta}$ minimizing $\mathcal{J}(\theta)$ stands for the estimate of the true value of θ^0. Generally, in practical situations only a subset of all parameters can be used for diagnosis. If such a situation takes place, then with no loss of generality the parameters of interest can be distinguished by partitioning the parameter vector into

$$\theta^{\mathsf{T}} = \left[\theta_1 \ \ldots \ \theta_s \vdots \theta_{s+1} \ \ldots \ \theta_m\right] = \left[\alpha^{\mathsf{T}} \vdots \beta^{\mathsf{T}}\right], \tag{8.1}$$

where α is a vector of s parameters, which are essential for effective fault detection, and β is the vector of some parameters, which are a part of the model but are not significant for fault detection (although they can be used for fault localization or are some representation of model uncertainty). Based on the observations, it is possible to test the simple null hypothesis

$$H_0 : \alpha = \alpha^0, \tag{8.2}$$

where α^0 is the nominal value for the vector α corresponding to the normal system performance.

The generalization of the likelihood function for the experiment considered takes the following form, cf.(2.81) [88, 152]:

$$\begin{aligned} \mathcal{L}(\theta|z) = &\left(\frac{1}{2\pi \det(C)}\right)^{N/2} \\ &\times \exp\left(-\frac{1}{2}\sum_{j=1}^{N}\int_T \|z^j(t) - h(\hat{y}(x^j(t),t;\theta),x^j(t),t)\|_C^2 \, \mathrm{d}t\right), \end{aligned} \tag{8.3}$$

where $C = C(x^j(t),t)$.

Setting $\Theta_0 = \{\theta \in \Theta : \alpha = \alpha^0\}$, we can define the following generalized log-likelihood ratio:

$$\lambda(z) = 2\ln\left(\frac{\sup_{\theta\in\Theta}\mathcal{L}(\theta|z)}{\sup_{\theta\in\Theta_0}\mathcal{L}(\theta|z)}\right) = \mathcal{J}(\tilde{\theta}) - \mathcal{J}(\hat{\theta}), \tag{8.4}$$

where

$$\hat{\theta} = \arg\min_{\theta\in\Theta}\mathcal{J}(\theta), \qquad \tilde{\theta} = \arg\min_{\theta\in\Theta_0}\mathcal{J}(\theta). \tag{8.5}$$

It should be pointed out that a number of different variations in the maximum likelihood ratio exist. One variation is to swap the numerator and the denominator. Another is to calculate the supremum over Θ_0' (A' denotes a complement of the set A) in the denominator of (8.4) instead of Θ [152].

The likelihood ratio test is widely used in statistics. The reason is partly that it is an optimal test in the case when both the null hypothesis H_0 and its alternative H_1 are simple, i.e., Θ_0 and Θ_0' reduce to one-element sets (cf. the Neyman–Pearson lemma [82]). The proofs of optimality also exist for many other cases where H_0 is simple [82]. In many situations where theoretical justification is missing, the likelihood ratio can still be shown to be very good in practice [152]. Nevertheless, there also exist a number of situations for which the likelihood ratio is not adequate. Nevertheless, the generalized log-likelihood ratio is commonly used, because it can be shown that, assuming the validity of the null hypothesis H_0, the sequence $\{\lambda(z)\}$ for $N \to \infty$ is weakly convergent to a χ^2 random variable on s degrees of freedom [88, Th. 3.6.1, p. 55].

The meaning of this fact is that we can compare the observed value of $\lambda(z)$ with some threshold k_γ obtained from the cumulative χ^2 distribution on s degrees of freedom where k_γ is such that $100(1-\gamma)\%$ of the distribution lies to the left of k_γ. The decision rule for a given significance level γ, which represents a fixed range of model uncertainty, takes the following form:

$$S = \begin{cases} \text{reject } H_0 & \text{if } \lambda(z) \geqslant k_\gamma \quad \text{(faulty system state)} \\ \text{accept } H_0 & \text{if } \lambda(z) < k_\gamma \quad \text{(nominal system state)} \end{cases} \tag{8.6}$$

The potential rejection of H_0 indicates an essential deviation of the vector α from the nominal value of this parameter and is a base for detection of abnormal states in the system. When the null hypothesis H_0 is true, we do not want to reject H_0. The mistake in rejecting H_0 when it is true is called the Type I error. Similarly, accepting H_0 when the alternative hypothesis of the form $H_1 : \alpha = \alpha^1 \neq \alpha_{\text{nom}}$ is true is called the Type II error [88]. In fault diagnosis, there is a direct connection between these errors and the probability of a false alarm, missed detection or missed isolation (localization).

8.2 Ψ_s-Optimum Sensor Location Problem

At this point, it is at least clear that, in order to achieve low probabilities of false alarms, missed detection, and missed isolation, we have to keep the probabilities of Type I and II errors low. Thus, the probabilities of these errors are a kind of performance measure for a single hypothesis test. From a practical point of view, it is not possible to minimize both the errors simultaneously. Therefore, a classical approach is to minimize the Type II error with a prescribed level of the Type I error.

From the point of view of the Type II error, it can be shown that, assuming the validity of the alternative hypothesis H_1, the sequence $\{\lambda(z)\}$ for $N \to \infty$ weakly converges to a noncentral χ^2 distribution on s degrees of freedom and with the noncentrality parameter [88, Th. 3.6.2, p. 58]

$$h_\alpha = N(\alpha - \alpha^0)^\mathsf{T} \left[M_{\alpha\alpha} - M_{\alpha\beta} M_{\beta\beta}^{-1} M_{\alpha\beta}^\mathsf{T}\right](\alpha - \alpha^0), \tag{8.7}$$

where

$$M_{\alpha\alpha} \in \mathbb{R}^{s\times s}, \quad M_{\alpha\beta} \in \mathbb{R}^{s\times(m-s)}, \quad M_{\beta\beta}^{-1} \in \mathbb{R}^{(m-s)\times(m-s)}$$

are the components of $M \in \mathbb{R}^{m\times m}$ according to the following block partitioning:

$$M = \left[\begin{array}{c:c} M_{\alpha\alpha} & M_{\alpha\beta} \\ \hdashline M_{\alpha\beta}^\mathsf{T} & M_{\beta\beta} \end{array}\right], \tag{8.8}$$

which stands for the FIM corresponding to the vector θ and defined by (2.85).

If only a subset of s parameters is of interest, with the parameter vector partitioning given by (8.1), and for a fixed significance level (i.e., fixed probability of false alarms), the power of the log-likelihood ratio test for the alternative hypothesis of the form $H_1 : \alpha \neq \alpha^0$, i.e.,

$$1 - \text{the probability of accepting } H_0 \text{ when } H_1 \text{ is true}, \tag{8.9}$$

can be made large by increasing the noncentrality parameter (8.7). This purpose can be achieved by taking a large number of sensors N or by trying to influence the value of h_α via the information matrix. In real world problems, the number of available sensors is always limited. Thus, in order to improve the experimental conditions, the second strategy should be used, in such a way as to obtain a 'high powered' test which leads directly to the low probability of missed detection. Introducing the scalar measure Ψ_s operating over the FIM components, we can state our sensor location problem as an optimization one:

$$\Psi_s[M] \longrightarrow \min. \tag{8.10}$$

Various choices for Ψ_s exist [9, 73]:

- D_s-optimality criterion:

$$\Psi_s(M) = -\ln\det[M_{\alpha\alpha} - M_{\alpha\beta}M_{\beta\beta}^{-1}M_{\alpha\beta}^{\mathsf{T}}], \tag{8.11}$$

- A_s-optimality criterion:

$$\Psi_s(M) = \operatorname{trace}[(M_{\alpha\alpha} - M_{\alpha\beta}M_{\beta\beta}^{-1}M_{\alpha\beta}^{\mathsf{T}})^{-1}]. \tag{8.12}$$

- s-sensitivity criterion:

$$\Psi_s(M) = -\operatorname{trace}[M_{\alpha\alpha} - M_{\alpha\beta}M_{\beta\beta}^{-1}M_{\alpha\beta}^{\mathsf{T}}]. \tag{8.13}$$

Under the validity of assumptions (A3)–(A5) and (A9), a number of characterizations of the optimal solutions can be provided by analogy to the results given in Section 2.1.4. In particular, the equivalence theorem (cf. Corr. 4.1) holds with components of the sensitivity function,

$$\psi_s(x,t,\xi) = \varsigma_s(\xi) - \phi_s(x,t,\xi), \tag{8.14}$$

$$\varsigma_s(\xi) = -\operatorname{trace}\Big[\mathring{\Psi}_s[M(\xi)]M(\xi)\Big], \tag{8.15}$$

$$\phi_s(x,t,\xi) = -\operatorname{trace}\Big[\mathring{\Psi}_s[M(\xi)]\Upsilon(x,t)\Big], \tag{8.16}$$

where

$$\mathring{\Psi}_s[M(\xi)] = \left.\frac{\partial \Psi_s(M)}{\partial M}\right|_{M=M(\xi)}. \tag{8.17}$$

The most popular result is the equivalence theorem defined for the case of the D_s-optimality [70], formulated first for a static regression model. Observe that for the partition

$$M^{-1} = \left[\begin{array}{c:c} D_{\alpha\alpha} & D_{\alpha\beta} \\ \hdashline D_{\alpha\beta}^{\mathsf{T}} & D_{\beta\beta} \end{array}\right], \tag{8.18}$$

where $D_{\alpha\alpha} \in \mathbb{R}^{s\times s}$, $D_{\alpha\beta} \in \mathbb{R}^{s\times(m-s)}$, $D_{\beta\beta} \in \mathbb{R}^{(m-s)\times(m-s)}$, we have [21, Fact 2.8.7, p.44]

$$D_{\alpha\alpha} = \left(M_{\alpha\alpha} - M_{\alpha\beta}M_{\beta\beta}^{-1}M_{\alpha\beta}^{\mathsf{T}}\right)^{-1}, \tag{8.19}$$

and further [21, Fact 2.15.8, p.73]

$$\det(D_{\alpha\alpha}) = \frac{\det(M_{\beta\beta})}{\det(M)}. \tag{8.20}$$

Consequently, minimization of the D_s-optimality criterion amounts to minimization of $\det(D_{\alpha\alpha})$, which is proportional to the determinant of the covariance matrix for α. Additionally, $\varsigma_s(\xi) = s$ and

$$\mathring{\Psi}_s[M(\xi)] = \frac{\partial \Psi_s(M)}{\partial M} = \left[\begin{array}{c:c} D_{\alpha\alpha} & D_{\alpha\beta} \\ \hdashline D_{\alpha\beta}^{\mathsf{T}} & D_{\beta\beta} - M_{\beta\beta}^{-1} \end{array}\right]. \tag{8.21}$$

Applying some matrix differentiation results, it is possible to obtain analogous forms of the matrix derivative of function Ψ_s for the Λ_s-optimality:

$$\mathring{\Psi}_s[M(\xi)] = \frac{\partial \Psi_s(M)}{\partial M} = \left[\begin{array}{c:c} -D_{\alpha\alpha}^2 & -2D_{\alpha\alpha}D_{\alpha\beta} \\ \hdashline -2D_{\alpha\beta}^{\mathsf{T}}D_{\alpha\alpha} & D_{\alpha\beta}^{\mathsf{T}}D_{\alpha\beta} \end{array}\right], \tag{8.22}$$

and the s-sensitivity criterion:

$$\mathring{\Psi}_s[M(\xi)] = \frac{\partial \Psi_s(M)}{\partial M} = \left[\begin{array}{c:c} I & 0 \\ \hdashline 0 & -M_{\beta\beta}^{-1}M_{\alpha\beta}M_{\alpha\beta}^{\mathsf{T}}M_{\beta\beta}^{-1} \end{array}\right]. \tag{8.23}$$

At this point, it is obvious that those simple modifications of the sensitivity function ψ_s pave the way for direct applicability of all algorithms studied in the previous chapters to solve the optimum scheduling problem for fault detectability.

8.3 Adaptive Thresholding Techniques

Test quantities should be designed in such a way as to be insensitive to uncontrolled effects like changes in inputs, disturbances, modelling errors, etc. These problems are closely related to robustness. In general, methods to achieve and analyze robustness can be divided into two groups. The first group considers robustness as a feature integrated into the design process of test quantities. The second group of approaches design the test quantities first without robustness considerations and then perform an additional step by adjusting or compensating the originally constructed test quantity. Experimental results show the advantage of the second groups of methods, while very scarce literature exists on the first group of approaches. In terms of robustness, hypothesis testing based on an adaptive threshold can significantly improve the reliability of diagnosis.

According to the analysis carried out in Section 8.1, the log-likelihood ratio is weakly convergent to the χ^2 distribution on s degrees of freedom. At the beginning, we set a given significance level γ, which represents model uncertainty, noise, etc. The commonly used values are $\gamma = 0.05$ or $\gamma = 0.01$. The significance level γ corresponds to the probability that a log-likelihood ratio exceeds a random value χ^2_γ. Values of χ^2_γ are tabulated in most statistical books.

In practice, due to modelling uncertainty and measurement noise, it is necessary to set threshold k_γ to a larger value in order to avoid false alarms. This operation causes a reduction in the fault detection sensitivity. Therefore, the choice of the threshold is only a compromise between fault decision sensitivity and the false alarms rate. For that reason, in the case considered, it is recommended to apply adaptive thresholds whose main idea is that they should vary in time since disturbances and other uncontrolled effects can also vary in time. Unfortunately, in general, there is a lack of contributions related to this subject. However, the delineated approach creates a possibility for construction of such an adaptive threshold based on a suitable adjustment of the significance level γ in the decision rule (8.6). Obviously, proper modelling of the threshold adaptation process is a difficult task and detailed analysis exceeds the scope of this paper. Nevertheless, we can propose some simple and effective techniques based on the estimation of statistical parameters based on the past observations of a test function.

It is well known that for the random variable $X \sim \chi^2(s)$ we have $E(X) = s$ and $\mathrm{var}(X) = 2s$. Unfortunately, the test quantity in our case is an approximation to $\chi^2(s)$ distributed random variable. Along the past of the test function (time window of length t_p), one can calculate the estimated values of the mean value:

$$\bar{\lambda}(t) = \frac{1}{t_p} \int_{t-t_p}^{t} \lambda(z)\,\mathrm{d}\tau \tag{8.24}$$

and the variance:

$$S^2_{\lambda}(t) = \frac{1}{t_p} \int_{t-t_p}^{t} (\lambda(z) - \bar{\lambda}(\tau))^2\,\mathrm{d}\tau. \tag{8.25}$$

Using (8.24) and (8.25), a threshold can be calculated according to the adaptive heuristic formula:

$$k(t) = \bar{\lambda}(t) + \alpha S_{\lambda}(t). \tag{8.26}$$

where $\alpha > 0$ is some fixed a priori margin for threshold variability.

8.4 Adaptive Trajectory Planning Based on Sequential Designs

Trajectories providing the most informative observations for detection and localization of moving contamination source strongly depend on the true vector of parameters θ describing the source which is obviously unknown. In the case when initial information about the parameters is missing, the only solution is to conduct additional analysis for predetermination of the initial estimates of the parameters of interest. Nevertheless, it is impossible to construct an optimal design for such an experiment without a sufficient knowledge about parameters, because most often optimal sensor locations depend on θ. On the other hand, from a practical point of view, it is desirable to have the computations dynamic data-driven, i.e., the current measurements from the mobile sensors must serve as a basis for the update of parameter estimates and these, in turn, can be used to correct the sensor movements. In the proposed numerical scheme, a nonlinear model-predictive-control-like technique is applied for this purpose.

A very reasonable idea for introduction of adaptivity is repetition of the measurement collection, estimation, prediction and design steps several times according to the sequential designs introduced in Section 7.1. This is equivalent to the division of the resources (e.g., the time horizon) into small parts which are related to the corresponding consecutive stages of the experiment. More precisely, for the purpose of adaptive estimation and analysis, the total observation horizon $T = [0, t_f]$ is divided using time moments $0 = t_0 < t_1 < \cdots < t_K = t_f$ forming its arbitrary partition into a collection of consecutive subintervals $T_k = [t_{k-1}, t_k)$, $k = 1, \ldots, K$. Based on the observational data collected chronologically in previous subintervals, the sensor routing problem over the current stage is then formulated as an optimal-control one with state-variable inequality constraints representing geometric constraints induced by admissible measurement regions and allowable distances between the sensor nodes. Taking account of the dynamic models of the vehicles conveying the sensors, the problem is finally reduced to determination of the control forces exerting on the sensor nodes according to the approaches delineated in Chapter 4 and readily solved numerically using common computational tools for algorithmic optimal control, e.g., RIOTS_95.

Algorithm 8.1. Sequential detection procedure
Input variables: y_0, u, x are initial conditions, controls and initial sensor coordinates for the first subinterval, respectively, ΔT_p is a fixed prediction horizon, K is a prescribed upper limit of experimental phases, γ is a fixed probability of false detection alarms

```
procedure DetectionBlock(y_0, u, x, ΔT_p, K, γ)
    k ← 1                                        ▷ Initialization of sequential design counter
    Z_meas ← ∅                                   ▷ Initialize set of observations
    y_k ← y_0                                    ▷ Setting initial conditions for first subinterval
    repeat
        T ← T_k = (t_{k-1}, t_k]                 ▷ Current design subinterval
        T̃_k ← (t_{k-1}, t_{k-1} + ΔT_p]          ▷ Current prediction horizon
        Z_meas ← Z_meas ∪ CollectMeas(x,t_q)     ▷ EXPERIMENT PHASE
        if GLRTest(Z_meas, y_0, γ) then          ▷ Generalized log-likelihood ratio test
            return
        end if
        θ̂_k ← EstimateParams(Z_meas, y_0)        ▷ ESTIMATION PHASE
        y_pred ← SolvePDEs(θ̂_k, T̃_k, y_k)        ▷ Prediction of PDE system evolution
        [u, x] ← SolveOCP(u, x, y_pred)          ▷ TRAJECTORY DESIGN PHASE
        k ← k + 1
        y_k ← y_pred(t_k)                        ▷ Update the initial conditions for prediction
    until k ≤ K
end procedure
```

8.4.1 Detection Block

An algorithmic scheme of the detection block is provided in Algorithm 8.1. The operators involved in this implementation are as follows:

- CollectMeas(x,t_q) returns the system response measured by sensor nodes at specified time instant t_q.
- GLRTest($Z_{\text{meas}}, y_0, \gamma$) is an implementation of (8.6). It returns true only if the log-likelihood ratio (8.4) calculated for the observations included in Z_{meas} and initial conditions in y_0 exceeds the threshold for the probability of false alarms fixed at γ.
- EstimateParams(Z_{meas}, y_0) determines the estimate of system parameters for the current design interval. The GLRTest procedure also estimates the system parameters but for detection purposes, so their accuracy is not adequate for design purposes.
- SolvePDEs($\widehat{\theta}_k, \tilde{T}_k, y_k$) generates the forecast of the system response and its sensitivities on the interval $\tilde{T}_k$ based on the estimated system parameters and initial conditions y_k estimated in the previous design interval.
- SolveOCP(u, x, y_{pred}) solves the corresponding optimal sensor routing problem (cf. Problem 4.2) and determines updates of the controls for sensor nodes using the predicted solution y_{pred} of the PDE system.

8.4.1.1 Identification (Localization) Block

Once the source is detected, we are able to provide a similar sequential scheme for source parameter identification starting with some parameter estimates (from the detection stage or previous experiments). A heuristic scheme of the identification block is embodied in Algorithm 8.2. The operators used here have the same meaning as those from the detection block.

Algorithm 8.2. Sequential identification (localization) procedure
Input variables: y_0, u, x are initial conditions, controls and initial sensor coordinates for the first subinterval, respectively, ΔT_p is a fixed prediction horizon, K is a prescribed upper limit of experimental phases, $\widehat{\theta}_0$ is an initial estimate of system parameters

---

procedure IDENTIFICATIONBLOCK($y_0, u, x, \Delta T_p, K, \widehat{\theta}_0$)
  $k \leftarrow 1$ ▷ Initialization of sequential design counter
  $y_k \leftarrow y_0$ ▷ Setting initial conditions for first subinterval
  repeat
    $Z_{\text{meas}} \leftarrow \emptyset$ ▷ Initialize set of observations
    $T \leftarrow T_k = (t_{k-1}, t_k]$; $\tilde{T}_k \leftarrow (t_{k-1}, t_{k-1} + \Delta T_p]$ ▷ Current design subinterval and prediction horizon
    $Z_{\text{meas}} \leftarrow Z_{\text{meas}} \cup$ COLLECTMEAS(x,t_q) ▷ EXPERIMENT PHASE
    $\widehat{\theta}_k \leftarrow$ ESTIMATEPARAMS(Z_{meas}, y_k) ▷ ESTIMATION PHASE
    $y_{\text{pred}} \leftarrow$ SOLVEPDES($\widehat{\theta}_k, \tilde{T}_k, y_k$) ▷ Prediction of PDE system evolution
    $[u, x] \leftarrow$ SOLVEOCP(u, x, y_{pred}) ▷ TRAJECTORY DESIGN PHASE
    $k \leftarrow k + 1$
    $y_k \leftarrow y_{\text{pred}}(t_k)$ ▷ Update the initial conditions for prediction
  until $k \leq K$
end procedure

8.5 Applications Example: Identification of Moving Pollution Sources

8.5.1 Source Localization

As an illustration of the proposed approach, consider the following example which consists in localization of a moving contamination source within a bounded area using a sensor network with mobile nodes equipped with concentration sensors.

8.5.1.1 Process Description

The spatial domain considered in our simulations is a square with a side length of 1 km (see Fig. 8.1). An active source of a toxic pollutant moves within this domain and emits a polluting substance to the atmosphere. Emission is assumed to start at the initial time of the simulation. The size of the domain is closely related to a physical model of the mesoscale atmospheric motion, see [109]. Therefore, the changes in the spatial concentration $y(x,t)$ of this substance over the observation interval $T = [0, 600]$ (in seconds) can be mathematically described by the following advection-diffusion equation:

$$\frac{\partial y(x,t)}{\partial t} + \nabla \cdot \big(v(x,t)y(x,t)\big) = \nabla \cdot \big(\kappa \nabla y(x,t)\big) + f(x,t), \quad (x,t) \in \Omega \times T, \tag{8.27}$$

subject to the boundary and initial conditions

$$\begin{cases} \dfrac{\partial y(x,t)}{\partial n} = 0, & (x,t) \in \Gamma \times T, \\ y(x,0) = y_0, & x \in \Omega, \end{cases} \tag{8.28}$$

where the term $f(x,t) = \mu \exp(-100\|x - \zeta(t)\|^2)$ constitutes the model of an active source of the pollutant with the emission intensity coefficient μ and the instantaneous location given by the trajectory $\zeta(t)$. Furthermore, κ denotes a turbulent diffusion coefficient and $\partial y/\partial n$ stands for the derivative of y in the direction of the outward normal to the boundary Γ. As for prior estimates of the unknown parameters μ and κ, the values of 10 kg/s and 60 m^2/s were used, respectively.

8.5.1.2 Experiment Settings

In our simulation, the pollution source is assumed to start at $t = 0$ from the point $(0.5, 0.75)$, then move at a constant pace along the arc of the circle with radius $\rho = \sqrt{5}/4$ and centre at $(0.0, 0.5)$, and finally terminate at point $(0.5, 0.25)$ at $t = 600$, i.e., its trajectory is given by

$$\begin{aligned} \zeta_1 &= \rho \sin\left(\frac{\omega_1 t}{600} + \omega_0\right), \\ \zeta_2 &= 0.5 + \rho \cos\left(\frac{\omega_1 t}{600} + \omega_0\right), \end{aligned} \tag{8.29}$$

where $\omega_1 = 2\arcsin(0.2\sqrt{5})$ and $\omega_0 = \arcsin(0.4\sqrt{5})$. The velocity field $v(x,t)$ varies in space and time according to the following model (cf. Fig. 8.1):

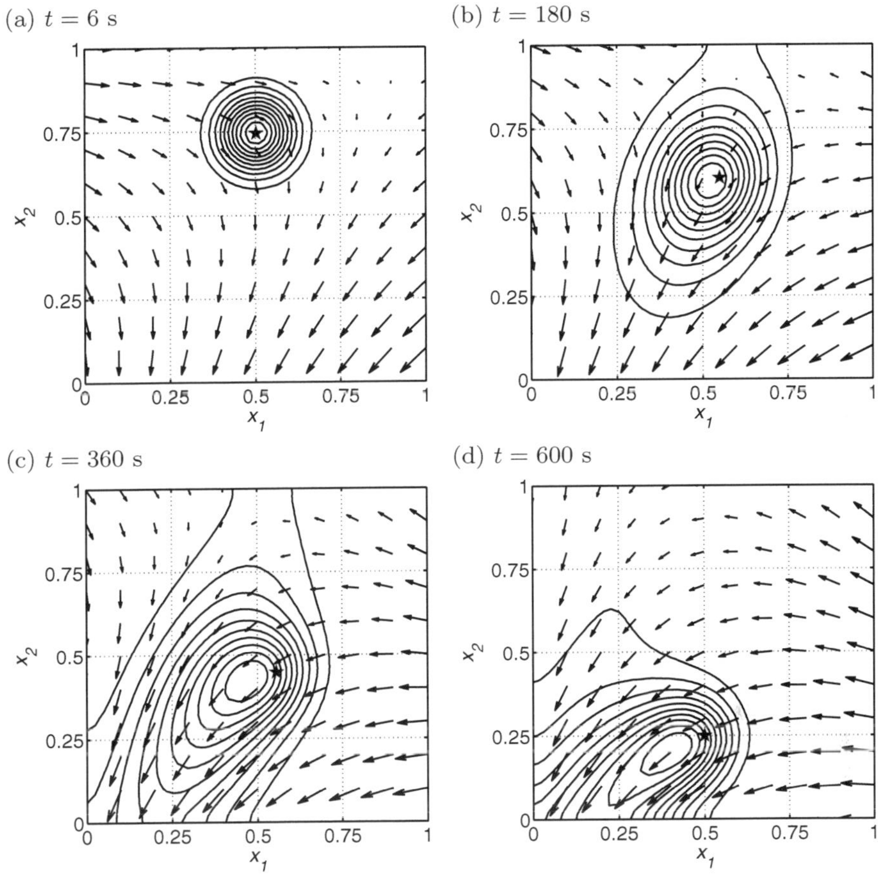

Fig. 8.1 Evolution of the wind velocity field versus the pollutant concentration and the actual position of the pollutant source (marked with asterisks)

$$v(x,t) = \Big(12\Big(x_2 - x_1 - \frac{t}{600}\Big), 12\Big(\frac{t(2x_1-1)}{600} + x_2 - 1\Big)\Big) \text{ [km/h]}. \quad (8.30)$$

The solution to (8.27)–(8.28) with simulated additive measurement noise with standard deviation $\sigma = 10^{-4}$ kg/m^3 constitutes the reference output signal for our simulations. Our purpose is to determine the evolution of the source, i.e., an estimate of its trajectory $\zeta(t)$ over the time interval T using a sensor network with mobile nodes.

Before proceeding with solution, we need to address the issue of reducing the problem of source identification to a finite-dimensional form. As Zhao and Nehorai [316] did, here we apply the parametrization approach commonly used in algorithmic optimal control, see [214] and [225]. It corresponds to explicit discretization of the source trajectories ζ_i. We can represent them, e.g., as linear combinations of canonical Lagrange interpolation polynomials or B-splines. To a great extent, the limitation of admissible source trajectories to a finite-dimensional subspace becomes a necessity if we wish to make the source identification problem tractable. Clearly, the selection of the subspace affects both the accuracy of numerical integration and the accuracy with which the solutions of the original problem are approximated. The reader interested in this complex problem is referred to specialized literature, e.g., [214, 248].

Since vapor measurements cannot be taken continuously due to technical limitations, we assume that the consecutive observations are taken at discrete times every $\varDelta t = 15$ s. Then, we divide the observation interval into 20 equal subintervals,

$$T_k = \big(t_k, t_k + 2\varDelta t\big], \quad t_k = 2(k-1)\varDelta t, \quad k = 1,\dots,20, \tag{8.31}$$

and approximate the source trajectory $\zeta(t)$ by linear splines,

$$\zeta(t) \approx \zeta_k + v_k(t - t_k), \quad t \in T_k, \quad k = 1,\dots,20, \tag{8.32}$$

where $v_k = (x_{k+1} - x_k)/(2\varDelta t)$ is the average source velocity in the interval T_k and ζ_k denotes the location of the source at time t_k. The ζ_ks become the parameters of interest in our source localization problem.

Since, usually, it is not possible to freely deploy sensors within the considered domain, in our simulation scenario the sensors are assumed to approach the contaminated area starting from arbitrary points on the left boundary and their initial positions are fixed (i.e., they are not optimized). For the first 150 seconds, i.e., within the first five time subintervals, the sensors move with the maximum speed to the center of the area simultaneously collecting the measurements. Based on those observations, the first estimate of the system parameters is determined and is used as an *a-priori* value to design the sensor trajectories for the next consecutive observation subinterval.

Then, for every T_k, $k = 6,\dots,20$, we have to update the controls for the sensor network nodes in order to accurately find the current location of the source. From among the system parameters $\theta_k = (\zeta_{k,1}, \zeta_{k,2}, v_{k,1}, v_{k,2}, \mu, \kappa)$, the localization of the pollution source is based on the knowledge of the first four. Thus, maximization of the reliability for localization of the contamination source is equivalent to the determination of D_s-optimal sensor trajectories, see [194]. In order to somewhat simplify the identification process of source coordinates and solve this problem in an on-line manner, we applied a computational scheme similar to that used by the optimal predictive control technique. Roughly speaking, for each consecutive time subinterval T_k the following steps are applied:

- First, an estimate $\hat{\theta}^k$ of the current parameter vector θ^k is determined based on measurements collected on subinterval T_k. The position ζ_k is updated and stored as the source position at time t_k and v_k is used for the

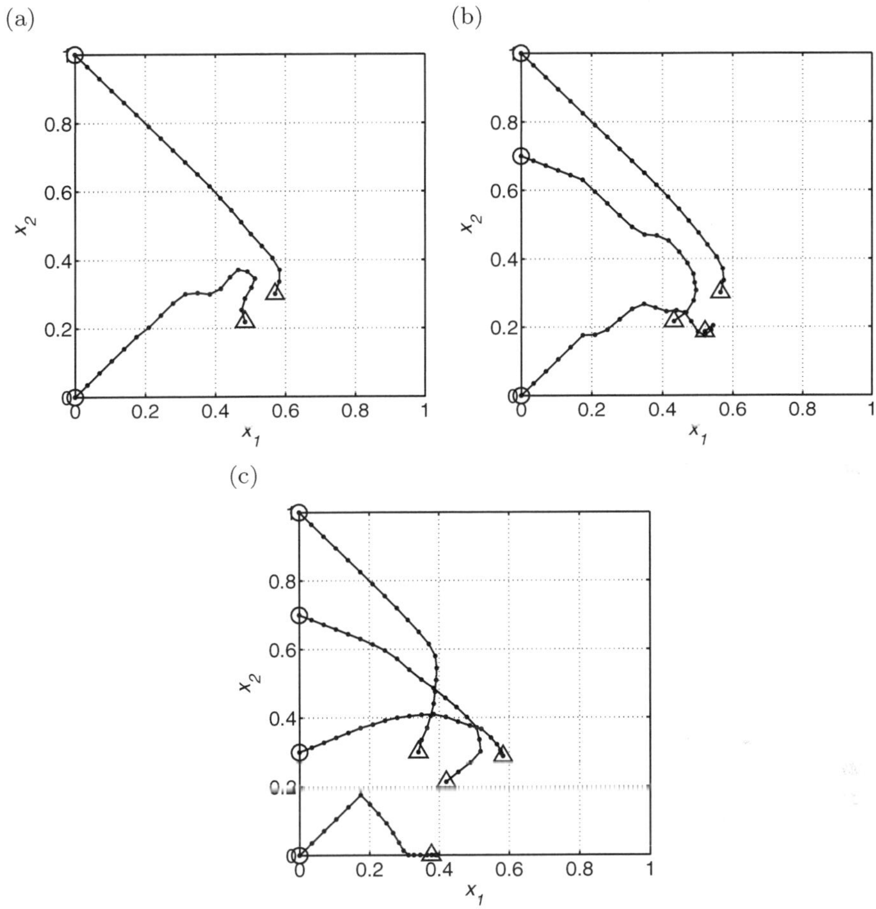

Fig. 8.2 Optimal sensor trajectories (starting and final positions are marked with open circles and triangles, respectively) for two (a), three (b) and four (c) sensors

prediction of source location and initial conditions for the system state at the next time instant t_{k+1}.
- The model equations (8.27) and (8.28) are solved using (8.32) as well as the current vector $\hat{\theta}^k$ on the interval $\tilde{T}_k = (t_k, t_k + 600]$ and the initial conditions estimated from the solution for $\tilde{T}_{k-1}$.
- Once we obtain the predicted response of our system, the D_s optimal controls for the sensor nodes on the interval $\tilde{T}_k$ are determined and applied on the interval T_k.

As regards the sensors dynamics, the model

$$\dot{q}(t) = u(t), \quad q(0) = q_0, \tag{8.33}$$

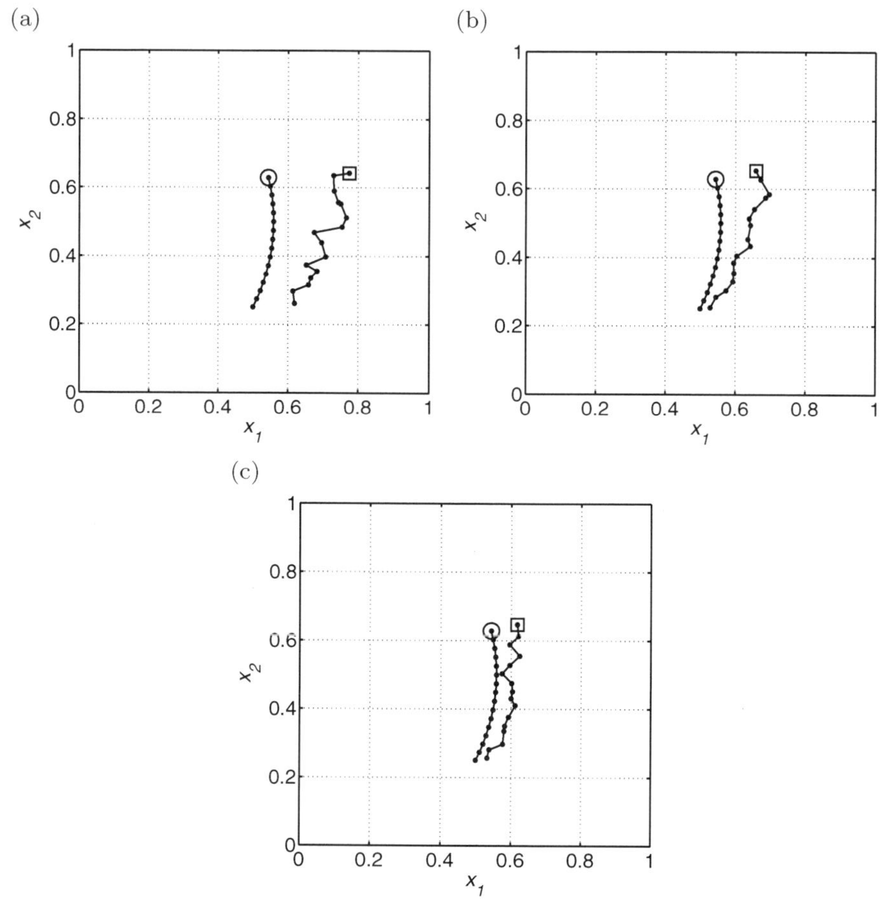

Fig. 8.3 Estimated trajectory of the pollution source for two (a), three (b) and four (c) sensors (starting from time $t = 150$ s indicated with a square). For comparison, the actual trajectory of the moving source is marked with an open circle

was applied, focusing on the direct control of sensor velocities. The following bounds for u were used:

$$-4.2 \leq u_i(t) \leq 4.2 \text{ [km/h]} , \quad \forall t \in T. \tag{8.34}$$

To slightly simplify computations, the constraints regarding collision avoidance were neglected. However, in practice any sensor network platform possesses some amount of autonomy allowing avoiding such critical situations.

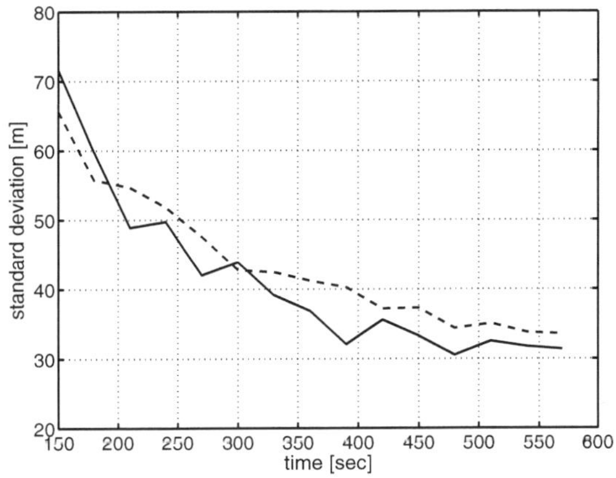

Fig. 8.4 Estimated standard deviation for the source coordinates in the case of four sensors (solid and dashed lines for $\zeta_{k,1}$ and $\zeta_{k,2}$, respectively)

8.5.1.3 Implementation Details

In order to verify the proposed approach, a MATLAB program was written using a PC equipped with Pentium M740 processor (1.73GHz, 1 GB RAM) running Windows XP and MATLAB 7 (R14). First, the system of PDEs was solved using the UMFPACK solver from the COMSOL 3.4 environment based on the finite-element method. Each time calculations were performed for a spatial mesh composed of 3912 triangles and 2017 nodes, and an evenly partitioned time interval (100 subintervals) using quadratic Lagrange elements. The sensitivity coefficients were then linearly interpolated and stored. In each consecutive time subinterval, the system parameters were estimated using the `nlinfit` function from the MATLAB Statistics Toolbox. Finally, for determining the optimal trajectories, the package RIOTS_95[247] was employed. From among its three main optimization procedures, the routine `riots` was used, which is based on the SQP algorithm.

In order to avoid convergence to local optima, the simulations at each time step were restarted several times from different initial starting points. Each simulation run took on average about 60 minutes.

8.5.1.4 Results

The optimal trajectories are presented in Fig. 8.2, where the sensor paths are shown for different numbers of nodes. The behavior of sensors is not that intuitive, since the concentration changes are quite complex due to a combination of different mass transport processes. It is clear that sensors attempt to reach the areas with high pollutant concentrations as quickly as possible using their maximum speed in the first part of trajectories. Then the trajectories strongly depend on the number of sensors, since the vehicles tend to cover the area trying to provide the most informative measurements and converge to source location to terminate in its vicinity. The effects of the presented localization technique are shown in Fig. 8.3, where we can easily compare the quality of localization for different numbers of sensors (the last point in the estimated source trajectory is obtained via velocity projection from the starting point from the last subinterval). It becomes clear that an increased number of sensors substantially improves the quality of localization. However, we can also observe the influence of the trajectory modelling error arising from the linearization of the source movements, as the estimated trajectory is located clearly on the right-hand side of the real trajectory. This is due to the linear velocity projection leading to an overestimated pollutant concentration on the right-hand side of the domain.

Figure 8.4 shows the slow improvement in the quality of source localization with the consecutive time steps. This can be explained in terms of better sensor locations at the end of the observation interval (i.e., in a closer vicinity of the moving pollution source) providing more informative measurements.

8.5.2 Source Detection and Identification

In this example, detection and identification of a moving contamination source using information collected with a mobile sensor network will be illustrated via a simulated scenario of pollution expansion over a two-dimensional spatial domain.

8.5.2.1 Process Description

The physical model in our example consists of one source of a pollutant moving inside the spatial domain being a unit square (1 km × 1 km). The changes in the spatial concentration $y(x,t)$ of the emitted pollutant over the time interval $T = [0, 400]$ (in seconds) are described by the advection-diffusion equation given by the system (8.27)–(8.28). This time the source term has the form

$$f(x,t) = H_{t_s}(t)\mu \exp(-100\|x - \zeta(t)\|^2), \tag{8.35}$$

where $H_\tau(t)$ denotes the Heaviside function,

$$H_\tau(t) = \begin{cases} 0 & \text{if } t < \tau, \\ 1 & \text{otherwise.} \end{cases} \tag{8.36}$$

This means that the source is starting to release a substance at time instant t_s with intensity μ and the temporal location given by trajectory $\zeta(t)$. As for prior estimates of the unknown parameters μ and κ, the values 10 kg/s and 90 m^2/s were used, respectively.

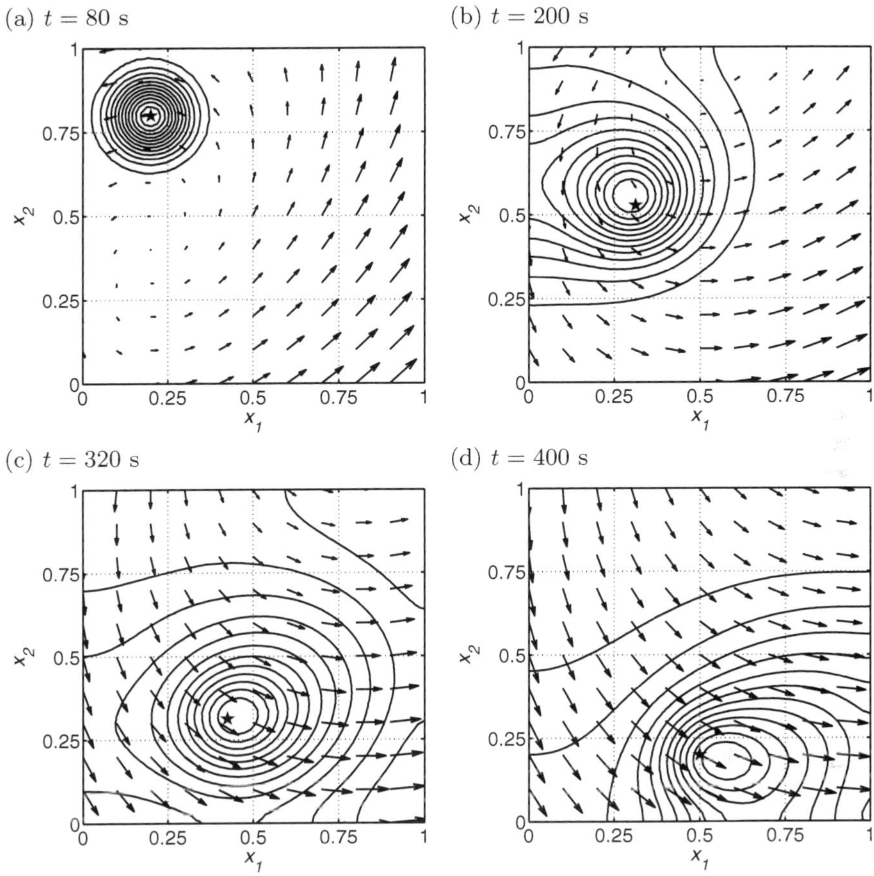

Fig. 8.5 Evolution of the wind velocity field versus the pollutant concentration and the actual position of the pollutant source (marked with asterisks)

8.5.2.2 Experiment Settings

In our simulation scenario, the pollution source is assumed to start the emission at $t_s = 80$ s from the point $(0.2, 0.8)$ and then move along a curve at a time-varying speed to the point $(0.5, 0.2)$ attained at $t = 400$ s. More precisely, its trajectory is given by

$$\begin{aligned} \zeta_{s,1} &= \frac{1}{3200}(3t + 400) \quad [\text{km}], \\ \zeta_{s,2} &= 5\left(\frac{t - 720}{1600}\right)^2 \quad [\text{km}]. \end{aligned} \tag{8.37}$$

The wind velocity field $v(x,t)$ is given by the following model (cf. Fig. 8.5):

$$v(x,t) = \Big(9\big(x_1 - x_2 + \frac{t}{400}\big), 9\big(x_1 - \frac{t}{400}\big)\Big) \ \text{[km/h]}. \qquad (8.38)$$

The solution to (8.27)–(8.28) with simulated additive measurement noise with standard deviation $\sigma = 10^{-2}$ kg/m^3 constitutes the reference output signal for our simulations. Our primary purpose here is to detect the presence of the pollution source using a sensor network with $N = 4$ mobile nodes. Then, after the detection, we are interested in accurate identification of crucial parameters characterizing the source, e.g., the initial time of the emission, as well as the intensity and current location of the source. Therefore, we have to design the sensor trajectories in such a way as to provide the most informative observations in this context.

Due to technical limitations, the measurements are assumed to be taken at multiples of $\Delta t = 10$ s. Taking this into account, the observation interval is divided into 20 equal design subintervals,

$$T_k = \big(t_k, t_k + 2\Delta t\big], \quad t_k = 2(k-1)\Delta t, \quad k = 1, \dots, 20. \qquad (8.39)$$

In such a way, two new observations per sensor are collected in each subinterval.

The detection problem considered here comprises the verification of the null hypothesis $H_0 : \mu = 0$ related to the zero intensity of the source, which is physically equivalent to its absence. To this end, a log-likelihood ratio technique can be applied. In our simulation scenario, the sensors are assumed to approach the contaminated area starting from the corners of the square domain with half of their maximum velocities in the direction of the center of the area (this time their initial positions are not optimized). Based on the collected measurements, the hypothesis H_0 is verified. Until the source is undetected, the sensors follow paths given a priori.

Under the validity of the alternative hypothesis $H_1 : \mu > 0$ (i.e., the presence of a source), the other source parameters become identifiable. Thus, once the source is detected, our goal is to properly design the sensor network node movements in order to accurately estimate the source characteristics. In order to slightly simplify parameter estimation, the source trajectory $\zeta(t)$ is approximated using linear splines,

$$\zeta(t) \approx \zeta_k + v_k(t - t_k), \qquad t \in T_k, \quad k = r, \dots, 20, \qquad (8.40)$$

where $v_k = (\zeta_{k+1} - \zeta_k)/(2\Delta t)$ is the average velocity of the source in the interval T_k, ζ_k denotes the source location at time t_k and $t_r \geq t_s$. Then, for each T_k, $k = r, \dots, 20$, we have to update the controls for the sensor network nodes in order to accurately find the current parameters of the source. From among the system parameters $\theta^k = (\mu, t_s, \zeta_{k,1}, \zeta_{k,2}, v_{k,1}, v_{k,2}, \kappa)$, we are interested in the first four, describing the current source location and intensity.

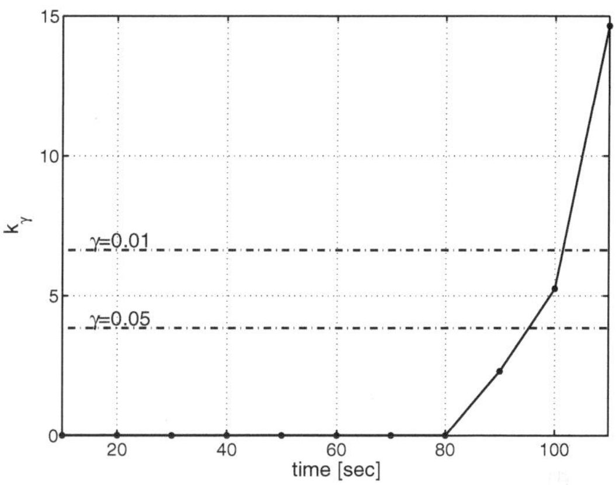

Fig. 8.6 Log-likelihood test for source detection (indicated thresholds for 5% and 1% false alarm levels)

Thus, maximization of the estimation accuracy for those parameters can be achieved via determination of D_s-optimal sensor trajectories.

To reduce the complexity of parameter estimation, the same sequential technique as delineated in Section 8.5.1.2 was applied to determine the estimates. As regards the sensors dynamics, again, the simple model (8.33) was employed, focusing on the direct control of sensor velocities with the following bounds for u:

$$-11.0 \leq u_l(t) \leq 11.0 \text{ [km/h]}\,, \quad \forall t \in T. \tag{8.41}$$

8.5.2.3 Implementation Details

In order to verify the proposed approach, a MATLAB program was written using a PC equipped with Pentium M740 processor (1.73GHz, 1 GB RAM) running Windows XP and MATLAB 7 (R14). First, the system of PDEs was solved using the `UMFPACK` solver from the COMSOL 3.4 environment exploiting the finite-element method. Each time calculations were performed for a spatial mesh composed of 3912 triangles and 2017 nodes and an evenly partitioned time interval (100 subintervals) using quintic Lagrange elements. The sensitivity coefficients were then linearly interpolated and stored. In each consecutive time subinterval, the system parameters were estimated using the `nlinfit` function from the MATLAB Statistics Toolbox. Finally, for determining the optimal trajectories, the package RIOTS 95 was employed. From among its three main optimization procedures, the routine `riots` was used, which is based on the SQP algorithm. In order to avoid getting stuck in a local optimum, the simulations were restarted several times from different starting points.

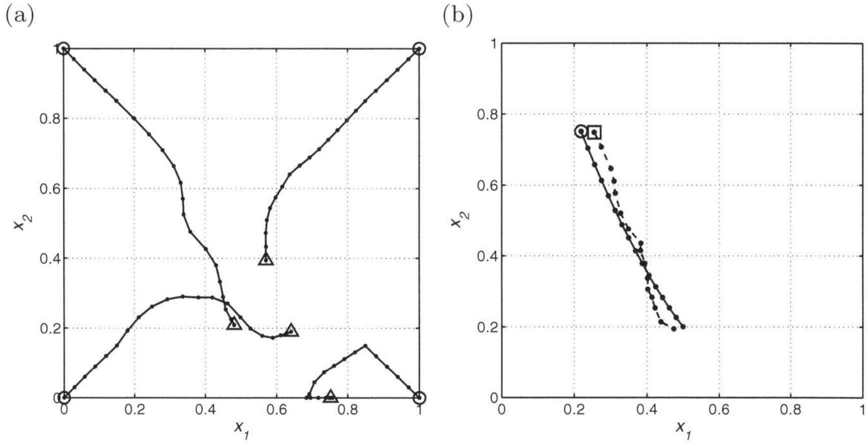

Fig. 8.7 Optimal sensor trajectories (starting and final positions marked with open circles and triangles, respectively) (a) and estimated trajectory of the pollution source (dashed line) and the actual trajectory (solid line). Their initial positions are indicated with a square and a circle, respectively (b)

8.5.2.4 Results

The phase of source detection is presented in Fig. 8.6, where the test statistics are calculated for each consecutive observation from sensors and compared with a threshold representing a constant false-alarm probability. Because in our scenario we assumed that there is no significant concentration of the pollutant before source activation, a likelihood ratio increases in a rather abrupt manner, as soon as one of the sensors measures a reasonably high concentration of the toxic substance. Therefore, for a 5% false alarm level, we observe almost immediate detection of the source after 20 s, i.e., within two consecutive measurements. Thus, starting from $t = 100$ s, the identification phase begins when the sensors trajectories are designed sequentially to accurately estimate the source characteristic parameters.

The optimal sensor trajectories are shown in Fig. 8.7(a). In order to explain the behavior of the sensor network in question, we have to compare the sensor paths with Fig. 8.5 to see that sensors attempt to follow the areas with greatest temporal concentration changes, which seems to be quite complex due to the combination of different mass transport processes.

The identification results are presented in Figs. 8.7(b) and 8.8, where we can easily compare the estimation quality of source parameters as well as localization effects.

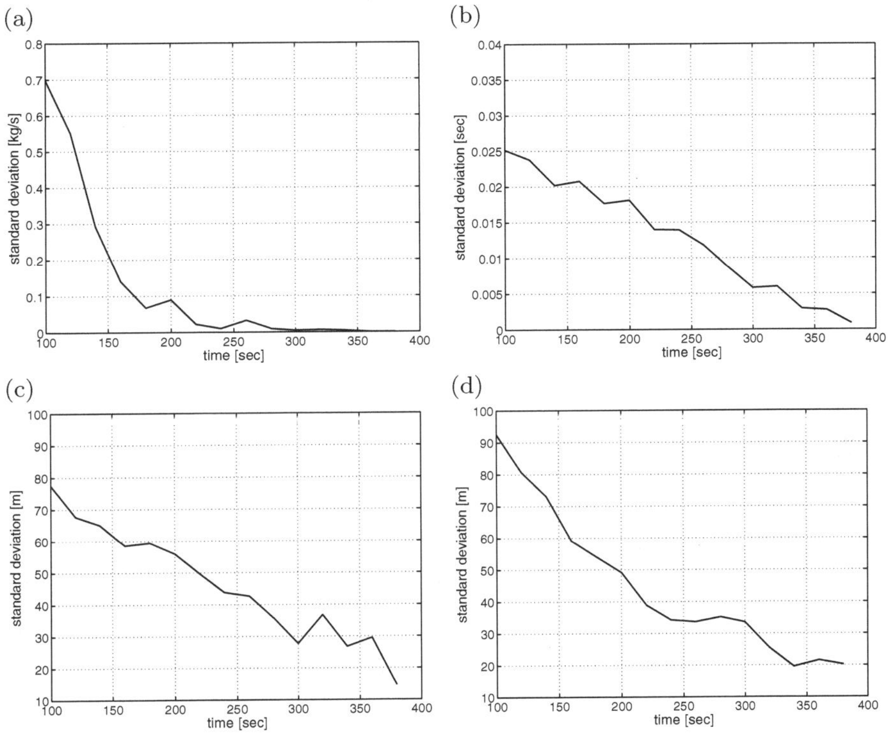

Fig. 8.8 Estimated standard deviation for the source parameters: intensity μ (a), emission starting time t_s (b), coordinate ζ_1 (c), coordinate ζ_2 (d)

8.6 Summary

In this chapter, some results related to the interpretation and use of data in DPS diagnostics using sensor networks with mobile nodes were presented. Specifically, theoretical research was carried out and computationally efficient methods to determine optimal configurations of mobile sensor networks for contaminating source detection and estimation were developed. What is more, proper implementation and testing of the computer code for verification of the proposed approach were done based on simulations. The key assumption made here was that the paths of the moving sources were unknown, but they were sufficiently smooth to be approximated by combinations of given basis functions. Parametrization of the source trajectories made it possible to reduce the source detection and estimation problem to that of parameter identification. In order to estimate the source and medium parameters, the maximum-likelihood estimation was used. Based on a scalar measure of fault detection reliability defined on the FIM, the problem was tailored to the framework of an optimal control one with state-variable inequality constraints representing geometric constraints induced by the admissible measurement regions and allowable distances between the sensors. Accordingly, numerical solvers of algorithmic optimal control can be employed. In our approach, RIOTS_95, a very efficient solver implemented in MATLAB has been applied.

The above approach was primarily aimed at off-line design of sensor trajectories. From a practical point of view, it is desirable to have the computations dynamic data-driven, i.e., the current measurements from the mobile sensors must serve as a basis for the update of parameter estimates and these, in turn, can be used to correct the sensor movements. This, however, gives rise to a highly nonlinear dynamic optimization problem. Thus, in the proposed research, an attempt was also made at applying a nonlinear receding-horizon approach to attack this issue.

The aim of the present work was to develop computationally efficient methods and algorithms to determine optimal trajectories of mobile sensor nodes for source identification in distributed parameter systems. In contrast to conventional approaches commonly used in distributed sensor networks, the knowledge of the mathematical model of the DPS in question is incorporated to form a basis for optimization. Consequently, the important information about the model governing the physical phenomenon in question is not lost and will be to the profit of early detection of potential system faults (e.g., chemical or biological threats).

Intuitively, we feel that the accuracy of the source identification problem must depend in a way on sensor locations. Numerical experience indicates that this influence is even more dramatic in the case of moving sources and sensors. This fact suggests that we may attempt to select these locations so as to obtain the best estimates of the source positions and intensities. But to form a basis for the comparison of different locations, a quantitative measure of the goodness of particular locations is required. Unfortunately, no measure of this type has been proposed yet. A principal aim of the work presented here was to demonstrate that criteria extensively applied in optimum experimental design and defined on the Fisher information matrix associated with the unknown parameters governing the process behavior can be employed for that purpose. Specifically, a Ψ_s-optimality criterion was used.

An online version of the proposed method for the design of optimal sensor trajectories could be of paramount importance in the fields of security, environmental and industrial monitoring, pollution control, etc. Motivating examples here include the detection of potential biochemical attacks and detecting leakages of dangerous biochemical materials. To deal with this issue, we have made an attempt at applying a nonlinear approach similar to that used in model predictive control. Simulation results demonstrate that the resulting scheme behaves well in practice, but a more thorough analysis is required to formally prove its convergence. Unfortunately, this problem is highly nontrivial and seeking its solution is left for future research.

Chapter 9
Extensions toward Challenging Problems of Network Scheduling

So far, numerous application-driven developments for the experimental design for DPSs related to the sensor scheduling in monitoring networks have been carefully studied. Nevertheless, all problems considered by no means exhaust all potential situations motivated by practical settings, and many difficult issues still remain open, posing challenges to researchers concerned with distributed measurement systems. In this chapter, two important extensions of the foregoing results to other experimental settings related to difficult design problems encountered in identification of real-world processes are discussed. The first one is the sensor scheduling problem for observations collected in a series for different realizations of processes with random parameters. Both the theoretical background and an algorithm for calculating optimum group experimental designs are provided to address this issue. The theory is applicable to those practical situations in which a dynamic system is sensitive to sampling or gives a different response at each run of the experiment. Together with the definition of group designs that is also introduced, this structure leads to a practical and numerically tractable representation of optimum designs for estimation of the mean values of the parameters. The second setting of great practical relevance which is investigated in this chapter is the problem of realization of the observational process under the presence of spatially correlated measurements. The task is extremely difficult, since information from different sensor nodes cannot be separated during the data fusion, leading to a far higher level of complexity compared to the uncorrelated setting.

9.1 Sensor Scheduling for Group Observations

An inherent characteristic of many dynamic processes is that any measurement of the process variables disturbs the experimental environment and

M. Patan: Optimal Sensor Networks Scheduling in Identification of DPSs, LNCIS 425, pp. 231–260.
springerlink.com

may lead to unpredictable changes in the parameter values. Although such disturbances are usually relatively small, they may significantly alter the system response and should not be neglected. This applies especially to distributed parameter processes, which are very sensitive to small perturbations in their parameters. In extreme situations, a perturbation can cause the process to transfer into an undesired state (e.g., instability and oscillations of electrical circuits or chemical reactions) or even result in its failure or destruction (e.g., in case of a plant in biological studies or quality testing of industrial devices). Applications include computer network monitoring and parallel computing [3, 123], computer-assisted impedance tomography [101, 252], ecosystem monitoring [241], thermal systems [10, 257], chemical reactions engineering [7, 91, 192], clinical trials [190] or pharmaco-kinetic modelling [26, 27, 314].

Here, we consider examples of identification for distributed systems where each of the repeated experiments may result in different parameter values or where sampling may cause disturbances in further runs of the process. In many cases, the nature of such perturbations is unpredictable. Therefore, the model parameters are treated as random variables. Interest is not in the individual parameters but in their mean value over a defined range. The question then is how to design a series of experiments to obtain a precise estimate of such a mean.

In the context of research on optimum design for nonlinear regression models, a similar problem has been addressed based on the notion of the so-called *population designs* [81, 166], with a majority of works stimulated by pharmaco-kinetic and pharmaco-dynamic problems in drug development studies, where it is assumed that an individual unit's (patient's) reaction to a treatment (dose) is constant, while the reaction over a population of patients is variable. However, the underlying models neglect an important factor of response sensitivity to sampling. Although some attempts to address this issue have been proposed within the framework of lumped systems [190, 192], in the context of DPSs this problem has received very little attention so far [189].

In the approach proposed here, an experimental unit's (device, object, process, etc.) response can vary randomly over space; for example, it may be affected by changes in the experimental setting, nonstationarity of parameters, etc. Another substantial difference is in the definition of the population design. In the standard approach, this is a collection of discrete designs grouped according to their size [166]. In this work, groups of individual measurement schedules are considered which form a continuous probability measure. Such a formulation offers appreciable flexibility in creating groups as well as individual experimental designs, while maintaining optimum statistical properties. This is achieved by the new idea of reformulating the optimization of a complex group design as the optimization of its simpler (aggregated) form, and then recovering the information on the group structure having optimized the aggregated design. As a result, we obtain a set of optimum solutions to choose from according to additional experimental requirements.

9.1.1 Group Scheduling Problem

9.1.1.1 Statistical Model

Suppose that there is a collection of N individual experimental units described by (2.66)–(2.68) (e.g., chemical reactors, thermal systems, electrical circuits etc.) for each of which n_j measurements are gathered, possibly according to different sensor spatial configuration. Due to a mutual interaction between measurement transducers and examined system each observation may result in a slight change in the experimental conditions and cause an unpredictable perturbation in the individual process parameters. Taking into account both sources of parameter variability, namely, the individual units and spatial locations, the 'group' observation strategy considered can be formally represented as (for the sake of clarity, the state has been assumed to be directly measurable)

$$z_j^\ell(t) = y(x_j^\ell, t; \theta_j^\ell) + \varepsilon(x_j^\ell, t), \quad t \in T, \quad \ell = 1, \dots, n_j, \quad k = 1, \dots N, \quad (9.1)$$

where $z_j^\ell(t)$ is the scalar output, $x_j^\ell \in X$ stands for the location of the ℓ-th sensor for the j-th experimental unit, X signifies the part of the spatial domain Ω where the measurements can be made and $\varepsilon(x_j^\ell, t)$ denotes zero-mean, Gaussian, spatially uncorrelated and white measurement noise.

Denote by $g(z_j^\ell(t)|\theta_j^\ell)$ the conditional probability density of observations z_j^ℓ given the value θ_j^ℓ. Under the assumption of the additivity of measurement noise, the density g has the same functional form as the pdf of $\varepsilon(\,\cdot\,,\,\cdot\,)$ but translated by the value of the model response vector y. Vectors $\theta_j^\ell \in \Theta$ are assumed to be realizations of the random vector $\theta = (\theta_1, \dots, \theta_m)^\mathsf{T}$ with probability density $p(\theta; \lambda)$. The function h is entirely determined by the hyperparameter vector $\lambda = (\lambda_1, \dots, \lambda_{\tilde{m}})^\mathsf{T}$. The primary interest is in efficient estimation of this vector of constant parameters λ (or of part of it). Then, the instantaneous log-likelihood function for λ given collection of observations z takes the form

$$\mathcal{L}(\lambda|z(t)) = \log \prod_{k=1}^{N} \prod_{\ell=1}^{n_j} \int_\Theta g(z_j^\ell(t)|\theta_j^\ell) p(\theta; \lambda)\,\mathrm{d}\theta, \quad (9.2)$$

where $z = [z_1, \dots, z_N]$ and $z_k = [z_j^1(t), \dots, z_j^{n_j}(t)]^\mathsf{T}$.

In applications, direct transformation of a priori knowledge about the parameters into a statistical distribution in closed form can be difficult. However, some typical examples for an approximation of $p(\,\cdot\,; \lambda)$ can be delivered by analogy to the robust designs in the average sense discussed in Section 7.3.1.

9.1.1.2 Experimental Design

Assume that a collection of N experimental units consists of G groups, each of size N_k, $k = 1, \ldots, G$, and that the individual units in the same group follow the same spatio-temporal schedule of measurements (design). Grouping can be interpreted as an additional ordering factor helping to solve the design problem. It may be directly related to economic constraints imposed on the total cost of the experiment or to technological constraints when the number of observations from some experimental units is limited (due to its destruction, state change, etc.). There is no loss of generality in assuming a group structure, since the number of groups can equal the number of individual units. The experimental design is constructed in two stages.

Individual level. The experimental design for an individual unit in the k-th group is a list of n_k spatial locations at which measurements are to be made, namely,

$$\xi_k = \{x_k^1, \ldots, x_k^{n_k}\}. \tag{9.3}$$

Since, in general, we allow replications of observations, i.e., some of the spatial locations may be the same, it is convenient to apply the concept of continuous designs introduced in Section 2.1.3. Using proportions p_k^i of the numbers of repeated observations to the number of all observations taken for the individual unit at x_k^i, we have

$$\xi_k = \left\{ \begin{matrix} x_k^1 & \ldots & x_k^{\ell_k} \\ p_k^1 & \ldots & p_k^{\ell_k} \end{matrix} \right\}, \quad p_k^i \in [0, 1], \quad \sum_{i=1}^{\ell_k} p_k^i = 1. \tag{9.4}$$

The design $\xi_k \in \Xi$, where Ξ denotes a set of admissible designs defined by (9.4), is a continuous measure on a set of ℓ_k distinct points in a design region X. To preserve the information about the number of measurements n_k for each individual unit, we denote the whole experimental system by the pair (ξ_k, n_k) for each individual unit.

Group level. Under the assumption that all N_k individual units in group k, $k = 1, \ldots, G$, follow the same sampling schedule, we can provide the following definition of group design [190, 192].

Definition 9.1. The *group design* is the collection of variables

$$\zeta = \left\{ \begin{matrix} (\xi_1, n_1) & \ldots & (\xi_G, n_G) \\ \alpha_1 & \ldots & \alpha_G \end{matrix} \right\}, \tag{9.5}$$

where $\alpha_k = \frac{N_k}{N}$ is the proportion of individual units in the whole population which follow individual design (ξ_k, n_k).

Obviously, we have $\sum_{k=1}^{G} \alpha_k = 1$; therefore, it is convenient to further relax the constraints in (9.5) by allowing the proportions α_k to be any real numbers from the interval (0,1]. A design $\zeta \in Z_{\mathrm{ad}}$, where Z_{ad} denotes a set of admissible designs defined by (9.5), is a continuous measure on a set of G distinct sensor allocations.

9.1.1.3 Estimation Accuracy Measure

Owing to the character of noise in (9.1), the FIM for a group design ζ is given by [189, 192]

$$M(\zeta, N) = N \sum_{k=1}^{G} \alpha_k M(\xi_k, n_k) = N \sum_{k=1}^{G} \alpha_k n_k \sum_{i=1}^{\ell_k} p_k^i M(x_k^i), \tag{9.6}$$

where

$$M(x_k^i) = \mathrm{E}\left\{ -\frac{\partial^2 \mathcal{L}_T(\lambda | \{z_k^i(t)\}_{t\in T})}{\partial\lambda\partial\lambda^\mathsf{T}} \right\}$$

is the elementary FIM for the observation made at location x_k^i and $\mathcal{L}_T(\,\cdot\,|\,\cdot\,)$ is a log-likehood based on the continuous data record $\{z_k^i(t)\}_{t\in T}$ defined by respective Radon-Nikodym derivative, cf. (2.12).

Since, for nonlinear models, the above formula is usually analytically intractable, approximation procedures are required for evaluation of the FIM. In the statistical literature, there exist a variety of methods including numerical integration and Monte-Carlo approximation. In the special case where p and g are normal density functions, linearization of the model around the expected value of the random-effect is most commonly used [81, 114, 240] as it leads to a relatively simple closed form of the FIM. Otherwise, we have to employ Monte-Carlo techniques [239] or resort to numerical integration.

Once the FIM is defined, by analogy to the design problems considered so far, we are interested in minimizing a functional Ψ operating on the set of admissible information matrices $\mathbb{M}$ established by (9.6):

$$\Psi[M(\zeta, N)] \longrightarrow \min. \tag{9.7}$$

That is, we look for a design $\zeta^\star$ which gives the optimum FIM for initially chosen values of the parameters λ.

Due to the assumption of independent observations, every new measurement provides some amount of information. Therefore, if the number of measurements is unlimited, the expected solution to the sensor selection problem is unbounded. Thus, it is necessary to constrain the total number of observations to be no greater than N_0, i.e.,

$$N \sum_{k=1}^{G} \alpha_k n_k \leqslant N_0. \tag{9.8}$$

If the number of all individual units in the design is not predetermined a priori and has to be estimated, it is convenient to relax the restriction of N being a positive integer and allow it to take any positive real value (this is not restrictive assertion if only the number of measurements is reasonably large). Then, we can formulate the following useful result.

Proposition 9.1. *Let the assumptions (A1)–(A3) hold. Then Ψ attains its minimum at some pair $(\zeta^\star, N^\star)$ for which the inequality in (9.8) becomes an equality.*

Proof. First, note that, since $M(x)$ is by definition a nonnegative definite matrix, then for any ζ and any positive N, the matrix $M(\zeta, N)$, being a sum of nonnegative definite matrices, is also a nonnegative definite matrix.

Further, let us assume that Ψ does not have its minimum on the boundary (9.8). Then the minimum occurs for some pair $N^\star$ and $\zeta^\star = \{(\xi_k^\star, n_k^\star, \alpha_k^\star)\}_{k=1}^G$ which strictly satisfy (9.8), i.e., $N^\star \sum_{k=1}^G \alpha_k^\star n_k^\star < N_0$. Then, there exists $\beta = \frac{N}{N^\star} > 1$ such that $M(\zeta^\star, N) = \beta M(\zeta^\star, N^\star)$ and $N \sum_{k=1}^G \alpha_k^\star n_k^\star = N_0$. But then, according to the monotonicity of Ψ (A3), we have

$$
\begin{aligned}
\Psi[M(\zeta^\star, N)] &- \Psi[\beta M(\zeta^\star, N^\star)] \\
&= \Psi[M(\zeta^\star, N^\star) + (\beta - 1)M(\zeta^\star, N^\star)] \leq \Psi[M(\zeta^\star, N^\star)].
\end{aligned}
\tag{9.9}
$$

Hence, the pair $(\zeta^\star, N)$ must be a minimum of Ψ. This contradiction finishes the proof. ■

Remark 9.1. Although the property established in Proposition 9.1 is easily shown, it is not obvious, since the constraint (9.8) is not convex with respect to the set of parameters N, α_k and n_k. Its significance, however, should not be underestimated. It allows restatement of the problem as the one of convex optimization, as will be shown below.

Remark 9.2. Existence of the optimal solution follows directly from the Weierstrass extreme value theorem, since the constraint (9.8) supplemented with the assumption of N being some positive and bounded real value assures that the set of all potential solutions (ζ, N) belong to a compact set, and criterion Ψ is convex and hence continuous (c.f. (A4)).

Hence, the optimum group sensor scheduling problem has the following form.

Problem 9.1. Find a pair (ζ, N) to minimize

$$
\mathcal{P}(\zeta, N) = \Psi(M(\zeta, N)) \tag{9.10}
$$

subject to (9.5) and (9.8).

Trying to solve Problem 9.1 directly is an extremely difficult task. Since it is necessary to simultaneously determine locations x_k^ℓs, proportions α_ks, numbers of observations in groups n_ks, the number of groups G and the number of units N, we encounter severe complexity. A convenient reformulation of the problem leading to a simpler computational scheme constitutes the subject of next section.

9.1.2 Solution via Aggregated Design

In order to obtain independence of the solution of the total cost of the experiment (i.e., to obtain the solution in terms of optimal proportions of the cost or numbers of repeated observations), we introduce, without any loss of generality, the *average* per total cost (normalized) FIM,

$$M(v) = \frac{N}{N_0}\sum_{k=1}^{G}\alpha_k n_k \sum_{i=1}^{\ell_k} p_k^i M(x_k^i) = \sum_{k=1}^{G}\beta_k M(\xi_k), \tag{9.11}$$

where

$$\beta_k = \frac{N}{N_0}\alpha_k n_k\,, \quad M(\xi_k) = \sum_{i=1}^{\ell_k} p_k^i M(x_k^i)$$

and

$$v = \begin{Bmatrix} \xi_1 & \dots & \xi_G \\ \beta_1 & \dots & \beta_G \end{Bmatrix}, \quad \beta_k \in (0,1], \quad \sum_{k=1}^{G}\beta_k = 1. \tag{9.12}$$

Due to Proposition 9.1, instead of solving the problem of minimizing $\Psi[M(\zeta, N)]$ subject to (9.8), we can equivalently solve the problem of minimization of $\Psi(v)$ subject to $\sum_{k=1}^{G}\beta_k = 1$. This significantly simplifies the very complex procedure of finding an optimum group design, as will be shown in the following.

Moreover, a number of characterizations for designs v can be generalized from the results given in Section 2.1.4 in a rather straightforward manner. Particularly, we have the following existence result.

Corollary 9.1. *Suppose that the assumptions (A1)–(A6) hold. Then an optimal design $v^\star$ exists and it consists of no more than $\widetilde{m}(\widetilde{m}+1)/2$ support points. Moreover, the set of optimal designs is convex.*

Furthermore, in order to establish the form of the general equivalence theorem for the group design, we introduce the *group sensitivity function* ψ_G, which exists due to the assumption (A6):

$$\psi_G(\xi, v) = \varsigma_G(v) - \phi_G(\xi, v), \tag{9.13}$$

$$\varsigma_G(v) = -\operatorname{trace}\Big[\mathring{\Psi}[M(v)]M(v)\Big], \tag{9.14}$$

$$\phi_G(\xi, v) = -\operatorname{trace}\Big[\mathring{\Psi}[M(v)]M(\xi)\Big], \tag{9.15}$$

where

$$\mathring{\Psi}[M(\upsilon)] = \left.\frac{\partial \Psi(M)}{\partial M}\right|_{M=M(\upsilon)}. \tag{9.16}$$

Corollary 9.2 (Generalized equivalence theorem for group design). *Suppose that the assumptions (A1)–(A6) hold. The following conditions are equivalent:*

(i) design $\upsilon^\star$ *minimizes* $\Psi[M(\upsilon)]$,
(ii) design $\upsilon^\star$ *minimizes* $\max_{\xi\in\Xi} \phi_G(\xi,\upsilon) - \varsigma_G(\upsilon)$,
(iii) $\max_{\xi\in\Xi} \phi_G(\xi,\upsilon^\star) = \varsigma_G(\upsilon^\star)$.

If the criterion $\Psi[\,\cdot\,]$ *is additionally strictly convex, all the designs which satisfy (i)–(iii) and their convex combinations have the same information matrices equal to* $M(\upsilon^\star)$.

These properties of the optimum designs $\upsilon^\star$ were derived assuming only the independence of observations among different groups. Therefore, direct application of Theorem (9.2) to build an efficient algorithm for calculating such optimal designs is not straightforward. In fact, there is a lack of such procedures in the literature.

However, taking advantage of the independence of the observations on the individual level, it is possible to strengthen this result. First, observe that the average FIM (9.11) may be rewritten in the form

$$M(\upsilon) = \sum_{k=1}^{G}\sum_{i=1}^{\ell_k} \frac{N}{N_0}\alpha_k n_k p_k^i M(x_k^i) = \sum_{k=1}^{G}\sum_{i=1}^{\ell_k} \gamma_k^i M(x_k^i), \tag{9.17}$$

where

$$\gamma_k^i = \frac{N}{N_0}\alpha_k n_k p_k^i = \beta_k p_k^i\,; \quad \sum_{k=1}^{G}\sum_{i=1}^{\ell_k} \gamma_k^i = 1.$$

Different groups do not have to have all support points different, that is, some points x_k^i may be the same for different k. Consequently, it is sensible to reformulate the problem to work with the locations $x_1,\dots,x_\ell$ (relabeled different time instants) in lieu of the x_k^is. Here, we introduce weights $\gamma_1,\dots,\gamma_\ell$ which are sums of the γ_k^is for the repeated time instants. This allows us to rewrite (9.17) as

$$M(\upsilon) = \sum_{j=1}^{\ell} \gamma_\ell M(x^j) = M(\omega), \tag{9.18}$$

where

$$\omega = \left\{\begin{matrix} x^1 & \dots & x^\ell \\ \gamma_1 & \dots & \gamma_\ell \end{matrix}\right\}, \quad \sum_{j=1}^{\ell}\gamma_j = 1, \quad \gamma_j = \sum_{k=1}^{G}\gamma_k^i. \tag{9.19}$$

Finally, we can reformulate the original optimization problem into the following form.

Problem 9.2. Find ω to minimize

$$\mathcal{P}(\omega) = \Psi[M(\omega)] \tag{9.20}$$

subject to (9.19).

This reformulation leads to solving the problem of finding a two level hierarchical optimal group design as that of finding an equivalent one level design. Note that $\omega \in \Xi$. We call ω an *aggregated design.*

The information about groups is included in γ_k^i and hence in γ_k. This information is later recovered after an optimum design ω has been found. Another advantage of such formulation of the group design is that it significantly reduces the dimension of the optimization problem. For every $\upsilon \in \Upsilon$ there exists $\omega \in \Xi$ such that $M(\upsilon) = M(\omega)$, and we have the analogous properties of these designs, in particular, the analogue of the existence result and the equivalence theorem can be easily provided [192].

Such reformulation of the original group scheduling problem makes it possible to find an optimal group design through solving the conventional problem of approximate nonhierarchical designs considered so far in the previous sections. It is then possible to apply existing numerical procedures. A substantial difficulty in determining the group designs arises from the fact that they are not unique. Indeed, the criterion Ψ is most often strictly convex on $\mathcal{M}(\Xi)$, and this guarantees that the optimal FIM is unique, but it does not necessarily mean that $(\zeta, N) \mapsto \Psi[M(\zeta, N)]$ is strictly convex in (ζ, N). Hence, there is no guarantee that the optimum group design is unique. Multiple global solutions $(\zeta^\star, N^\star)$ may yield the same minimizing value of $M(\zeta, N)$. Furthermore, there may be multiple local minima of $\Psi(\,\cdot\,)$ that can seriously complicate the optimization process.

9.1.3 Numerical Scheme for Group Sensor Scheduling

Problem 9.2 is a classical problem of optimum experimental design theory and has been studied thoroughly in the literature since the early 1970s, resulting in many effective algorithms for its efficient solution (in particular, all algorithms studied in Chapter 3 can be adopted to its efficient solution).

Here we propose a new technique for finding optimum group designs based on the reformulation of the optimization problem. Instead of solving the original Problem 9.1, we first solve the equivalent Problem 9.2, which is far simpler. The solution to the latter requires a method to transform the optimal $\omega^\star$ into the original group design pair $(\zeta^\star, N^\star)$. This can be achieved in three steps.

Step 1. Solve the optimization task defined by Problem 9.2 and as a result determine the aggregated design $\omega^\star$.

Step 2. Transform $\omega^\star$ into an equivalent continuous design $\upsilon^\star$, which satisfies $M(\upsilon) = M(\omega)$.

Step 3. Transform $\upsilon^\star$ into an equivalent design pair $(\zeta^\star, N^\star)$.

After solving the minimization problem in Step 1, we have

$$\omega^\star = \left\{ \begin{matrix} x^{1\star} & \dots & x^{\ell\star} \\ \gamma_1^\star & \dots & \gamma_\ell^\star \end{matrix} \right\}, \quad \sum_{j=1}^{\ell} \gamma_j^\star = 1, \quad \gamma_k^\star > 0. \tag{9.21}$$

Then, we have to retrieve the components of $\upsilon^\star$, i.e, $\{(\xi_k^\star, \beta_k^\star)\}_{k=1,\dots,G}$ or $\{(x_k^{i\star}, p_k^{i\star}, \beta_k^\star)\}_{k=1,\dots,G;\; i=1,\dots,\ell_k}$. Here we allow the weights p_k^i to take zero values. Then each individual design has the same general form:

$$\xi_k^\star = \left\{ \begin{matrix} x^{1\star} & \dots & x^{\ell\star} \\ p_k^{i\star} & \dots & p_k^{\ell\star} \end{matrix} \right\}, \quad \sum_{k=1}^{\ell} p_k^{i\star} = 1, \quad p_k^{i\star} \geq 0$$

and, technically, the problem of determining the design $\upsilon^\star$ simplifies to the determination of the weights $p_k^{i\star}$ and $\beta_k^\star$. This can be achieved by solving the system of equations

$$\begin{cases} \beta_k p_k^i - \gamma_k^i = 0,\; i = 1, \dots, \ell, \;\; k = 1, \dots, G & (\ell G \text{ nonlinear eqs., see (9.17)}), \\ \sum_{i=1}^{\ell} p_k^i = 1, \quad k = 1, \dots, G & (G \text{ linear equations}), \\ \sum_{k=1}^{G} \gamma_k^i = \gamma_i^\star,\; i = 1, \dots, \ell & (\ell \text{ linear equations}). \end{cases} \tag{9.22}$$

It is clear that this system of $\ell G + \ell + G$ equations with $G + 2\ell G$ variables is underdetermined if the number of groups $G > 1$. There are $\ell(G-1)$ more variables than equations. Although the first sG equations are nonlinear, it is easy to show that a solution with nonnegative values of all variables always exists. Treating $\ell(G-1)$ variables γ_k^i as nonnegative parameters (satisfying the condition that for any $k = 1, \dots, G$ there exists at least one positive value γ_k^i), the solution becomes simple:

$$\begin{cases} \beta_k^\star = \sum_{i=1}^{\ell} \gamma_k^i, \; k = 1, \dots, G \\ p_k^{i\star} = \gamma_k^i / \beta_k^\star, \quad i = 1, \dots, \ell, \;\; k = 1, \dots, G. \end{cases} \tag{9.23}$$

These values are further used in Step 3. In the special case when replications of observations are impossible, additional binary constraints to the system (9.22) have to be added, namely,

$$p_k^i \in \{0, 1/n_k\}, \quad k = 1, \dots, G. \tag{9.24}$$

The optimal values of the parameters $\alpha_k^\star, n_k^\star$, $k = 1, \dots, G$, and $N^\star$ can be retrieved solving the system of equations

$$\begin{cases} \frac{N}{N_0}\alpha_k n_k = \beta_k^\star,\ k = 1, \dots, G & (G \text{ nonlinear equations}) \\ \sum_{k=1}^{G} \alpha_k = 1 & (1 \text{ linear equation}). \end{cases} \tag{9.25}$$

Here, again, we have an underdetermined system of nonlinear equations, since we have $G + 1$ equations and $2G + 1$ variables, which can be easily solved numerically. Such a formulation of the optimal group design problem does not lead to unique solutions. This, however, gives room for tailoring optimum designs to practical requirements and makes it possible to use any additional information the experimenter might have. It is an advantage of the method that it gives the experimenter some freedom to impose additional constraints on the design variables.

For example, if the numbers of observations per individual unit in each group, n_k, $k = 1, \dots, G$, are known or can be chosen arbitrarily, then the optimal solution exists and the optimum total number of units and the optimum weights are, respectively,

$$N^\star = N_0 \sum_{k=1}^{G} \frac{\beta_k^\star}{n_k} \quad \text{and} \quad \alpha_k^\star = \frac{N_0}{N^\star}\frac{\beta_k^\star}{n_k}, \quad k = 1, \dots, G. \tag{9.26}$$

Two special forms of 'simple' group designs are given in the following result

Proposition 9.2. *If $\omega^\star$ is a solution to Problem 9.2 given by (9.21), then*

(i) the design $\zeta^\star = \left\{ \begin{smallmatrix} \omega^\star \\ 1 \end{smallmatrix} \right\}$ and $N^\star = N_0/n_1$ for any $n_1 > 0$ minimizes $\mathcal{P}(\zeta^\star, N^\star)$;

(ii) the design $\zeta^\star = \left\{ \begin{smallmatrix} \omega_1^\star & \dots & \omega_\ell^\star \\ \gamma_1^\star & \dots & \gamma_\ell^\star \end{smallmatrix} \right\}$ and $N^\star = N_0 \sum_{k=1}^{\ell} \gamma_k^\star / n_k$, where $\omega_k^\star = \left\{ \begin{smallmatrix} x_k^\star \\ 1 \end{smallmatrix} \right\}$ for any $n_k > 0$, $k = 1, \dots, \ell$, minimizes $\mathcal{P}(\zeta^\star, N^\star)$.

Proof. To prove Proposition 9.2, it is sufficient to show that the information matrices $M(\zeta^\star, N^\star)$ and $M(\omega^\star)$ are the same, and this is a matter of simple calculation. Indeed, we have the following.

(*i*) In this case, from (9.11), we have

$$M(\zeta^\star, N^\star) = \frac{\frac{N_0}{n_1}}{N_0} \sum_{i=1}^{\ell} n_1 \gamma_i^\star M(x^{i^\star}) = \sum_{i=1}^{\ell} \gamma_i^\star M(x^{i\star}) = M(\omega^\star),$$

which proves the first part of the claim.

(*ii*) Here, using the result from (9.23) that $\beta_j^\star = \gamma_j^\star$, $j = 1, \dots, \ell$, and applying (9.26), we obtain

$$\alpha_j^\star = \frac{N_0}{N^\star}\frac{\beta_j^\star}{n_j}, \quad j = 1, \dots, \ell$$

and

$$M(\zeta^\star, N^\star) = \frac{N_0 \sum_{j=1}^{\ell} \frac{\gamma_j^\star}{n_j}}{N_0} \sum_{j=1}^{\ell} \frac{N_0}{N_0 \sum_{j=1}^{\ell} \frac{\gamma_j^\star}{n_j}} \frac{\gamma_j^\star}{n_j} n_j M(x^{j^\star})$$
$$= \sum_{j=1}^{\ell} \gamma_j^\star M(x^{j^\star}) = M(\omega^\star),$$

which completes the proof. ■

The above result indicates that when the observations are independent, there exist group designs of very simple forms: a single group design (in the sense of the same observation schedule for all $N^\star$ individuals) and the design with ℓ-groups, each with one support point. In the latter, a $N^\star$ individuals are divided into ℓ groups according to the weights $\gamma_k^\star$, and all individuals within the same group are sampled only once at the same single time point.

Despite the complex formulation of the problem, the approximations of optimal group designs are simple. To the best of the author's knowledge, this fact has not been known in the statistics literature.

9.2 Measurement Strategies for Correlated Observations

This section is dedicated to the subject of sensor network design in the presence of spatially correlated measurements, being one of the most characteristic properties of information fusion for spatial data [34, 46, 73, 169, 180, 196, 281]. Although such an experimental setting is of great practical importance, it is consequently neglected in most works on sensor location for parameter estimation in dynamic DPSs, which is on the one hand very convenient and leads to elegant theoretical results, but cannot be justified in a high number of applications. This is because the complexity of many real-world engineering problems cannot be fully explained in terms of dynamic relations between its responses since the interaction of the system with the experimental environment may be of very confound character. Environmental monitoring, incipient fault detection, surveillance, computer-assisted tomography and many industrial experiments are the most typical areas where this factor emerges, raising the issue of its practical importance. Moreover, it is often not possible to use replicated measurements and it is rather clear that in many situations several observations clustered in close vicinity one to another do not provide much more information than a single measurement. This fact is of crucial importance as the classical concept of design measures is not applicable to this case. The situation for dynamic multi-output systems may be even more complicated since the measurements are collected from different outputs according to some regular schedule. Thus, random errors can be correlated not only in space and time, but also some mutual influence among different system responses may occur. Obviously, such simultaneous correlations imply a very complex nature of the problem.

To deal with it, two main approaches can be distinguished. The first idea is to exploit some well-known numerical techniques of constrained optimization where suitably defined additional constraints representing admissible distances between observations in time and space are imposed. Direct application of optimization techniques by no means excludes the clusterization effect, but when the number of sensors is quite high, the problem complexity considerably increases. In terms of interpretation and applicability, such an approach is very attractive in the context of mobile sensors [281]. The second approach consists in taking into account mutual correlations between all measurements by appropriately modifying the information matrix [191, 283]. This alternative approach constitutes the subject of this section.

9.2.1 Correlated Measurement Errors

For the clarity of presentation, the deliberations are limited to the case of N stationary sensors with the observations taken continuously in time. Generalizations to more general observation strategies (i.e., networks equipped with scanning and mobile sensors) can be made with minor difficulties of technical nature. Let us assume that the random errors in the observation equation (2.75) are correlated and the covariance structure is known (either the covariance matrix C or the covariance kernel is given).

Observations. Assuming for simplicity that the system state y is directly measurable, this is equivalent to the output equations

$$z^j(t) = y(x^j, t; \theta) + \varepsilon(x^j, t), \quad j = 1, \dots, N, \tag{9.27}$$

where $z^j(\,\cdot\,)$ is an s-dimensional output, $t \in T = [0, t_f]$ and the Gaussian random field $\varepsilon(\,\cdot\,,\,\cdot\,)$ representing measurement errors satisfy

$$\mathrm{E}\big[\varepsilon(x,t)\big] = 0, \tag{9.28}$$

$$\mathrm{E}[\varepsilon(x,t)\varepsilon^{\mathsf{T}}(\chi,\tau)] = K(x,\chi,t;\theta)\delta(t-\tau). \tag{9.29}$$

where $K(x,\chi,t;\theta) \in \mathbb{R}^{s\times s}$ is a known continuous covariance kernel and the $Ns \times Ns$ covariance matrix takes the form

$$C(t;\theta) = \begin{bmatrix} K(x^1,x^1,t;\theta) & \cdots & K(x^1,x^N,t;\theta) \\ \vdots & \ddots & \vdots \\ K(x^N,x^1,t;\theta) & \cdots & K(x^N,x^N,t;\theta) \end{bmatrix}. \tag{9.30}$$

Notice that, if any two sensors are placed at the same location, then the corresponding columns (and rows) of C are identical, which leads to the singularity of the covariance matrix. Since the replications of the measurements are no longer justified, the following compact notation for the design of experiment will be subsequently used in the further analysis:

$$\xi = \{x^1, \dots, x^N\}. \tag{9.31}$$

Information matrix. Given the set of experimental observations $z(t) = [z^1(t)^\mathsf{T}, \dots, z^N(t)^\mathsf{T}]^\mathsf{T}$ measured continuously on T, the instantaneous log-likelihood function of the parameters θ takes the form

$$\begin{aligned}\mathcal{L}(\theta|z(t)) = &-\frac{1}{2}sN\ln 2\pi - \frac{1}{2}\ln\det C(\xi,t;\theta)\\ &-\frac{1}{2}\left(z-y(x,t;\theta)\right)^\mathrm{T}C^{-1}(\xi,t;\theta)\left(z-y(x,t;\theta)\right),\end{aligned} \tag{9.32}$$

with $C(\xi,t;\theta)$ denoting the covariance matrix for the set of observations taken at locations from ξ. For such an experimental setting it is possible to formulate the following result determining the form of the information matrix.

Proposition 9.3. *Let the Gaussian random field $\varepsilon(\,\cdot\,,\,\cdot\,)$ satisfy (9.28)–(9.29). Then the the elements of the temporary Fisher information matrix take the form*

$$M_{ij}(t) = \frac{\partial y^\mathsf{T}}{\partial\theta_i}C^{-1}\frac{\partial y}{\partial\theta_j} + \frac{1}{2}\operatorname{trace}\left(C^{-1}\frac{\partial C^{-1}}{\partial\theta_i}C^{-1}\frac{\partial C^{-1}}{\partial\theta_j}\right), i,j=1,\dots,m, \tag{9.33}$$

where $C = C(\xi,t;\theta)$ and $y = y(x,t;\theta)$.

Proof. First, in order to clarify the notation within the proof, we will write shortly C, y and ε instead of $C(\xi,t;\theta)$, $y = y(x,t;\theta)$ and $\varepsilon(x,t)$, respectively. Differentiating the log-likelihood function with respect to the individual parameter θ_i, we obtain

$$\begin{aligned}\frac{\partial\mathcal{L}(\theta|z)}{\partial\theta_i} &= -\frac{1}{2}\frac{\partial\ln\det C}{\partial\theta_i} - \frac{1}{2}\frac{\partial(\varepsilon^\mathsf{T}C^{-1}\varepsilon)}{\partial\theta_i}\\ &= -\frac{1}{2}\left(\operatorname{trace}\left(C^{-1}\frac{\partial C}{\partial\theta_i}\right) - 2\varepsilon^\mathsf{T}C^{-1}\frac{\partial y}{\partial\theta_i} - \varepsilon^\mathsf{T}C^{-1}\frac{\partial C}{\partial\theta_i}C^{-1}\varepsilon\right).\end{aligned} \tag{9.34}$$

Introducing notations $C_k = \partial C/\partial\theta_k$ and $y_k = \partial y/\partial\theta_k$, the individual (i,j)-th element of information matrix takes the form

$$\begin{aligned}M_{ij} &= \mathrm{E}\left\{\left(\frac{\partial\mathcal{L}(\theta|z)}{\partial\theta_i}\right)\left(\frac{\partial\mathcal{L}(\theta|z)}{\partial\theta_j}\right)\right\} \\ &= \mathrm{E}\Big\{\frac{1}{4}\big[\operatorname{trace}(C^{-1}C_i)\operatorname{trace}(C^{-1}C_j) \\ &+ 4y_i^\top C^{-1}\varepsilon\varepsilon^\top C^{-1}y_j + \varepsilon^\top C^{-1}C_iC^{-1}\varepsilon\varepsilon^\top C^{-1}C_jC^{-1}\varepsilon \\ &+ \operatorname{trace}(C^{-1}C_i)(-2\varepsilon^\top C^{-1}y_i) + \operatorname{trace}(C^{-1}C_j)(-2\varepsilon^\top C^{-1}y_j) \\ &- \operatorname{trace}(C^{-1}C_i)\varepsilon^\top C^{-1}C_jC^{-1}\varepsilon - \operatorname{trace}(C^{-1}C_j)\varepsilon^\top C^{-1}C_iC^{-1}\varepsilon \\ &+ 2\varepsilon^\top C^{-1}y_i\varepsilon^\top C^{-1}C_jC^{-1}\varepsilon + 2\varepsilon^\top C^{-1}y_j\varepsilon^\top C^{-1}C_iC^{-1}\varepsilon\big]\Big\}.\end{aligned} \tag{9.35}$$

From (9.28)–(9.29) we have

$$\mathrm{E}[2\varepsilon^\top C^{-1}y_i] = 2\,\mathrm{E}(\varepsilon^\top)C^{-1}y_i = 0 \tag{9.36}$$

$$\mathrm{E}[\varepsilon^\top C^{-1}C_iC^{-1}\varepsilon] = \operatorname{trace}(C^{-1}C_iC^{-1}\,\mathrm{E}(\varepsilon\varepsilon^\top)) = \operatorname{trace}(C^{-1}C_i). \tag{9.37}$$

Furthermore, from (9.36) and (9.37), under the normality of ε, we deduce [249, Th. 1, p. 55]

$$\begin{aligned}\mathrm{E}[2\varepsilon^\top C^{-1}y_i\varepsilon^\top C^{-1}C_jC^{-1}\varepsilon] &= \operatorname{cov}(2\varepsilon^\top C^{-1}y_i;\varepsilon^\top C^{-1}C_jC^{-1}\varepsilon) \\ &+ \mathrm{E}[2\varepsilon^\top C^{-1}y_i]\,\mathrm{E}[\varepsilon^\top C^{-1}C_jC^{-1}\varepsilon] = 0.\end{aligned} \tag{9.38}$$

And, finally, from (9.37), under the normality of ε, we have [249, Eq. 58, p. 66]

$$\begin{aligned}\mathrm{E}[\varepsilon^\top C^{-1}C_iC^{-1}\varepsilon\varepsilon^\top C^{-1}C_jC^{-1}\varepsilon] &= \operatorname{cov}(\varepsilon^\top C^{-1}C_iC^{-1}\varepsilon;\varepsilon^\top C^{-1}C_jC^{-1}\varepsilon) \\ &+ \mathrm{E}[\varepsilon^\top C^{-1}C_iC^{-1}\varepsilon]\,\mathrm{E}[\varepsilon^\top C^{-1}C_jC^{-1}\varepsilon] \\ &= \operatorname{trace}(C^{-1}C_iC^{-1}C_j) + \operatorname{trace}(C^{-1}C_i)\operatorname{trace}(C^{-1}C_j).\end{aligned} \tag{9.39}$$

Substituting the relations (9.36)–(9.39) to (9.35) and simplifying, we obtain

$$\begin{aligned}M_{ij} &= \mathrm{E}\Big\{\frac{1}{4}\big[\operatorname{trace}(C^{-1}C_i)\operatorname{trace}(C^{-1}C_j) + 4\operatorname{trace}(y_i^\top C^{-1}\varepsilon\varepsilon^\top C^{-1}y_j) \\ &+ \operatorname{trace}(C^{-1}C_iC^{-1}C_j) + \operatorname{trace}(C^{-1}C_i)\operatorname{trace}(C^{-1}C_j) \\ &- \operatorname{trace}(C^{-1}C_i)\operatorname{trace}(C^{-1}C_j) - \operatorname{trace}(C^{-1}C_j)\operatorname{trace}(C^{-1}C_i)\big]\Big\} \\ &= \operatorname{trace}(y_i^\top C^{-1}\,\mathrm{E}(\varepsilon\varepsilon^\top)C^{-1}y_j) + \mathrm{E}\left[\frac{1}{2}\operatorname{trace}(C^{-1}C_iC^{-1}C_j)\right] \\ &= y_i^\top C^{-1}y_j + \frac{1}{2}\operatorname{trace}(C^{-1}C_iC^{-1}C_j),\end{aligned} \tag{9.40}$$

which is the desired claim (cf. (9.33)). ■

Owing to the uncorrelated setting within time domain we have the FIM for the continuous data record $M(\xi) = \int_T M(t)\,dt$. From the point of view of numerical calculations, it is convenient to rewrite the FIM as

$$M(\xi) = P(\xi) + Q(\xi), \tag{9.41}$$

where

$$P(\xi) = \int_T F^\mathsf{T}(\xi, t) W(\xi, t) F(\xi, t)\, \mathrm{d}t \tag{9.42}$$

$$Q(\xi) = \int_T \left\{ \frac{1}{2} \operatorname{trace}\left(W(\xi, t) S_i(\xi, t) W(\xi, t) S_j(\xi, t) \right) \right\}_{i,j=1}^{m} \mathrm{d}t, \tag{9.43}$$

and $F^\mathsf{T}(\,\cdot\,,\,\cdot\,)$ is $(m \times Ns)$-dimensional matrix calculated at a prior value θ^0 of the parameter vector θ,

$$F(\xi, t) = \begin{bmatrix} F(x^1, t) \\ \vdots \\ F(x^N, t) \end{bmatrix}, \quad F(x, t) = \left(\frac{\partial y(x, t; \theta)}{\partial \theta} \right)_{\theta = \theta^0}, \tag{9.44}$$

the matrix $W(\xi, t)$ is the inverse of the variance-covariance matrix $C(\xi, t)$ at θ^0, and $S_i(\xi, t)$ is a matrix of the derivatives of the elements of $C(\xi, t)$ with respect to θ_i at θ^0, that is,

$$W(\xi, t) = [C(\xi, t; \theta^0)]^{-1}, \quad S_i(\xi, t) = \left(\frac{\partial C(\xi, t; \theta)}{\partial \theta_i} \right)_{\theta = \theta^0}, \quad i = 1, \ldots, m. \tag{9.45}$$

Here, in general, the matrix C depends on the model parameters θ and its structure depends on the assumed kind of the correlations, both among the responses and in space. In such a case, the information matrix (9.33) does not inherit the advantages of its counterpart (2.83) in the uncorrelated setting. Particularly, the valuable property of the additivity of information matrices corresponding to single observations no longer holds. Thus, the information pieces from individual observations cannot be separated and therefore direct application of the results from convex optimization is rather impossible. Furthermore, as has already been mentioned, the classical definition of design measures as the proportion of experimental effort is generally impractical. From among all designs, the one which minimizes the performance index Ψ has to be selected. This can be formulated as the optimization problem

$$\xi^\star = \arg \min_{\xi \in \Xi(X)} \Psi[M(\xi)], \tag{9.46}$$

which, besides a similar notation, does not have much in common with (2.23). For example, it does not necessarily have to be convex anymore and direct application of numerical algorithms known from convex optimization is impossible.

9.2.2 *Exchange Procedure for Computation of Optimal Designs*

Clearly, a practical determination of optimal designs in the context of correlated measurements causes extreme difficulties in building an efficient algorithm for finding optimum designs due to the character of the problem [6, 191, 283]. For that reason, there are few available results on this subject, in which the contributors try to imitate some iterative methods of optimal design construction for the uncorrelated case. For calculation of exact designs, it is possible to adapt the exchange-type numerical scheme proposed by Brimkulov et al. [34, 283], which was originally used for determination of D-optimum sampling points for parameter estimation in linear models of random fields.

9.2.2.1 General Scheme of the Algorithm

A generalized version of this exchange algorithm is outlined below in Algorithm 9.1. Starting from an arbitrary initial N-point design $\xi^{(0)}$, in each iteration this procedure generates a new support point which is included into the current design instead of an existing point which will be deleted, in such a way as to maximize the resulting decrease in the performance index Ψ of the FIM. For iteration r the following notation is used: $F^{(r)}$ is the current matrix of sensitivity coefficients in (9.44), $W^{(r)}$ stands for the inverse of the current covariance matrix, as in (9.45), $M^{(r)}$ means the resulting FIM and $P^{(r)}, Q^{(r)}$ are its components as in (9.43). Furthermore, $U = \{1, \dots, N\}$, and $\xi^{(r)}_{x^j \leftrightarrows x}$ denotes the design in which the support point x^j is replaced by x. As a convenient performance index expressing the quality of new candidate location x replacing the existing support point x^j the ratio

$$\Delta(x^j, x) = \{\Psi[M(\xi^{(r)}_{x^j \leftrightarrows x})] - \Psi[M(\xi^{(r)})]\}/\Psi[M(\xi^{(r)})] \qquad (9.47)$$

is used.

From the point of view of nonlinear programming, treating the design problem as an optimization one with a collection of decision variables $\xi = \{x^1, \dots, x^N\}$ and the performance index $\Psi[M(\xi)]$, the algorithm outlined above is very similar to the Gauss–Seidel relaxation algorithm (also known as the block coordinate descent method, cf. [22]). The only difference lies in the update which takes place only for the coordinate x^j for which the resulting decrease in $\Psi[M(\xi)]$ is the largest. In the classical Gauss–Seidel relaxation scheme, each iteration consists of cyclic repetitions of N one-dimensional search steps with respect to variables $x^1, \dots, x^N$ taken in cyclic order. If an increase in $\Psi[M(\xi)]$ is attained for some j after performing the search, then the corresponding variable x^j is immediately updated. Consequently, the presented simple exchange routine has similar properties regarding convergence

Algorithm 9.1. Exchange algorithm for correlated measurements
Input variables: $\xi^{(0)}$ is an initial design such that $x^{i(0)} \neq x^{j(0)}$ for $i, j \in U$, $i \neq j$ and $M(\xi^{(0)})$ is nonsingular, δ is some given positive tolerance

1: **procedure** DesignCorrelated($\xi^{(0)}, \delta$)
2: $\quad F^{(0)} \leftarrow F(\xi^{(0)}, t)$, $W^{(0)} \leftarrow W(\xi^{(0)}, t)$ ▷ Initialization
3: $\quad P^{(0)} \leftarrow P(\xi^{(0)})$, $Q^{(0)} \leftarrow Q(\xi^{(0)})$, $M^{(0)} \leftarrow P^{(0)} + Q^{(0)}$
4: $\quad r \leftarrow 0$
5: $\quad$ **repeat**
6: $\qquad (j^*, x^*) \leftarrow \arg\min_{(j,x) \in U \times X} \Delta(x^j, x)$ ▷ Searching for support candidate
7: $\qquad \xi^{(r+1)} \leftarrow \xi^{(r)}_{x^{j^*} \leftrightarrows x^*}$
8: $\qquad F^{(r+1)} \leftarrow F(\xi^{(r+1)}, t)$, $W^{(r+1)} \leftarrow W(\xi^{(r+1)}, t)$ ▷ Update of matrices
9: $\qquad P^{(r+1)} \leftarrow P(\xi^{(r+1)})$, $Q^{(r+1)} \leftarrow Q(\xi^{(r+1)})$, $M^{(r+1)} \leftarrow P^{(r+1)} + Q^{(r+1)}$
10: $\qquad r \leftarrow r + 1$
11: $\quad$ **until** $|\Delta(x^{j^*}, x^*)| \leq \delta$
12: **end procedure**

to the Gauss–Seidel algorithm and, in particular, only convergence to a local minimum is assured. In effect, the solution returned by the proposed routine may significantly depend on the choice of starting points. Therefore, it would be very difficult to give any general advice regarding the initial configuration design $\xi^{(0)}$ and parameters θ^0. Nevertheless, the starting designs based on the lack of correlation seem to be a rather poor choice since the correlation matrix is then singular, which leads to numerical problems. Based on simulation results, there is evidence that the designs with properly spread support points (according to the correlation strength) are much better as the initial solutions. From a practical point of view, in order to obtain an approximation to a global maximum, a multi-start procedure with several restarts of the algorithm from different initial designs is necessary.

In spite of the simplicity of Algorithm 9.1, its direct adaptation raises some difficulties, which stem from the fact that the form of the FIM is much more cumbersome than in the case of independent measurements and the computational effort of updating matrices in lines 6 and 9 is not acceptable in most practical situations, especially when the number of observations is large. Therefore, additional steps need to be introduced during its implementation in order to significantly reduce the computational cost, and this issue is considered in the next subsection.

9.2.2.2 Implementation Details

The performance of Algorithm 9.1 can be substantially improved by suitable optimization of the computations. This may be achieved through elimination of time-consuming matrix inversions appearing in searching for a new support candidate (line 6) which requires multiple repetition of the passage from design $\xi^{(r)}$ to $\xi^{(r)}_{x^j \leftrightarrows x}$. Such a task can be split into two stages: the removal of location x^j from $\xi^{(r)}$, and then the inclusion of a new point x in place of the deleted one. Obviously, both the stages imply suitable updates in all the matrices corresponding to the current design.

First, consider the inclusion of a new point x into the actual design. In order to determine the updated version of the FIM (for $N+1$ spatial locations), introduce the matrices

$$V(\xi, x, t) = \left[K(x, x^1, t), \dots, K(x, x^N, t)\right]^{\mathsf{T}}, \tag{9.48}$$

$$B(\xi, x, t) = -C^{-1}(\xi, t)V(\xi, x, t), \tag{9.49}$$

$$\Gamma(\xi, x, t) = [K(x, x, t) - V^{\mathsf{T}}(\xi, x, t)C^{-1}(\xi, x, t)V(\xi, x, t)]^{-1}, \tag{9.50}$$

$$G(\xi, x, t) = \left[F^{\mathsf{T}}(x, t) + F^{\mathsf{T}}(\xi, t)B(\xi, x, t)\right], \tag{9.51}$$

defined for the design $\xi = \{x^1, \dots, x^N\}$.

The following result creates a possibility for recursive calculation of the component P of the FIM (9.43).

Proposition 9.4. *The component P of the information matrix for the design $\bar{\xi} = \xi \cup \{x\}$ is given by*

$$P(\bar{\xi}) = P(\xi) + \int_T G(\xi, x, t)\Gamma(\xi, x, t)G^{\mathsf{T}}(\xi, x, t)\,\mathrm{d}t.$$

Proof. When including a point x into the design ξ so as to form the augmented design $\bar{\xi} = \{x^1, \dots, x^N, x\}$, the respective covariance matrix may be represented in partitioned form:

$$C(\bar{\xi}, t) = \left[\begin{array}{c|c} C(\xi, t) & V(\xi, x, t) \\ \hline V^{\mathsf{T}}(\xi, x, t) & K(x, x, t) \end{array}\right]. \tag{9.52}$$

Applying the Frobenius formula [70, Th. 1.1.4, p. 16] for inverting the block matrix $C(\bar{\xi}, t)$, we obtain

$$C^{-1}(\bar{\xi}, t) = \left[\begin{array}{c|c} C^{-1}(\xi, t) + B(\xi, x, t)\Gamma(\xi, x, t)B^{\mathsf{T}}(\xi, x, t) & B(\xi, x, t)\Gamma(\xi, x, t) \\ \hline \Gamma(\xi, x, t)B^{\mathsf{T}}(\xi, x, t) & \Gamma(\xi, x, t) \end{array}\right]. \tag{9.53}$$

Because $F^\mathsf{T}(\bar{\xi},t) = \left[\, F^\mathsf{T}(\xi,t) \,\vdots\, F^\mathsf{T}(x,t) \,\right]$, from (9.53) we get

$$
\begin{aligned}
P(\bar{\xi}) = \int_T F^\mathsf{T}(\bar{\xi},t)C^{-1}(\bar{\xi},t)F(\bar{\xi},t)\,\mathrm{d}t &= \int_T F(\xi,t)^\mathsf{T}C^{-1}(\xi,t)F(\xi,t)\,\mathrm{d}t \\
&+ \int_T F^\mathsf{T}(\xi,t)B(\xi,x,t)\Gamma(\xi,x,t)B^\mathsf{T}(\xi,x,t)F(\xi,t)\,\mathrm{d}t \\
&+ \int_T F^\mathsf{T}(x,t)\Gamma(\xi,x,t)B^\mathsf{T}(\xi,x,t)F(\xi,t)\,\mathrm{d}t \\
&+ \int_T F^\mathsf{T}(\xi,t)B(\xi,x,t)\Gamma(\xi,x,t)F(x,t)\,\mathrm{d}t \\
&+ \int_T F^\mathsf{T}(x,t)\Gamma(\xi,x,t)F(x,t)\,\mathrm{d}t \\
=& P(\xi) + \int_T \left[F^\mathsf{T}(x,t) + F^\mathsf{T}(\xi,t)B(\xi,x,t)\right]\Gamma(\xi,x,t) \\
&\quad \times \left[F(x,t) + B^\mathsf{T}(\xi,x,t)F(\xi,t)\right]\mathrm{d}t \\
=& P(\xi) + \int_T G(x)\Gamma(x)G^\mathsf{T}(x)\,\mathrm{d}t,
\end{aligned}
\tag{9.54}
$$

which is the desired conclusion. ■

In order to construct a numerical procedure for determining optimal experiment designs, it is necessary to establish relationships between the information matrices after removing a support point and adding a new one to the design. Assume that a point $x_\mathrm{r} \in \xi$ is to be replaced by a new point x_a. Then, without loss of generality, we can assume that $x_\mathrm{r} = x^N$. In fact, it is always possible to rearrange the points in the design and all matrices (by swapping appropriate rows and columns) in such a way as to obtain the location x_r in the N-th position. Such an interchange makes the resulting formulae clearer and more elegant, while simplifying the implementation. Denote by $\tilde{\xi} = \xi \setminus x^N$ and by $\bar{\xi} = \tilde{\xi} \cup x_\mathrm{a}$ the resulting designs. Then Proposition 9.4 yields the following relation between the information matrices $P(\bar{\xi})$ and $P(\xi)$, corresponding to the designs $\bar{\xi}$ and ξ:

$$
\begin{aligned}
P(\bar{\xi}) = P(\xi) &- \int_T G(\xi,x^N,t)\Gamma(\xi,x^N,t)G^\mathsf{T}(\xi,x^N,t)\,\mathrm{d}t \\
&+ \int_T G(\tilde{\xi},x_\mathrm{a},t)\Gamma(\tilde{\xi},x_\mathrm{a},t)G^\mathsf{T}(\tilde{\xi},x_\mathrm{a},t)\,\mathrm{d}t.
\end{aligned}
\tag{9.55}
$$

To improve the numerical efficiency of calculations of the FIM Q component, we exploit the symmetry of matrices S_j. We can apply the well-known result from linear matrix algebra.

Proposition 9.5. *Let A be a square matrix of size m and let B be a symmetric matrix of size m. Then*

$$\operatorname{trace}(AB) = \sum_{k=1}^{n}\sum_{\ell=1}^{n} a_{k\ell}b_{k\ell} = \sum_{k,\ell}(A \star B)_{k\ell}.$$

where the operator $\star$ denotes array multiplication, i.e., element by element.

From Proposition 9.5 it follows immediately that it is possible to reduce the computational cost of determining the matrix Q by taking, for each $i = 1,\dots,m$, $A_i = WS_iW$ and $B = S_j$. Once the inverse W of the covariance matrix C is established, the set of symmetric matrices A_i for $i = 1,\dots,m$ should be calculated and stored. Obviously, since they are symmetric, it is sufficient to calculate only their upper triangular elements and then reconstruct the rest. Finally, the elements of the matrix Q can be computed according to the formula in Proposition 9.5. In such a way, some cumbersome matrix multiplications in (9.43) can be omitted.

Based on the above results, we are capable of precisely establishing the necessary calculations during the passage from the design $\xi^{(r)}$ to $\xi^{(r)}_{x^N \leftrightarrows x}$, which guarantees essential reduction in the computational burden. A detailed scheme of the computations is as follows.

Stage 1: Deletion of x^N from $\xi^{(r)}$. Write $F^{(r)}(t)$ as

$$F^{(r)}(t) = \begin{bmatrix} F_{\mathrm{r}}(t) \\ \hdashline F(x^N,t) \end{bmatrix}, \tag{9.56}$$

where $F_{\mathrm{r}}(t) \in \mathbb{R}^{s(N-1)\times m}$ and $F(x^N,t) \in \mathbb{R}^{s\times m}$. Deletion of x_N from design $\xi^{(r)}$ implies removing $F(t)$ from $F^{(r)}(t)$, so that we have $F_{\mathrm{r}}(t)$ instead of $F^{(r)}(t)$. Consequently, some updates of matrices $W^{(r)}(t)$ and $P^{(r)}$ are also necessary. Namely, decomposing the symmetric matrix $W^{(r)}(t)$ into

$$W^{(r)}(t) = C^{-1\,(r)}(t) = \begin{bmatrix} V_{\mathrm{r}}(t) & L(x^N,t) \\ \hdashline L^{\mathsf{T}}(x^N,t) & \Gamma_{\mathrm{r}}(x^N,t) \end{bmatrix}, \tag{9.57}$$

where $V_{\mathrm{r}}(t) \in \mathbb{R}^{s(N-1)\times s(N-1)}$, $L(x^N,t) \in \mathbb{R}^{s(N-1)\times s}$, $\Gamma_{\mathrm{r}}(x^N,t) \in \mathbb{R}^{s\times s}$, and setting

$$G_{\mathrm{r}}(x^N,t) = F^{\mathsf{T}}(x^N,t) + F_{\mathrm{r}}^{\mathsf{T}}(t)L(x^N,t)\Gamma_{\mathrm{r}}^{-1}(x^N,t), \tag{9.58}$$

from (9.53) it is easy to deduce the following counterparts of $W^{(r)}(t)$ and $P^{(r)}$:

$$W_{\mathrm{r}}(t) = V_{\mathrm{r}} - L(x^N,t)\Gamma_{\mathrm{r}}^{-1}(x^N,t)L^{\mathsf{T}}(x^N,t), \tag{9.59}$$

$$P_{\mathrm{r}} = M^{(r)} - G_{\mathrm{r}}(x^N,t)\Gamma_{\mathrm{r}}(x^N,t)G_{\mathrm{r}}^{\mathsf{T}}(x^N,t). \tag{9.60}$$

Stage 2: Inclusion of x into the design resulting from Stage 1. At this stage, a new sensitivity matrix $F_{\mathrm{a}}(t)$ is constructed:

$$F_{\mathrm{a}}(t) = \begin{bmatrix} F_{\mathrm{r}}(t) \\ \hdashline F(x,t) \end{bmatrix}, \tag{9.61}$$

where $F(x,t)$ is defined according to (9.44). Such an augmentation influences the form of the matrices $W_{\mathrm{r}}(t)$ and P_{r} obtained at Stage 1. Analyzing the proof of Proposition 9.4, it is easy to deduce the respective updated versions $W_{\mathrm{a}}(t)$ and P_{a}. Rewritting (9.48)–(9.50) as follows:

$$V(x,t) = \left[K(x,x^1,t),\dots,K(x,x^{N-1},t)\right]^{\mathsf{T}}, \tag{9.62}$$

$$B(x,t) = -W_{\mathrm{r}}V(x,t), \tag{9.63}$$

$$\Gamma_{\mathrm{a}}(x,t) = [K(x,x,t) - V^{\mathsf{T}}(x,t)W_{\mathrm{r}}(t)V(x,t)]^{-1}, \tag{9.64}$$

$$G_{\mathrm{a}}(x,t) = \left[F^{\mathsf{T}}(x,t) + F_{\mathrm{r}}^{\mathsf{T}}(t)B(x,t)\right], \tag{9.65}$$

we obtain

$$W_{\mathrm{a}}(t) = \left[\begin{array}{c:c} W_{\mathrm{r}}(t) + B(x,t)\Gamma_{\mathrm{a}}(x,t)B^{\mathsf{T}}(x,t) & B(x,t)\Gamma_{\mathrm{a}}(x,t) \\ \hdashline \Gamma_{\mathrm{a}}(x,t)B^{\mathsf{T}}(x,t) & \Gamma_{\mathrm{a}}(x,t) \end{array}\right], \tag{9.66}$$

$$P_{\mathrm{a}} = P_{\mathrm{r}} + G_{\mathrm{a}}(x,t)\Gamma_{\mathrm{a}}(x,t)G_{\mathrm{a}}^{\mathsf{T}}(x,t). \tag{9.67}$$

Remark 9.3. It can be shown that $G_{\mathrm{a}}(x,t) = \mathbf{0}$ for any $x_{\mathrm{a}} = x_i$, $i = 1,\dots,N$, i.e., when x_{a} coincides with the point which already exists in the design. Moreover, the set of eigenvalues for any matrix $W(\xi,t)S_j(\xi,t)$, $j = 1,\dots,m$, where $\xi = \{x^1,\dots,x^{N-1}\} \cup x_{\mathrm{a}}$, consist of eigenvalues of $W(\tilde{\xi},t)S_j(\tilde{\xi},t)$ ($\tilde{\xi} = \{x^1,\dots,x^{N-1}\}$) and one additional zero eigenvalue if $x_{\mathrm{a}} \in \tilde{\xi}$ (the proofs of both facts are rather extensive but only of technical nature, hence omitted). The meaning of this fact is that replication of an observation does not alter the components $P(\tilde{\xi})$ and $Q(\tilde{\xi})$ of the information matrix. It can be interpreted in such a way that the additional measurement at the same time instant does not provide any new information on the estimated parameters. Therefore, there will be no replications in the optimal solution, and the proposed numerical algorithm should automatically avoid repetitions of the observations at the same time instant, since it does not lead to improvement in the criterion value.

9.3 Applications in Engineering

9.3.1 Identification of a Magnetic Brake System

As for demonstration of the approach to the group sensing strategy described in the first part of the chapter, consider a sensor scheduling problem related

Fig. 9.1 Scheme of a magnetic brake

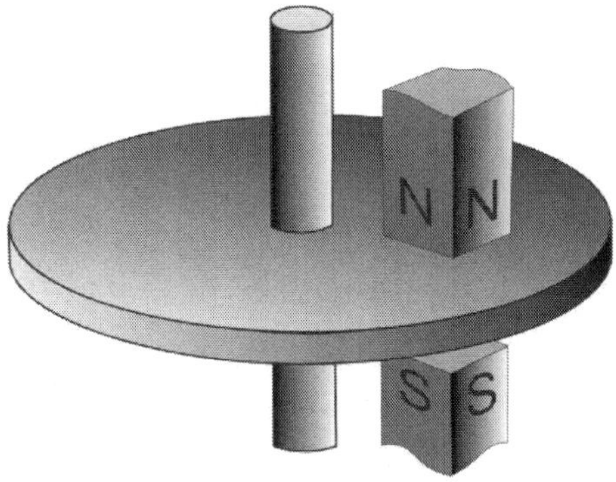

to parameter identification of a magnetic brake system. In its simplest form, a magnet brake consists of a disc of conductive material and a permanent magnet. The disc rotates within a constant magnetic field induced by the magnet (cf. Fig. 9.2). The magnetic field generates eddy currents within the conductor and the Lorentz forces slow down the disc. Assuming for simplicity that the magnetic flux has only a vertical component, while the currents and the electric field have no z-components, from Lorentz' equation we obtain the following PDE describing the vertical component $B = B(x, t)$ of the magnetic field within the area of a plate Ω at time t:

$$\nabla \cdot \left(\nabla B(x,t) + d\mu\sigma(x)\omega(t) \left(B(x,t) + B_0\right) \begin{pmatrix} x_2 \\ -x_1 \end{pmatrix} \right) = 0, \qquad (9.68)$$

where $B_0 = 0.1\mathrm{T}$ is a vertical component of the magnetic field induced by the permanent magnet, μ is a magnetic permeability, σ is an electrical conductivity, d denotes disk thickness and ω is angular velocity. The PDE is supplemented by the Dirichlet condition $B = 0$ on the disc boundary $\partial\Omega$.

The induced torque slows the disc down and the evolution of the system over time interval $(0, t_f]$ can be described by the following initial value problem:

$$J \frac{\mathrm{d}\omega(t)}{\mathrm{d}t} = T_q, \quad \omega(0) = \omega_0, \qquad (9.69)$$

where T_q is a total torque affecting the disc:

$$T_q = \int_\Omega \frac{1}{\mu} \left(x_2 \frac{\partial B}{\partial x_1} - x_1 \frac{\partial B}{\partial x_2} \right) (B + B_0)\, \mathrm{d}x_1 \mathrm{d}x_2, \qquad (9.70)$$

J being a disc moment of inertia.

In our example it is assumed that a copper disc with a radius of 10 cm and thickness $d = 2$ cm is used ($\mu = 4\pi \cdot 10^{-7}$ [H/m]) with the initial angular velocity of 140 rps. Due to the nonhomogeneity of the plate, conductivity coefficient vary slightly in space. It is assumed that these changes can be approximated by the following model:

$$\sigma(x) = \theta_1 + \theta_2 \rho + \theta_3 \rho^2, \tag{9.71}$$

where $\rho = \sqrt{x_1^2 + x_2^2}$. During the test experiments, measurements can be taken from many different devices and it is possible to run the tests many times for each examined magnetic brake. Thus, the parameters θ_1, θ_2 and θ_3 are subject to random fluctuations and require proper estimation based on measurement data taken according to a suitable designed group sensor schedule. The parameter values are treated as realizations of a random variable whose unknown means, variances and covariances are the focus of interest (nine hyperparameters). Further, we assume that the variables θ_i are normally distributed. The sensors can be located on the rectangular grid with a side length of 1 cm and are shown in Fig. 9.4, indicated with black points. Due to technical limitations, the sensors cannot be located in the vicinity of the magnet and the plate axis (there are 198 such locations).

In order to verify the proposed approach, a MATLAB program was written using a PC equipped with Intel Centrino T9300 processor (2.5GHz, 3 GB RAM) running Windows Vista and MATLAB 2007b. A PDE was solved using efficient solvers of the COMSOL environment based on the finite element method [43]. The nominal values of the expected values for the system parameters were assumed to be

$$\mathrm{e}(\theta)_0 = 5.99 \cdot 10^7 \cdot (1\ \mathrm{S/m}, -1.5\ \mathrm{S/m^2}, 8\ \mathrm{S/m^3})$$

and

$$\mathrm{cov}(\theta)_0 = 10^6 \cdot \begin{bmatrix} 0.1 & 0.05 & 0.02 \\ 0.05 & 0.12 & 0.04 \\ 0.02 & 0.04 & 0.1 \end{bmatrix}.$$

Calculations were performed for a spatial mesh composed of 1380 triangles, 718 nodes and an evenly partitioned time interval (51 subintervals). Finally, the initial value problem for the system dynamics was simulated using SIMULINK. The total number of measurements N_0 was assumed to be 100.

In order to calculate $\omega^\star$ in Step 1 of the algorithmic scheme, the MATLAB-based package CVX dedicated for solving the convex programming problems was used [90]. As a result, the following aggregated design was generated:

$$\omega^\star = \left\{ \begin{matrix} (0.0157, -0.0368), & 0.11 \\ (0.0157, -0.0263), & 0.06 \\ (0.0578, -0.0368), & 0.32 \\ (0.0684, 0.0368), & 0.14 \\ (0.0684, 0.0473), & 0.08 \\ (0.0894, -0.0157), & 0.29 \end{matrix} \right\}.$$

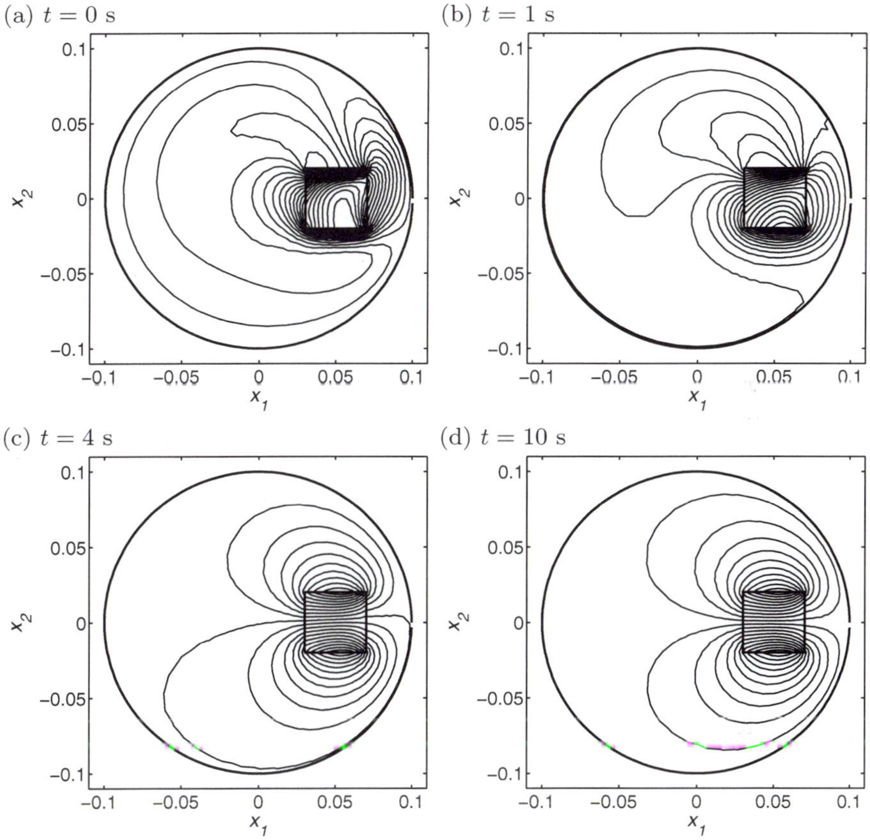

Fig. 9.2 Temporal changes of magnetic flux density B ($\omega_0 = 140$[rps], $B_0 = 0.1$[T])

Then, from Step 2 with the additional assumption that for each device we should use as many sensors as possible, it is possible to obtain the nontrivial group schedule illustrated in Fig. 9.4. The calculated optimal number of experimental units (e.g., devices used or replicated trials) was equal to $N^\star = 32$. It is clear that the locations of probes tend to form patterns reflecting the areas of the greatest changes in magnetic flux density, but due to the complexity of the system dynamics they also substantially depend on the values of input variables and, in general, it is difficult to predict their configuration.

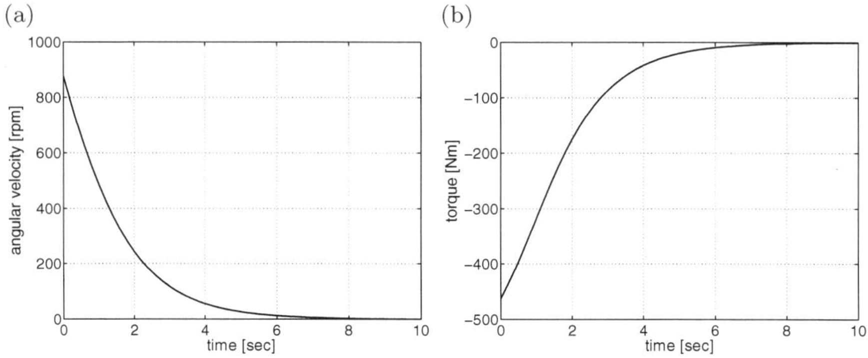

Fig. 9.3 Example plots of angular velocity (a) and braking torque (b) ($\omega_0 = 140$[rps], $B_0 = 0.1$[T])

9.3.2 Air Pollution Monitoring

As a representative example, illustrating the sensor array design problem in the presence of correlated measurements, consider an atmospheric pollutant transport-chemistry process over a unit circle. For simplicity, assume that the velocity of the transport medium is zero everywhere. Instead, we take into account an active source of pollution and reaction, which leads to the decay of pollutant concentration $y(x,t)$. The entire process over the observation interval $T = [0,10]$ is modelled with the following diffusion-reaction equation:

$$\frac{\partial y(x,t)}{\partial t} = \theta_1 \Big[\nabla^2 y(x,t) + 4y(x,t) - 4\theta_2 e^{-4\theta_1 t} \Big], \quad x \in \Omega = \{x : \|x\| < 1\}, \tag{9.72}$$

where the second term on the right-hand side is responsible for the reaction and the last term represents some exponential source of contamination over Ω. Equation (9.72) is supplemented by suitable boundary and initial conditions,

$$y(x,t) = 0, \qquad (x,t) \in \partial\Omega \times T, \tag{9.73}$$

$$y(x,0) = \theta_2(1 - x_1^2 - x_2^2), \qquad x \in \Omega. \tag{9.74}$$

This time the task consists in determining the locations of an arbitrary number of stationary sensors for estimation of the diffusion coefficient θ_1 and the relative amplitude of the source θ_2. The measurements are assumed to be corrupted by noise which is zero-mean and correlated in time and space with a covariance kernel $k(x,\chi,t,\tau) = \exp(-\rho_t|t-\tau|)\exp(-\rho_x\|x-\chi\|)$.

The problem (9.72)–(9.74) has the closed-form solution, i.e.,

$$y(x,t) = \theta_2(1 - x_1^2 - x_2^2)e^{-4\theta_1 t}. \tag{9.75}$$

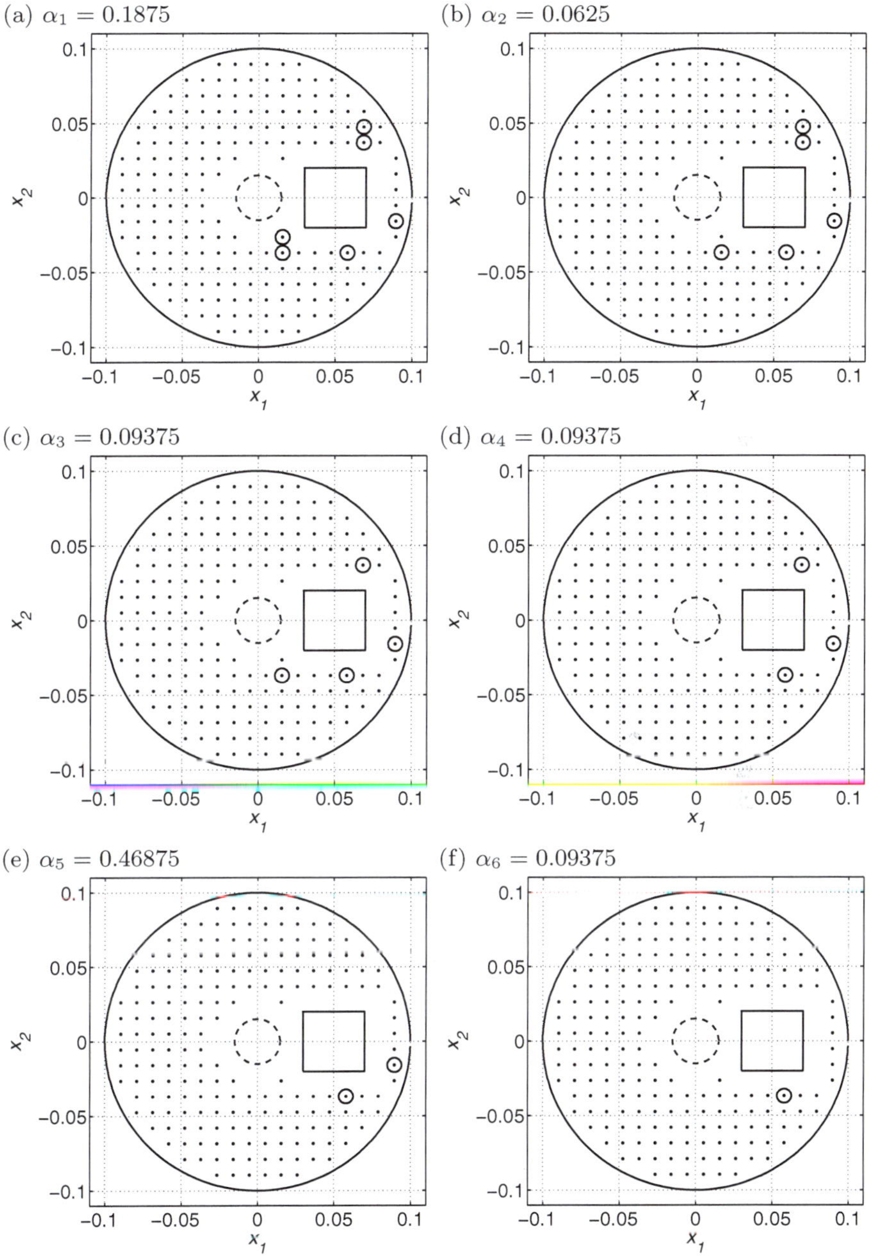

Fig. 9.4 D-optimal sensor locations (circles) for groups of experimental units (experiment trials)

The nominal parameter values $\theta_1 = 0.05$ and $\theta_2 = 0.02$ were used. The central symmetry of the problem and the fact that the sensitivities $\partial y/\partial\theta_1$ and $\partial y/\partial\theta_2$ take their maximal values at point $(x_1, x_2) = (0, 0)$ suggest that a one-point design

$$\xi^\star = \{(0,0)\} \tag{9.76}$$

is D-optimal for the stationary sensor strategy under the assumption of independent measurements. Indeed, passing the polar coordinates, it can be verified that the determinant of the FIM in the case of a one-point design takes the form

$$\begin{aligned}\det(M(\xi)) =& \frac{1}{4}e^{-8} - \frac{9}{2}e^{-4} + \frac{1}{4} - r^2 + 18e^{-4}r^2 - 27e^{-4}r^4 + \frac{3}{2}r^4 - r^6 + \frac{1}{4}r^8 \\ & - e^{-8}r^2 + \frac{3}{2}e^{-8}r^4 - e^{-8}r^6 + 18e^{-4}r^6 + \frac{1}{4}e^{-8}r^8 - \frac{9}{2}e^{-4}r^8,\end{aligned} \tag{9.77}$$

where $r = \|x\|$ and $0 \le r \le 1$. The function (9.77) takes its maximal value at $r = 0$, which corresponds to the centre of the circular domain, which is the most informative point for observations, and it can be easily checked that (9.76) is optimal.

In order to find optimal locations for taking measurements, a program based on the developed algorithm was written in Matlab v. 6.5. R12.1 and run on a PC equipped with Duron 900MHz CPU and running Windows 2000. The two-dimensional search for a candidate to include into the current design was implemented using the routine `fmincon` from the Optimization Toolbox.

The influence of the mutual correlations of observations on sensor allocation was tested by varying the coefficient ρ_x. For clarity, the correlation in time was assumed to be negligible (i.e., $\rho_t = 100$). The results obtained for relative small and medium numbers of measurements are presented in Fig. 9.5.

An analysis of the obtained results leads to the conclusion that the level of correlation directly affects the distances between the sensors, which increase when the correlation is more intense. If the correlation is small, then the measurements tend to cluster in the vicinity of the optimal design support for the case of independent measurements. Another important remark is that the higher the correlation, the lower the criterion value. This effect results from a more global character of the mutual influence between random error realizations in the case of higher correlation values. Due to the interference with the estimation of the unknown parameters, the valuable information in observations is reduced. Note that due to the problem symmetry every solution constructed by rotation of this presented in Fig. 9.5 by any angle constitute an equivalent approximation of the optimal sensor allocation. This phenomenon also affects the performance of the algorithm and, because of numerical reasons, in the case of $N = 15$ measurements the solutions are not as regular as in the simplest case when $N = 5$.

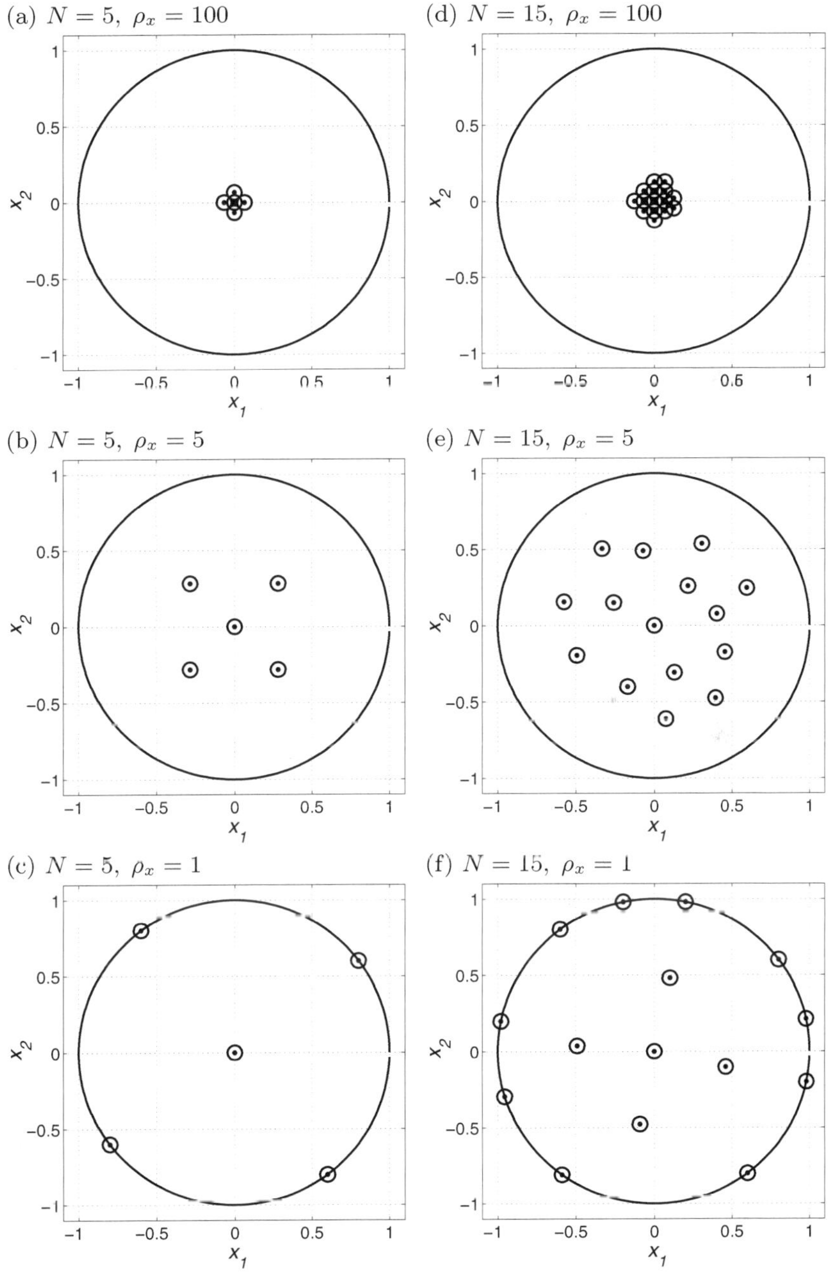

Fig. 9.5 D-optimum sensor allocation for $N = 5$ and 15 measurements for small ($\rho_x = 100$), medium ($\rho_x = 5$) and considerable ($\rho_x = 1$) spatial correlations, respectively

9.4 Summary

In the first part of the chapter, we dealt with a sensor scheduling problem for accurate parameter estimation for DPSs with random model parameters. The set-up is particularly relevant in practical situations where taking a sample may cause some disturbance to system conditions or where the conditions themselves are not completely stable in space or time. The disturbance or instability may affect the response and hence also the values of the model parameters. Estimation of the expectation rather than of the individual values of the parameters was of interest. Experiments in such situations will be large and some grouping may help to plan and perform them.

An essential contribution here is a novel definition of the so-called *group design* to suit the experimental conditions. Such designs, in general, have a complicated structure which causes serious optimization problems. Therefore, a reformulation of the design which greatly simplifies the calculations and also gives the flexibility to introduce additional experimental restrictions is proposed. In such a way, an approach is developed which is capable of solving scheduling tasks for large-scale sensor arrays. A particularly appealing in the delineated approach is the flexibility of obtaining the whole class of optimum designs. This allows tailoring them to specific practical conditions.

In the second part of the chapter, the question of correlated observations, being one of the most serious problems which may appear in applications, was addressed. The main contribution here is to provide appropriate formulation of the problem in terms of optimum experimental design and to develop an efficient exchange-type procedure dedicated to DPSs. The computational algebra derived here significantly increases the efficiency of the algorithm. In fact, the main effort lies not in the algorithm itself but in its effective implementation to make this complex problem numerically tractable. To our best knowledge, there is a lack of alternative approaches in the data analysis literature. The proposed routine constitutes a broad generalization of the algorithm proposed in [34], which can be further refined to take into account more sophisticated correlation structures [180, 196, 281, 283]. The crucial novelty introduced in this work is that the model, and so the optimum design, was extended from the standard error assumption of independence to allow for covariances dependent on the unknown model parameters. It covers a common situation in engineering practice, i.e., when the variance of measurements is nonhomogeneous. In practice, however, the covariance kernel or, alternatively, the explicit form of the measurement covariance matrix may be unknown. Thus, there exists a strong necessity for more systematic approaches for estimation of the covariance matrix.

Chapter 10
Conclusions and Further Research Directions

Undoubtedly, the optimization and control of DPSs are intensively expanding research areas with a high number of applications. The process of data acquisition, being an integral part of control design, is fundamental since in distributed systems it exerts a strong influence on the accuracy of estimation, the quality of control and prediction of system behavior. Therefore, sensor networks, being, in fact, modern observational systems which have recently emerged in the context of monitoring distributed processes, are becoming a very important field of research. Their advantageous features related to distributed sensing and data fusion, robustness to faults and failures of sensor nodes, as well as flexibility in realization of the measurement process facilitate wide applicability of such systems. Nevertheless, the problem of sensor scheduling for parameter estimation in DPSs is very difficult since its intrinsic nonlinearity rather excludes simple methods. Furthermore, most often, the dependence between the observations and the system performance is not intuitive and has obscure or confounding nature. Despite these facts, the problem has been considered by many authors and a number of relevant contributions have already been reported in the literature. In addition to this, some new approaches to determining optimal sensor locations are still proposed in order to provide a more general context, a wider range of potential criteria and better understanding of the interconnection between optimal sensor routing and the quality of the identifiers. On the other hand, engineers expect efficient techniques which are easy to implement and provide benefits overcoming the cost of the application.

In such a way, although various results for the sensor scheduling problem exist, engineers seem to be reluctant to apply them in practice. This is a direct consequence of the complexity of most sensor location approaches, which lead to sophisticated and inefficient algorithms. Moreover, the existing methods are often intricate and difficult to implement. However, taking into account that the progress in computational mathematics combined with the rapidly increasing computer power steadily extend the range of potential applications, there is a strong necessity to develop more effective systematic

M. Patan: Optimal Sensor Networks Scheduling in Identification of DPSs, LNCIS 425, pp. 261–266.
springerlink.com

approaches which would pave the way to algorithms of great efficiency and reasonable complexity.

Bearing this in mind, the original objective of the research reported in this monograph was to develop effective and reliable methods to solve the sensor scheduling problems encountered in a wide class of practical applications for monitoring DPSs using sensor networks or sensor arrays. In order to accomplish this task the appropriate theoretical substantiation for the adopted approach was developed. Furthermore, some known methods were generalized and several new algorithms constructed. The following is a final concise summary of the contributions provided by this work to the state-of-the-art in optimal sensor scheduling for parameter estimation in DPSs:

- Systematizes and generalizes the classical results of optimal experiment design for stationary sensors to MIMO systems with possible output correlation.
- Develops expeditious methods of activating scanning sensors, and in particular:
  - extends the theory and adapts some algorithms of nonlinear programming and optimal experiment design to solve sensor activation problems;
  - introduces an approach based on clusterization-free designs for an arbitrarily fixed switching schedule and develops an extremely efficient exchange algorithm, which is very easy to implement;
  - transforms the problem to the equivalent Mayer problem of optimal control in the case of an optimized switching schedule and adapts some recently developed techniques of discrete-valued optimal control (the CPET transform) for its solution.
- Formulates and solves the problems of trajectory planning for sensor networks with mobile nodes:
  - refines Rafajłowicz's approach to constructing optimal trajectories of moving sensors; derives optimality conditions for nonparameterized and parameterized trajectories;
  - extends Uciński's approach to sensor path planning towards design of resource aware sensor routing strategies; specifically, it is shown how to reduce the problem to a state-constrained optimal-control problem including alternative objectives related to the network resources or behavior (i.e. energy limits, path lengths, coverage, etc.);
  - develops the method of sensor routing for heterogeneous sensor networks adapting the concept of continuous experimental designs and applies the two-phase first-order algorithm to construct optimal trajectories of mobile nodes.

It is also demonstrated via simulations that the proposed approaches can be efficiently applied to wide spectrum of challenging engineering problems providing the evidence that the mobility of nodes lead to significant improvement of the observational system performance.
- Devises a new approach to decentralized scheduling for sensor networks identifying DPSs. A mathematical formulation and decomposition of the problem in the general settings of scanning and mobile observations is provided together with the decentralized procedures for it solution dedicated for scalable sensor networks. The main contribution here consists in the construction of the specialized protocol of information exchange for the simultaneous optimization and estimation of the global optimality criterion at each sensor node. Consequently, the proposed method leads to the fully distributed scheme of calculations which can be implemented with great ease and is robust wrt individual nodes failures.
- Addresses the issue of selecting optimal observation points for scanning networks in view of a combinatorial optimization approach. The solution of this scheduling problem being, NP-hard one, is achieved via branch-and-bound technique enhanced with simplicial decomposition scheme perfectly suited for large-scale problems. A parallel version of the branch-and-bound application is also developed together with the multi-objective formulation of the problem paying attention to resource-aware system maintenance and providing the evidence for the great flexibility of this approach for significant refinements and generalizations.
- Develops some approaches to optimal sensor allocation in the presence of the model or design variables uncertainty:

  - enhances a sequential design techniques; introduces the additional design loop which makes the sequential approach more flexible, leading to a seamless design and estimation phase;
  - provides a characterization of robust designs in the minimax and average sense with the appropriate discussion; an original application of statistical learning theory is presented, which paves the way for efficient use of the randomized algorithms to determine the sensor schedules in an average sense;
  - formulates the sensor scheduling problem under the sensor location uncertainty and adapts it to the framework of randomized design for determining fine approximations to optimal solutions.

- Reports and enhances methods of optimal sensor placement to model-based fault diagnosis. It adapts the parametric approach to describe the DPS state and develops methods of reliability maximization of fault detection and isolation based on structured hypothesis testing. A theoretical research has been delivered and computationally efficient algorithms developed to determine optimal configurations of sensor networks for fault detection and identification. In the proposed approach, an attempt was also made at applying a nonlinear receding-horizon technique to provide

the dynamic data-driven computation scheme, i.e., the current measurements from the sensors must serve as a basis for the update of parameter estimates and these, in turn, can be used to correct the sensor movements.
- Deals with a sensor scheduling problem for accurate parameter estimation for DPSs with random model parameters based on the novel definition of the so-called *group design* to suit the experimental conditions. Furthermore, a proper reformulation of the design which greatly simplifies the calculations and also gives the flexibility to introduce additional experimental restrictions is proposed. This leads to the computational scheme which is capable of solving scheduling tasks for large-scale sensor arrays. A particularly appealing in the delineated approach is the flexibility of obtaining the whole class of optimum designs. This allows tailoring them to specific practical conditions.
- Develops an approach to solve the sensor location problem in the case of correlated observations. Specifically, a method which includes mutual correlations between measurements directly into the measurement covariance matrix is proposed, and a relaxation algorithm is developed for solving the resulting computational problem. Additional contribution is also that the model was extended to allow for covariances dependent on the unknown model parameters.
- Validates the concepts and methods considered through tangible application examples close to real-world engineering problems. In particular, the performance of the algorithms is illustrated with numerous examples of DPSs in one, two or three-dimensional spatial domains including transmission lines, computer-assisted tomography, air and groundwater pollution, heat transfer, magnetic brake and identification of the contamination and heating sources.

From the point of view of the potential applications, the main quality of the proposed approach to sensor network scheduling is the transparency of the solutions leading to many efficient and easy-to-implement numerical procedures. The author strongly believes that these advantages establish a firm position of such a methodology regarding applications in engineering. In addition to this, note that few alternatives exist in the literature which takes into account the factor of qualitative monitoring of distributed process together with maintenance of the resources of sensor network. Nevertheless, due to the complexity of the task, there still remain open problems which require closer attention and indicates potential significant extensions. In what follows, a brief list of possible areas for further investigation is presented.

Further development of parallel and decentralized algorithms for sensor scheduling. Computing on clusters of PCs has become as common as computing on scientific workstations had been a decade ago, therefore the parallel versions of sensor scheduling/routing algorithms aimed at solving large-scale problems create a very attractive direction of research. Furthermore, since the wireless sensor network is, in fact, a computer network with distributed resources, a natural direction of research is towards fully decentralized algorithms providing the efficiency and robustness of data acquisition and processing. Such aspects of sensor scheduling have great potential in applications and are very promising research areas for the near future.

Extension of the considered class of DPSs. The great advantage of the delineated approach is that it is independent of a specific form of PDEs used as a mathematical model of considered DPS. In such a manner the presented approach can be rather easily adopted to the more general classes of multi-output DPSs since only the formula for calculating the elements of FIM will be properly changed. One of the interesting examples can be the case of DPSs with possible time delays. However, the main problem which has to be addressed while trying to derive such a generalization is to answer the question whether or not the approach based on the FIM is still applicable.

Development of the approaches robust to the structural model uncertainty. In model-based systems analysis, proper calibration of the model involves two stages, namely, an appropriate selection of the model structure and then the estimation of its parameters. The problem of the model quality assessment is usually omitted due to its difficulty. Therefore, effective sensor scheduling procedures for verification of model structure are of great practical relevance.

Further development of approaches for correlated measurements. It is clear that in real-world problems the independence of the observations often cannot be sufficiently justified and mutual correlations between measurements should be taken into account. Since the classical concept of design measures is not applicable in this case, this fact is of crucial importance as it significantly increases the problem complexity. In Section 9.2, an exchange-type relaxation algorithm was adopted. Attempts to extend those concepts towards distributed and parallel algorithms would be quite natural. The close connection of correlated measurements with practical problems makes such results very desirable.

Establishing the links between sensor scheduling problems for parameter and state estimation. The state estimation of the dynamic system is one of the fundamental problems in engineering applications. Most common approaches to this task rely on definition of suitable criterion operating on the estimate error covariance matrix of Kalman-Bucy filter or some observability measure. Although generalizations to DPSs have been extensively investigated since the 1970s, there still are no effective implementations dedicated to systems with two or three spatial dimensions. The reason is a large scale of the problem because, e.g., for three-dimensional domains, after discretization its dimensionality can reach even the order of few millions variables. Recent works [57, 286] indicate that some algorithms of sensor location for parameter estimation can be effectively employed for the purpose of state estimation. This suggests that solutions to the problem of sensor scheduling for parameter estimation can possibly be successfully employed in the setting of state estimation, which in the literature was considered to be completely different. The potential of this research area cannot be overestimated as the state estimation problem is of great practical relevance.

Simultaneous estimation and experimental design. Most of the approaches considered within the scope of this dissertation concern implementation of sensor allocation strategies before conducting the experiment, i.e., they work off-line. The only exception are sequential design techniques discussed within the framework of parameter uncertainties. From a practical point of view, it would be interesting to investigate the problem of simultaneously taking measurements, estimating the parameters and updating the locations of scanning or movable sensors.

References

1. Agrawal, D.P., Biswas, R., Jain, N., Mukherjee, A., Sekhar, S., Gupta, A.: Sensor systems: State of the art and future challenges. In: Wu, J. (ed.) Handbook on Theoretical and Algorithmic Aspects of Sensor, Ad Hoc Wireless, and Peer-to-Peer Networks, ch. 20. Auerbach Publications/Taylor & Francis Group, Boca Raton (2006)
2. Aihara, S.I.: Consistency of extended least square parameter estimate for stochastic distributed parameter systems. In: Bastin, G., Gevers, M. (eds.) Proc. 4th European Control Conf., EUCA, Brussels, Belgium, July 1-4 (1997), Published on CD-ROM
3. Akl, S.G., Yao, W.: Parallel computation and measurement uncertainty in nonlinear dynamical systems. Journal of Mathematical Modelling and Algorithms 4(1), 5–15 (2005)
4. Amouroux, M., Babary, J.P.: Sensor and control location problems. In: Singh, M.G. (ed.) Systems & Control Encyclopedia, vol. 6, pp. 4238–4245. Pergamon Press, Oxford (1988)
5. Andó, B., Cammarata, G., Fichera, A., Graziani, S., Pitrone, N.: A procedure for the optimization of air quality monitoring networks. IEEE Transactions on Systems, Man, and Cybernetics – Part C: Applications and Reviews 29(1), 157–163 (1999)
6. Angelis, L., Bora-Senta, E., Moyssiadis, C.: Optimal exact experimental designs with correlated errors through a simulated annealing algorithm. Computational Statistics & Data Analysis 37(3), 275–296 (2001)
7. Atkins, P.W.: Physical Chemistry. Oxford University Press, Oxford (1998)
8. Atkinson, A.C., Donev, A.N.: Optimum Experimental Designs. Clarendon Press, Oxford (1992)
9. Atkinson, A.C., Donev, A.N., Tobias, R.: Optimum Experimental Design, with SAS. Oxford University Press, Oxford (2007)
10. Autrique, L., Perez, L., Scheer, E.: On the use of periodic photothermal methods for materials diagnosis. Sensors and Actuators B 135, 478–487 (2009)
11. Azhogin, V.V., Zgurovski, M.Z., Korbicz, J.: Filtration and Control Methods for Stochastic Distributed-Parameter Processes. Vysha Shkola, Kiev (1988) (in Russian)

12. Banks, H.T.: Computational issues in parameter estimation and feedback control problems for partial differential equation systems. Physica D 60, 226–238 (1992)
13. Banks, H.T., Fitzpatrick, B.G.: Statistical methods for model comparison in parameter estimation problems for distributed systems. Journal of Mathematical Biology 28, 501–527 (1990)
14. Banks, H.T., Kunisch, K.: Estimation Techniques for Distributed Parameter Systems. In: Systems & Control: Foundations & Applications. Birkhäuser, Boston (1989)
15. Banks, H.T., Smith, R.C., Wang, Y.: Smart Material Structures: Modeling, Estimation and Control. In: Research in Applied Mathematics. Masson, Paris (1996)
16. Bard, Y.: Nonlinear Parameter Estimation. Academic Press, New York (1974)
17. Basseville, M., Nikiforov, I.: Detection of Abrupt Changes. Prentice-Hall, New York (1993)
18. Bennett, A.F.: Inverse Methods in Physical Oceanography. In: Cambridge Monographs on Mechanics and Applied Mathematics. Cambridge University Press, Cambridge (1992)
19. Bensoussan, A., Prato, G.D., Delfour, M.C., Mitter, S.: Representation and Control of Infinite Dimensional Systems, 2nd edn. Birkhäuser, Basel (2007)
20. Berliner, M.L., Nychka, D., Hoar, T. (eds.): Studies in the Atmospheric Sciences. Lecture Notes in Statistics, vol. 144. Springer, Berlin (2000)
21. Bernstein, D.S.: Matrix Mathematics. Theory, Facts, and Formulas with Application to Linear Systems Theory. Princeton University Press, Princeton (2005)
22. Bertsekas, D.P.: Nonlinear Programming, 2nd edn. Optimization and Computation Series. Athena Scientific, Belmont (1999)
23. Bertsekas, D.P., Tsitsiklis, J.N.: Parallel and Distributed Computation: Numerical Methods. Athena Scientific, Belmont (1997)
24. Billingsley, P.: Statistical Inference for Markov Processes. University of Chicago Press, Chicago (1961)
25. Boer, E.P.J., Hendrix, E.M.T., Rasch, D.A.M.K.: Optimization of monitoring networks for estimation of the semivariance function. In: Atkinson, A.C., Hackl, P., Müller, W. (eds.) Proc. 6th Int. Workshop on Model-Oriented Data Analysis, mODa 6, Puchberg/Schneeberg, Austria, pp. 21–28. Physica-Verlag, Heidelberg (2001)
26. Bogacka, B., Atkinson, A., Patan, M.: Designs for discriminating between models by testing the equality of parameters. In: 6th St. Petersburg Workshop on Simulation, vol. 2, pp. 589–594. St. Petersburg, VVM com. Ltd, Russia (2009)
27. Bogacka, B., Patan, M., Johnson, P.J., Youdim, K., Atkinson, A.C.: Optimum design of experiments for enzyme inhibition kinetic models. Journal of Biopharmaceutical Statistics 21(3), 555–572 (2011)
28. Boukas, E., Liu, Z.: Deterministic and stochastic time delay systems. In: Control Engineering. Birkhäuser, Basel (2003)
29. Boukerche, A. (ed.): Handbook of Algorithms for Wireless Networking and Mobile Computing. Chapman & Hall/CRC, Boca Raton (2006)

30. Boyd, S., Ghaoui, L.E., Feron, E., Balakrishnan, V.: Linear Matrix Inequalities in System and Control Theory. SIAM Studies in Applied Mathematics, vol. 15. SIAM, Philadelphia (1994)
31. Boyd, S., Ghosh, A., Prabhakar, B., Shah, D.: Randomized gossip algorithms. IEEE Transactions on Information Theory 52(6), 2508–2530 (2006)
32. Boyd, S., Vandenberghe, L.: Convex Optimization. Cambridge University Press, Cambridge (2004)
33. Braca, P., Marano, S., Matta, V.: Running consensus in wireless sensor network. In: Proc. 9th Int. Conf. Information Fusion, Cologne, Germany, June 30-July 3 (2008), Published on CD-ROM
34. Brimkulov, U.N., Krug, G.K., Savanov, V.L.: Design of Experiments in Investigating Random Fields and Processes. Nauka, Moscow (1986) (in Russian)
35. Bullo, F., Cortés, J., Martínez, S.: Distributed Control of Robotic Networks. Applied Mathematics Series. Princeton University Press (2009), `http://coordinationbook.info`
36. Butkovskiy, A.G., Pustylnikov, A.M.: Mobile Control of Distributed Parameter Systems. John Wiley & Sons, New York (1987)
37. Cassandras, C.G., Li, W.: Sensor networks and cooperative control. European Journal of Control 11(4–5), 436–463 (2005)
38. Chavent, G.: On the theory and practice of non-linear least-squares. Advances in Water Resources 14(2), 55–63 (1991)
39. Chen, J., Patton, R.J.: Robust Model-Based Fault Diagnosis for Dynamic Systems. Kluwer Academic Publishers, Berlin (1999)
40. Chen, W.H., Seinfeld, J.H.: Optimal location of process measurements. International Journal of Control 21(6), 1003–1014 (1975)
41. Chiang, L.H., Russel, E.L., Braatz, R.D.: Fault Detection and Diagnosis in Industrial Systems. Springer, London (2001)
42. Chong, C.Y., Kumar, S.P.: Sensor networks: Evolution, opportunities, and challenges. Proceedings of the IEEE 91(8), 1247–1256 (2003)
43. COMSOL AB: COMSOL Multiphysics Modelling Guide, ver. 3.4 (2007)
44. Cook, D., Fedorov, V.: Constrained optimization of experimental design. Statistics 26, 129–178 (1995)
45. Cottle, R.W.: Manifestations of the Schur complement. Linear Algebra and Its Applications 8, 189–211 (1974)
46. Cressie, N.A.C.: Statistics for Spatial Data, revised edn. John Wiley & Sons, New York (1993)
47. Curtain, R.F., Zwart, H.: An Introduction to Infinite-Dimensional Linear Systems Theory. In: Texts in Applied Mathematics. Springer, New York (1995)
48. Daley, R.: Atmospheric Data Analysis. Cambridge University Press, Cambridge (1991)
49. de Cogan, D., de Cogan, A.: Applied Numerical Modelling for Engineers. Oxford University Press, New York (1997)
50. Demetriou, M.A.: Activation policy of smart controllers for flexible structures with multiple actuator/sensor pairs. In: El Jai, A., Fliess, M. (eds.) Proc. 14th Int. Symp. Mathematical Theory of Networks and Systems, Perpignan, France, June 19-23 (2000), Published on CD-ROM

51. Demetriou, M.A.: Detection and containment policy of moving source in 2D diffusion processes using sensor/actuator network. In: Proceedings of the European Control Conference, Kos, Greece, July 2-5 (2006), Published on CD-ROM
52. Demetriou, M.A.: Power management of sensor networks for detection of a moving source in 2-D spatial domains. In: Proceedings of the 2006 American Control Conference, Minneapolis, MN, June 14-16 (2006), Published on CD-ROM
53. Demetriou, M.A.: Process estimation and moving source detection in 2-D diffusion processes by scheduling of sensor networks. In: Proceedings of the 2007 American Control Conference, New York City, USA, July 11-13 (2007), Published on CD-ROM
54. Demetriou, M.A.: Natural consensus filters for second order infinite dimensional systems. Systems & Control Letters 58(12), 826–833 (2009)
55. Demetriou, M.A., Hussein, I.I.: Estimation of spatially distributed processes using mobile spatially distributed sensor network. SIAM Journal on Control and Optimization 48(1), 266–291 (2009), doi:10.1137/060677884
56. Demetriou, M.A., Paskaleva, A., Vayena, O., Doumanidis, H.: Scanning actuator guidance scheme in a 1-D thermal manufacturing process. IEEE Transactions on Control Systems Technology 11(5), 757–764 (2003)
57. Demetriou, M.A., Uciński, D.: State estimation of spatially distributed processes using mobile sensing agents. In: Proc. American Control Conference, San Francisco, USA (2011), Published on CD-ROM
58. Demkowicz, L., Kurtz, J., Pardo, D., Paszynski, M., Rachowicz, W., Zdunek, A.: Computing with hp-Adaptive Finite Elements. Frontiers: Three Dimensional Elliptic and Maxwell Problems with Applications, vol. 2. Chapman and Hall/CRC, Boca Raton (2007)
59. Du, C., Xie, L.: H_∞ Control and Filtering of Two-dimensional Systems. LNCIS, vol. 278. Springer, Heidelberg (2002)
60. Duch, W., Korbicz, J., Rutkowski, L., Tadeusiewicz, R. (eds.): Biocybernetics and Biomedical Engineering 2000. Neural Networks. Akademicka Oficyna Wydawnicza, PLJ (2000) (in Polish)
61. Dullerud, G.E., Paganini, F.: A Course in Robust Control Theory. A Convex Approach. In: Texts in Applied Mathematics, vol. 360. Springer, New York (2000)
62. El-Farra, N., Christofides, P.D.: Switching for control of spatially-distributed processes. Computer & Chemical Engineering 28, 111–128 (2004)
63. El Jai, A.: Distributed systems analysis via sensors and actuators. Sensors and Actuators A 29, 1–11 (1991)
64. El Jai, A., Amouroux, M. (eds.) Proceedings of the First International Workshop on Sensors and Actuators in Distributed Parameter Systems, Perpignan, France, December 16–18. IFAC, Perpignan (1987)
65. El Jai, A., Pritchard, A.J.: Sensors and Controls in the Analysis of Distributed Systems. John Wiley & Sons, New York (1988)
66. Emirsajłow, Z.: The Linear Quadratic Control Problem for Infinite Dimensional Systems with Terminal Targets. Technical University Publishers, Szczecin (1991)

67. Emirsajłow, Z., Townley, S.: On application of the implemented semigroup to a problem arising in optimal control. International Journal of Control 78(4), 298–310 (2005)
68. Ermakov, S.M. (ed.): Mathematical Theory of Experimental Design. Nauka, Moscow (1983) (in Russian)
69. Ermakov, S.M., Zhigljavsky, A.A.: Mathematical Theory of Optimal Experiments. Nauka, Moscow (1987) (in Russian)
70. Fedorov, V.V.: Theory of Optimal Experiments. Academic Press, New York (1972)
71. Fedorov, V.V.: Optimal design with bounded density: Optimization algorithms of the exchange type. Journal of Statistical Planning and Inference 22, 1–13 (1989)
72. Fedorov, V.V.: Design of spatial experiments: Model fitting and prediction. Tech. Rep. TM-13152, Oak Ridge National Laboratory, Oak Ridge, TN (1996)
73. Fedorov, V.V., Hackl, P.: Model-Oriented Design of Experiments. Lecture Notes in Statistics. Springer, New York (1997)
74. Fitzpatrick, B.G.: Bayesian analysis in inverse problems. Inverse Problems 7, 675–702 (1991)
75. Fitzpatrick, B.G.: Large sample behavior in Bayesian analysis of nonlinear regression models. Journal of Mathematical Analysis and Applications 192, 607–626 (1995)
76. Fitzpatrick, B.G., Yin, G.: Empirical distributions in least squares estimation for distributed parameter systems. Journal of Mathematical Systems, Estimation, and Control 5(1), 37–57 (1995)
77. Flatau, A.B., Chong, K.P.: Dynamic smart material and structural systems. Engineering Structures 24, 261–270 (2002)
78. Fleming, W.H., Rishel, R.W.: Deterministic and Stochastic Optimal Control. In: Applications of Mathematics. Springer, New York (1975)
79. Floudas, C.A.: Mixed integer nonlinear programming, MINLP. In: Floudas, C.A., Pardalos, P.M. (eds.) Encyclopedia of Optimization, vol. 3, pp. 401–414. Kluwer Academic Publishers, Dordrecht (2001)
80. Ford, I., Titterington, D.M., Kitsos, C.P.: Recent advances in nonlinear experimental design. Technometrics 31(1), 49–60 (1989)
81. Gagnon, R.C., Leonov, S.L.: Optimum population designs for pk models with serial sampling. Journal of Biopharmaceutical Statistics 15, 143–163 (2005)
82. Garthwaite, I.T., Jolliffe, B.J.: Statistical Inference. Prentice-Hall, London (1995)
83. Gendron, B., Crainic, T.: Parallel branch-and-bound algorithms: Survey and synthesis. Operational Research 42(6), 1042–1066 (1994)
84. Gendron, B., Crainic, T.: A parallel branch-and-bound algorithm for multicommodity location with balancing requirements. Computers and Operation Research 24(1), 829–847 (1997)
85. Gerdts, M.: Solving mixed-integer optimal control problems by branch&bound: A case study from automobile test-driving with gear shift. Journal of Optimization Theory and Applications 26, 1–18 (2005)
86. Gibson, J.S., George, H., Wu, C.F.: Least-squares estimation of input/output models for distributed linear systems in the presence of noise. Automatica 36, 1427–1442 (2000)

87. Gil, M.I.: Stability of Finite and Infinite Dimensional Systems. Kluwer Academic Publishers, Boston (1998)
88. Goodwin, G.C., Payne, R.L.: Dynamic System Identification. In: Mathematics in Science and Engineering. Academic Press, New York (1977)
89. Grabowski, P.: Lecture Notes on Optimal Control Systems. University of Mining and Metallurgy Publishers, Cracow (1999)
90. Grant, M., Boyd, S.: Graph implementations for nonsmooth convex programs. In: Blondel, V., Boyd, S., Kimura, H. (eds.) Recent Advances in Learning and Control (A Tribute to M. Vidyasagar). LNCIS, pp. 99–110. Springer, Heidelberg (2008)
91. Gray, P., Scott, S.: Chemical Ooscillations and Iinstabilities: Nonlinear chemical kinetics. Oxford University Press, New York (1993)
92. Grenander, U.: Stochastic processes and statistical inference. Arkiv för Matematik 1(17), 195–277 (1951)
93. Hall, D.L., Llinas, J. (eds.): Handbook of Multisensor Data Fusion. CRC Press, Boca Raton (2001)
94. Hartl, R.F., Sethi, S.P., Vickson, R.G.: A survey of the maximum principles for optimal control problems with state constraints. SIAM Review 37(2), 181–218 (1995)
95. Haug, E.J., Choi, K.K., Komkov, V.: Design Sensitivity Analysis of Structural Systems. In: Mathematics in Science and Engineering. Academic Press, Orlando (1986)
96. Hearn, D.W., Lawphongpanich, S., Ventura, J.A.: Finiteness in restricted simplicial decomposition. Operations Research Letters 4(3), 125–130 (1985)
97. Hearn, D.W., Lawphongpanich, S., Ventura, J.A.: Restricted simplicial decomposition: Computation and extensions. Mathematical Programming Study 31, 99–118 (1987)
98. Hirsch, M.J., Pardalos, P.M., Murphey, R., Grundel, D. (eds.): Proceedings of the 7th International Conference on Cooperative Control and Optimization. Advances in Cooperative Control and Optimization. Springer, Berlin (2008)
99. Hogg, N.G.: Oceanographic data for parameter estimation. In: Malanotte-Rizzoli, P. (ed.) Modern Approaches to Data Assimilation in Ocean Modeling, Elsevier Oceanography, pp. 57–76. Elsevier, Amsterdam (1996)
100. von Hohenbalken, B.: Simplicial decomposition in nonlinear programming algorithms. Mathematical Programming 13, 49–68 (1977)
101. Holder, D.S. (ed.): Electrical Impedance Tomography: Methods, History and Applications. Taylor & Francis, Philadelphia (2004)
102. Holder, D.S., Tidswell, T.: Electrical impedance tomography on brain function. In: Holder, D.S. (ed.) Electrical Impedance Tomography: Methods, History and Applications, pp. 127–166. Taylor & Francis, Philadelphia (2004)
103. Horn, R.A., Johnson, C.R.: Matrix Analysis. Cambridge University Press, Cambridge (1986)
104. Hussein, I.I., Demetriou, M.A.: Estimation of distributed processes using mobile spatially distributed sensors. In: Proceedings of the 2007 American Control Conference, New York, USA, July 11-13 (2007), Published on CD-ROM
105. Isakov, V.: Inverse Problems for Partial Differential Equations. In: Applied Mathematical Sciences. Springer, New York (1998)
106. ISO: International vocabulary of metrology — Basic and general concepts and associated terms (VIM). International Organisation of Standarization (2008)

107. Iyengar, S., Tandon, A., Wu, Q., Cho, E., Rao, N.S., Vaishnavi, V.K.: Deployment of sensors: An overview. In: Iyengar, S.S., Brooks, R.R. (eds.) Distributed Sensor Networks, pp. 483–504. Chapman & Hall/CRC, Boca Raton (2005)
108. Iyengar, S.S., Brooks, R.R. (eds.): Distributed Sensor Networks. Chapman & Hall/CRC, Boca Raton (2005)
109. Jacobson, M.Z.: Fundamentals of Atmospheric Modeling. Cambridge University Press, Cambridge (1999)
110. Jain, N., Agrawal, D.P.: Current trends in wireless sensor network design. International Journal of Distributed Sensor Networks 1, 101–122 (2005)
111. Jennings, L.S., Fisher, M.E., Teo, K.L., Goh, C.J.: MISER 3: Optimal Control Software, Version 2.0. Theory and User Manual. Department of Mathematics, University of Western Australia, Nedlands (2002), `http://www.cado.uwa.edu.au/miser/`
112. Jeremić, A., Nehorai, A.: Design of chemical sensor arrays for monitoring disposal sites on the ocean floor. IEEE Transactions on Oceanic Engineering 23(4), 334–343 (1998)
113. Jeremić, A., Nehorai, A.: Landmine detection and localization using chemical sensor array processing. IEEE Transactions on Signal Processing 48(5), 1295–1305 (2000)
114. Jones, B., Wang, J.: Constructing optimal designs for fitting pharmacokinetic models. Statistics and Computing 9, 209–218 (1999)
115. Karatzas, I., Shreve, S.: Brownian Motion and Stochastic Calculus. Springer, New York (1988)
116. Kazimierczyk, P.: Optimal experiment design; Vibrating beam under random loading. Tech. Rep. 7/1989, Institute of Fundamental Technological Research, Polish Academy of Sciences, Warsaw (1989)
117. Kempe, D., Dobra, A., Gehrke, J.: Gossip-based computation of aggregate information. In: Proc. Conf. Foundations of Computer Science, pp. 482–491 (2003)
118. Kiefer, J., Wolfowitz, J.: Optimum designs in regression problems. The Annals of Mathematical Statistics 30, 271–294 (1959)
119. Kim, M.H., Kim, S., Lee, H.J., Lee, Y.J., Kim, K.Y.: An experimental study of electrical impedance tomography for the two phase flow visualization. International Communications in Heat Mass Transfer 29(2), 193–202 (2002)
120. Klamka, J.: Controllability of Dynamical Systems. In: Mathematics and Its Applications. Kluwer Academic Publishers, Dordrecht (1991)
121. Klamka, J.: Constrained controllability of semilinear systems with delays. Nonlinear Dynamics 56(1-2), 169–177 (2009)
122. Klamka, J., Wyrwał, J.: Controllability of second-order infinite-dimensional systems. Systems and Control Letters 57(2), 386–391 (2008)
123. Kontoghiorghes, E.J. (ed.): Handbook of Parallel Computing and Statistics. CRC Press, Boca Raton (2006)
124. Korbicz, J., Kościelny, J. (eds.): Modeling, diagnostics and process control: implementation in the DiaSter system. Springer, Berlin (2010)
125. Korbicz, J., Kościelny, J., Kowalczuk, Z., Cholewa, W.: Fault Diagnosis. Models, Artificial Intelligence, Applications. Springer, Heidelberg (2004)

126. Korbicz, J., Uciński, D.: Sensors allocation for state and parameter estimation of distributed systems. In: Gutkowski, W., Bauer, J. (eds.) Proc. IUTAM Symposium, Zakopane, Poland, August 31- September 3, pp. 178–189. Springer, Berlin (1994)
127. Korbicz, J., Uciński, D., Pieczyński, A., Marczewska, G.: Knowledge-based fault detection and isolation system for power plant. Applied Mathematics and Computer Science 3(3), 613–630 (1993)
128. Korbicz, J., Zgurovsky, M.Z., Novikov, A.N.: Suboptimal sensors location in the state estimation problem for stochastic non-linear distributed parameter systems. International Journal of Systems Science 19(9), 1871–1882 (1988)
129. Korbicz, J., Zgurowski, M.Z.: Estimation and Control of Stochastic Distributed-Parameter Systems. Państwowe Wydawnictwo Naukowe, Warsaw (1991) (in Polish)
130. Kovarik, K.: Numerical Models in Grounwater Pollution. Springer, Berlin (2000)
131. Kowalewski, A.: Optimal Control Problems of Distributed Parameter Systems with Boundary Conditions Involving Time Delays. In: Scientific Bulletins of the University of Mining and Metallurgy, Automatics Series. University of Mining and Metallurgy Publishers, Cracow (1991) (in Polish)
132. Kowalewski, A.: Optimal Control of Infinite Dimensional Distributed Parameter Systems with Delays. University of Mining and Metallurgy Publishers, Cracow (2001) (in Polish)
133. Kowalewski, A.: Optimal Control of Infinite Dimensional Distributed Parameter Systems with Delays. University of Mining and Metallurgy Press, Cracow (2001)
134. Kowalewski, A., Lasiecka, I., Sokołowski, J.: Sensitivity analysis of hyperbolic optimal control problems. Computational Optimization and Applications (2011) (in print)
135. Królikowski, A.: Design of input sequence for linear dynamic system identification. IEEE Transactions on Automatic Control 28(1), 95–97 (1983)
136. Królikowski, A., Eykhoff, P.: Input signals design for system identification: A comparative analysis. In: Prep. 7th IFAC/IFORS Symp. Identification and System Parameter Estimation, York, pp. 915–920 (1985)
137. Kubrusly, C.S.: Distributed parameter system identification: A survey. International Journal of Control 26(4), 509–535 (1977)
138. Kubrusly, C.S., Malebranche, H.: Sensors and controllers location in distributed systems – A survey. Automatica 21(2), 117–128 (1985)
139. Kuczewski, B.: Computational Aspects of Discrimination between Models of Dynamic Systems. University of Zielona Góra Press, Zielona Góra (2006), `http://zbc.uz.zgora.pl`
140. Kunisch, K.: A review of some recent results on the output least squares formulation of parameter estimation problems. Automatica 24(4), 531–539 (1988)
141. Lakshmi, S., Aronson, J.: A parallel branch-and-bound method for cluster analysis. Annals of Operation Research 90, 65–86 (1999)
142. Lam, R.L.H., Welch, W.J., Young, S.S.: Uniform coverage designs for molecule selection. Technometrics 44(2), 99–109 (2002)

143. Landaw, E.M.: Optimal experiment design for biologic compartmental systems with applications to pharmacokinetics. Ph.D. thesis, University of California, Los Angeles, USA (1980)
144. Lasiecka, I.: Active noise control in an acoustic chamber: Mathematical theory. In: Domek, S., Kaszyński, R., Tarasiejski, L. (eds.) Proc. 5th Int. Symp. Methods and Models in Automation and Robotics, Międzyzdroje, Poland, August 25-29, vol. 1, pp. 13–22. Szczecin University of Technology Press, Szczecin (1998)
145. Lasiecka, I., Triggiani, R.: Control Theory for Partial Differential Equations: Continuous and Approximation Theories. In: Encyclopedia of Mathematics and Its Applications, vol. I, II, Cambridge University Press, Cambridge (2000)
146. Lee, H.W.J., Ali, M., Wong, K.H.: Global optimization for a class of optimal discrete-valued control problems. DCDIS Series B 11, 735–756 (2004)
147. Lee, H.W.J., Cai, X.Q., Teo, K.L.: Optimal control approach to manpower planning problem. Mathematical Problems in Engineering 7, 155–175 (2001)
148. Lee, H.W.J., Lee, W.R., Wang, S., Teo, K.L.: Construction of sub-optimal feedback control for chaotic systems using B-splines with optimally chosen knot points. International Journal of Bifurcation and Chaos in Applied Science and Engineering 11, 2375–2387 (2001)
149. Lee, H.W.J., Teo, K.L.: Control parametrization enhancing technique for solving a special class of ode with state dependent switch. Journal of Optimization Theory and Applications 118, 55–66 (2003)
150. Lee, H.W.J., Teo, K.L., Lim, A.E.B.: Sensor scheduling in continuous time. Automatica 37, 2017–2023 (2001)
151. Lee, H.W.J., Teo, K.L., Rehbock, V., Jennings, L.S.: Control parametrization enhancing technique for optimal discrete-valued control problems. Automatica 35, 1401–1407 (1999)
152. Lehmann, E.L., Romano, J.P.: Testing Statistical Hypotheses, 3rd edn. Springer (2005)
153. Li, X., Yong, J., Li, H.C.: Optimal Control Theory for Infinite Dimensional Systems (Systems and Control). Springer, Berlin (1995)
154. Liu, C.Q., Ding, Y., Chen, Y.: Optimal coordinate sensor placements for estimating mean and variance components of variation sources. IEE Transactions 37, 877–889 (2005)
155. van Loon, M.: Numerical smog prediction, I: The physical and chemical model. Tech. Rep. NM-R9411, Centrum voor Wiskunde en Informatica, Amsterdam (1994)
156. van Loon, M.: Numerical smog prediction, II: Grid refinement and its application to the Dutch smog prediction model. Tech. Rep. NM-R9523, Centrum voor Wiskunde en Informatica, Amsterdam (1995)
157. Lou, Y., Christofides, P.D.: Optimal actuator/sensor placement for nonlinear control of the Kuramoto-Sivashinsky equation. IEEE Transactions on Control Systems Technology 11(5), 737–745 (2003)
158. Luo, Z.H., Guo, B.Z., Morgul, O.: Stability and Stabilization of Infinite Dimensional Systems with Applications. Springer, Heidelberg (1999)
159. Malanotte-Rizzoli, P. (ed.): Modern Approaches to Data Assimilation in Ocean Modeling. Elsevier Oceanography. Elsevier, Amsterdam (1996)

160. Malanowski, K., Nahorski, Z., Peszyńska, M. (eds.): Modelling and Optimization of Distributed Parameter Systems. IFIP. Kluwer Academic Publishers, Boston (1996)
161. Malebranche, H.: Simultaneous state and parameter estimation and location of sensors for distributed systems. International Journal of Systems Science 19(8), 1387–1405 (1988)
162. Mao, G.Y., Petzold, L.R.: Efficient integration over discontinuities for differential-algebraic systems. Computers and Mathematics with Applications 43(1-2), 65–79 (2002)
163. Martínez, S., Bullo, F.: Optimal sensor placement and motion coordination for target tracking. Automatica 42, 661–668 (2006)
164. MathWorks: Optimization Toolbox for Use with Matlab. User's Guide, Version 2. The MathWorks, Inc., Natick, MA (2000)
165. Mehra, R.K.: Optimization of measurement schedules and sensor designs for linear dynamic systems. IEEE Transactions on Automatic Control 21(1), 55–64 (1976)
166. Mentré, F., Mallet, A., Baccar, D.: Optimal design in random-effects regression models. Biometrika 84(2), 429–442 (1997)
167. Mitkowski, W.: Stabilization of Dynamic Systems. Wydawnictwa Naukowo-Techniczne, Warsaw (1991) (in Polish)
168. Morton, K.W.: Numerical Solution of Convection-Diffusion Problems. Chapman and Hall, London (1996)
169. Müller, W.G.: Collecting Spatial Data. Optimum Design of Experiments for Random Fields, 2nd revised edn. Contributions to Statistics. Physica-Verlag, Heidelberg (2001)
170. Müller, W.G.: Collecting Spatial Data. Optimum Design of Experiments for Random Fields, 3rd revised edn. Physica-Verlag, Heidelberg (2007)
171. Nakano, K., Sagara, S.: Optimal scanning measurement problem for a stochastic distributed-parameter system. International Journal of Systems Science 19(7), 1069–1083 (1988)
172. de Nevers, N.: Air Pollution Control Engineering, 2nd edn. McGraw-Hill, New York (2000)
173. Nocedal, J., Wright, S.J.: Numerical Opimization. Springer, New York (1999)
174. Noor, J.: Electrical Impedance Tomography. A Low Frequence Approach. Lambert Academic Publishers, Cologne (2010)
175. Nychka, D., Piegorsch, W.W., Cox, L.H. (eds.): Case Studies in Environmental Statistics. Lecture Notes in Statistics, vol. 132. Springer, New York (1998)
176. Ögren, P., Fiorelli, E., Leonard, N.E.: Cooperative control of mobile sensor networks: Adaptive gradient climbing in a distributed environment. IEEE Transactions on Automatic Control 49(8), 1292–1302 (2004)
177. Omatu, S., Seinfeld, J.H.: Distributed Parameter Systems: Theory and Applications. Oxford Mathematical Monographs. Oxford University Press, New York (1989)
178. Patan, K., Patan, M.: Optimal Training Sequences for Locally Recurrent Neural Networks. In: Alippi, C., Polycarpou, M., Panayiotou, C., Ellinas, G. (eds.) ICANN 2009, Part I. LNCS, vol. 5768, pp. 80–89. Springer, Heidelberg (2009)
179. Patan, K., Patan, M.: Selection of Training Data for Locally Recurrent Neural Network. In: Diamantaras, K., Duch, W., Iliadis, L.S. (eds.) ICANN 2010, Part II. LNCS, vol. 6353, pp. 134–137. Springer, Heidelberg (2010)

180. Patan, M.: Optimal Observation Strategies for Parameter Estimation of Distributed Systems. University of Zielona Góra Press (2004), http://zbc.uz.zgora.pl
181. Patan, M.: Control-oriented observation strategies for parameter estimation in distributed-parameter systems. In: Proc. 12th IEEE Int. Conf. Methods and Models in Automation and Robotics, MMAR 2006, Międzyzdroje, Poland, pp. 139–144. PPH Zapol, Dmochowski, Sobczyk, Inc, Szczecin (2006), Published on CD-ROM
182. Patan, M.: Optimal activation policies for continous scanning observations in parameter estimation of distributed systems. International Journal of Systems Science 37(11), 763–775 (2006)
183. Patan, M.: A Parallel Sensor Scheduling Technique for Fault Detection in Distributed Parameter Systems. In: Luque, E., Margalef, T., Benítez, D. (eds.) Euro-Par 2008. LNCS, vol. 5168, pp. 833–843. Springer, Heidelberg (2008)
184. Patan, M.: Resource-limited trajectory design for mobile sensors in parameter estimation of distributed systems. In: Malinowski, K., Rutkowski, L. (eds.) Recent Advances in Control and Automation, pp. 160–171. Academic Publishing House EXIT, Warsaw (2008)
185. Patan, M.: Configuring sensor network for parameter estimation of distributed parameter systems under the location uncertainty. In: Computer Methods in Mechanics - CMM 2009: 18th International Conference: Short Papers, Committee on Mechanics, Departament of Technical Sciences, Polish Academy of Sciences, Polish Association for Computational Mechanics, University of Zielona Góra, Zielona Góra, pp. 359–360. The University of Zielona Góra Press, Zielona Góra (2009) ISBN: 978-83-7481-245-0
186. Patan, M.: Decentralized mobile sensor routing for parameter estimation of distributed systems. In: Proc. 1st IFAC Workshop on Estimation and Control of Networked Systems - NecSys 2009, Venice, Italy, September 24-26, pp. 210–215 (2009), Published on CD-ROM
187. Patan, M.: Distributed configuration of sensor networks for parameter estimation in spatio-temporal systems. In: Proc. European Control Conference, ECC 2009, Budapest, Hungary, August 23-26, pp. 4871–4876 (2009), Published on CD-ROM
188. Patan, M.: Decentralized scheduling of sensor network for parameter estimation of distributed systems. In: Proc. 13th Int. Conf. Information Fusion, Edinburgh, UK, July 26-29 (2010), Published on CD-ROM
189. Patan, M.: Group sensor scheduling for parameter estimation of random-effects distributed systems. In: Grzech, A., witek, P., Brzostowski, K. (eds.) Applications of Systems Science, pp. 23–32. Academic Publishing House EXIT, Warsaw (2010), Published on CD-ROM
190. Patan, M., Bogacka, B.: Efficient sampling windows for parameter estimation in population models. In: Atkinson, A.C., Hackl, P., Müller, W. (eds.) Proc. 8th Int. Workshop on Model Oriented Data Analysis, mODa 8, Almagro, Spain, pp. 100–108. Physica-Verlag, Heidelberg (2007)
191. Patan, M., Bogacka, B.: Optimum experimental designs for dynamic systems in the presence of correlated errors. Computational Statistics & Data Analysis 51(12), 5644–5661 (2007)

192. Patan, M., Bogacka, B.: Optimum group designs for random-effects nonlinear dynamic processes. Chemometrics and Intelligent Laboratory Systems 101, 73–86 (2010)
193. Patan, M., Chen, Y., Tricaud, C.: Resource-constrained sensor routing for parameter estimation of distributed systems. In: Proc. 17th IFAC World Congress, Seoul, South Korea, July 4–8 (2008), Published on DVD-ROM
194. Patan, M., Patan, K.: Optimal observation strategies for model-based fault detection in distributed systems. International Journal of Control 78(18), 1497–1510 (2005)
195. Patan, M., Uciński, D.: Optimal strategies of scanning sensors for parameter estimation of distributed systems. In: Kaszyński, R. (ed.) Proc. 9th IEEE Int. Conf. Methods and Models in Automation and Robotics, Międzyzdroje, Poland, August 25-28, vol. 1, pp. 115–120. University of Technology Press, Szczecin (2003)
196. Patan, M., Uciński, D.: Optimal sensor location for parameter estimation of distributed systems in the presence of correlated measurement errors. In: Proc. 10-th IEEE Int. Conf. Methods and Models in Automation and Robotics, MMAR 2004, Control theory: Control Engineering: Modeling and Simulation, Międzyzdroje, Poland, vol. 1, pp. 51–56. University of Technology Press, Szczecin (2004), Published on CD-ROM
197. Patan, M., Uciński, D.: Robust Activation Strategy of Scanning Sensors via Sequential Design in Parameter Estimation of Distributed Systems. In: Wyrzykowski, R., Dongarra, J., Paprzycki, M., Waśniewski, J. (eds.) PPAM 2004. LNCS, vol. 3019, pp. 770–778. Springer, Heidelberg (2004)
198. Patan, M., Uciński, D.: Optimal activation strategy of discrete scanning sensors for fault detection in distributed-parameter systems. In: Proc. 16th IFAC World Congress, Prague, Czech Republic, July 4–8 (2005), Published on CD-ROM
199. Patan, M., Uciński, D.: Configuring a sensor network for fault detection in distributed parameter system – Part II: Solution by branch-and bound. In: Proc. 8th Conf. on Diagnostics of Processes and Systems, Subice, Poland, September 10-12 (2007)
200. Patan, M., Uciński, D.: Configuring a sensor network for fault detection in distributed parameter systems. International Journal of Applied Mathematics and Computer Science 18(4), 513–524 (2008)
201. Patan, M., Uciński, D.: Optimal scheduling of mobile sensor networks for detection and localization of stationary contamination sources. In: Proc. 9th Int. Conf. Information Fusion, Cologne, Germany, June 30-July 3 (2008), Published on CD-ROM
202. Patan, M., Uciński, D.: Configuration of sensor network with uncertain location of nodes for parameter estimation in dostributed parameter systems. In: Proc. 14th IEEE Int. Conf. Methods and Models in Automation and Robotics, MMAR 2009, Międzyzdroje, Poland, West Pomeranian University of Technology Press, Szczecin (2009), Published on CD-ROM
203. Patan, M., Uciński, D.: Mobile sensor routing for detection of moving contamination sources - part 2: algorithms and results. In: Diagnosis of Processes and Systems, pp. 203–210. Pomeranian Science and Technology Publishers PWNT (2009)

204. Patan, M., Uciński, D.: Sensor scheduling with selection of input experimental conditions for identification of distributed systems. In: Proc. 15th Int. Conf. Methods and Models in Automation and Robotics, MMAR 2010, Międzyzdroje, Poland, pp. 148–153. West Pomeranian University of Technology Press, Szczecin (2010), Published on CD-ROM
205. Patan, M., Uciński, D.: Time-constrained sensor scheduling for parameter estimation of distributed systems. In: Proc. 49th IEEE Conference on Decision and Control, Atlanta, USA, December 16-19, pp. 7–12 (2010), Published on CD-ROM
206. Patan, M., Uciński, D.: Resource-aware sensor activity scheduling for parameter estimation of distributed systems. In: Proc. 18th IFAC World Congress, Milano, Italy, August 28-September 2 (2011), Published on CD-ROM
207. Patan, M., Uciński, D., Baranowski, P.: Optimal observation strategies for fault detection in distributed-parameter systems. Pomiary Automatyka Kontrola 1(9), 71–73 (2005) (in Polish)
208. Patriksson, M.: Simplicial decomposition algorithms. In: Floudas, C.A., Pardalos, P.M. (eds.) Encyclopedia of Optimization, vol. 5, pp. 205–212. Kluwer Academic Publishers, Dordrecht (2001)
209. Patton, R.J., Frank, P.M., Clark, R.: Issues of Fault Diagnosis for Dynamic Systems. Springer, Berlin (2000)
210. Pázman, A.: Foundations of Optimum Experimental Design. In: Mathematics and Its Applications. D. Reidel Publishing Company, Dordrecht (1986)
211. Pázman, A.: Nonlinear Statistical Models. Kluwer, Dordrecht (1993)
212. Pflug, G.C.: Optimization of Stochastic Models. The Interface Between Simulation and Optimization. In: Engineering and Computer Science: Discrete Event Dynamic Systems. Kluwer Academic Publishers, Boston (1996)
213. Pierre, D.A.: Optimization Theory with Applications, Series in Decision and Control. Series in Decision and Control. John Wiley & Sons, New York (1969)
214. Polak, E.: Optimization. Algorithms and Consistent Approximations. Applied Mathematical Sciences. Springer, New York (1997)
215. Polis, M.P.: The distributed system parameter identification problem: A survey of recent results. In: Proc. 3rd IFAC Symp. Control of Distributed Parameter Systems, Toulouse, France, pp. 45–58 (1982)
216. Polyanin, A.D.: Linear Partial Differential Equations for Engineers and Scientists. Chapman and Hall/CRC, New York (2002)
217. Porat, B., Nehorai, A.: Localizing vapor-emitting sources by moving sensors. IEEE Transactions on Signal Processing 44(4), 1018–1021 (1996)
218. Pronzato, L.: Removing non-optimal support points in D-optimum design algorithms. Statistics & Probability Letters 63, 223–228 (2003)
219. Pronzato, L.: A minimax equivalence theorem for optimum bounded design measures. Statistics & Probability Letters 68, 325–331 (2004)
220. Pronzato, L., Walter, É.: Robust experiment design via stochastic approximation. Mathematical Biosciences 75, 103–120 (1985)
221. Pronzato, L., Walter, É.: Robust experiment design via maximin optimization. Mathematical Biosciences 89, 161–176 (1988)
222. Pukelsheim, F.: Optimal Design of Experiments. In: Probability and Mathematical Statistics, John Wiley & Sons, New York (1993)
223. Pukelsheim, F., Rieder, S.: Efficient rounding of approximate designs. Biometrika 79(4), 763–770 (1992)

224. Pukelsheim, F., Torsney, B.: Optimal weights for experimental design on linearly independent support points. Annals of Statistics 19(3), 1614–1625 (1991)
225. Pytlak, R.: Numerical Methods for Optimal Control Problems with State Constraints. Springer, Berlin (1999)
226. Pytlak, R., Vinter, R.B.: A feasible directions algorithm for optimal control problems with state and control constraints: Convergence analysis. SIAM Journal on Control and Optimization 36, 1999–2019 (1998)
227. Quereshi, Z.H., Ng, T.S., Goodwin, G.C.: Optimum experimental design for identification of distributed parameter systems. International Journal of Control 31(1), 21–29 (1980)
228. Rafajłowicz, E.: Design of experiments for parameter identification of the static distributed systems. Systems Science 4(4), 349–361 (1978)
229. Rafajłowicz, E.: Design of experiments for eigenvalue identification in distributed-parameter systems. International Journal of Control 34(6), 1079–1094 (1981)
230. Rafajłowicz, E.: Optimal experiment design for identification of linear distributed-parameter systems: Frequency domain approach. IEEE Transactions on Automatic Control 28(7), 806–808 (1983)
231. Rafajłowicz, E.: Choice of Optimum Input Signals in Linear Distributed-Parameter Systems Identification. In: Monographs, Technical University Press, Wrocław (1986) (in Polish)
232. Rafajłowicz, E.: Optimum choice of moving sensor trajectories for distributed parameter system identification. International Journal of Control 43(5), 1441–1451 (1986)
233. Rafajłowicz, E.: Algorithms of Experimental Design with Implementations in MATHEMATICA. Akademicka Oficyna Wydawnicza PLJ, Warsaw (1996) (in Polish)
234. Rafajłowicz, E.: Optimization of Experiments with Applications to Production Quality Monitoring. Wrocław University of Technology Press, Wrocław (2005) (in Polish)
235. Rao, M.M.: Measure Theory and Integration. John Wiley & Sons, New York (1987)
236. Rao, S.S.: Engineering Optimization. John Wiley & Sons, New Jersey (2009)
237. Reemtsen, R.: Semi-infinite programming: Discretization methods. In: Floudas, C.A., Pardalos, P.M. (eds.) Encyclopedia of Optimization, pp. 100–107. Kluwer Academic Publishers, Dordrecht (2001)
238. Reinefeld, A.: Heuristic search. In: Floudas, C.A., Pardalos, P.M. (eds.) Encyclopedia of Optimization, vol. 2, pp. 409–411. Kluwer Academic Publishers, Dordrecht (2001)
239. Retout, S., Mentré, F.: Further developments of the Fisher information matrix in nonlinear mixed effects models with evaluation in population pharmacokinetics. Journal of Biopharmaceutical Statistics 13(2), 209–227 (2003)
240. Retout, S., Mentré, F., Bruno, R.: Fisher information matrix for non-linear mixed-effect models: evaluation and application for optimal design of enoxaparin population pharmacokinetics. Statistics in Medicine 21, 2623–2639 (2002)
241. Richter, O., Söndgerath, D.: Parameter Estimation in Ecology. The Link between Data and Models. VCH, Weinheim (1990)

242. Rockafellar, R.T.: Convex Analysis. Princeton University Press, Princeton (1970)
243. Rudin, W.: Real and Complex Analysis. McGraw-Hill, New York (1986)
244. Russell, S.J., Norvig, P.: Artificial Intelligence: A Modern Approach, 2nd edn. Pearson Education International, Upper Saddle River (2003)
245. Sastry, S., Iyengar, S.S.: Real-time sensor-actuator networks. International Journal of Distributed Sensor Networks 1, 17–34 (2005)
246. Schittkowski, K.: Numerical Data Fitting in Dynamical Systems — A Practical Introduction with Applications and Software. Kluwer Academic Publishers, Dordrecht (2002)
247. Schwartz, A.L.: Theory and implementation of numerical methods based on Runge-Kutta integration for solving optimal control problems. Ph.D. thesis, University of California, Berkeley (1996)
248. Schwartz, A.L., Polak, E., Chen, Y.: A Matlab Toolbox for Solving Optimal Control Problems. Version 1.0 for Windows (1997), `http://www.schwartz-home.com/~adam/RIOTS/`
249. Searle, S.R.: Linear Models. John Wiley & Sons, New York (1971)
250. Seber, G.A.F., Wild, C.J.: Nonlinear Regression. John Wiley & Sons, New York (1989)
251. Shimizu, K., Aiyoshi, E.: Necessary conditions for min-max problems and algorithms by a relaxation procedure. IEEE Transactions on Automatic Control AC-25(1), 62–66 (1980)
252. Sikora, J.: Numerical Approaches to the Impedance and Eddy-Current Tomographies. Technical University Press, Warsaw (2000) (in Polish)
253. Silvey, S.D.: Optimal Design. An Introduction to the Theory for Parameter Estimation. Chapman and Hall, London (1980)
254. Silvey, S.D., Titterington, D.M., Torsney, B.: An algorithm for optimal designs on a finite design space. Communications in Statistics — Theory and Methods 14, 1379–1389 (1978)
255. Sinopoli, B., Sharp, C., Schenato, L., Schaffert, S., Sastry, S.S.: Distributed control applications within sensor networks. Proceedings of the IEEE 91(8), 1235–1246 (2003)
256. Sivergina, I.F., Polis, M.P.: Comments on "Model based solution techniques for the source localization problem". IEEE Transactions on Control Systems Technology 10(4), 633–633 (2002)
257. Skrbic, Z., Divjakovic, V.: Temperature influence on changes on parameters of the unit cell of biopolymer phb. Polymer 37(3), 505–507 (1996)
258. Sokołowski, J., Zolesio, J.P.: Introduction to Shape Optimization: Shape Sensitivity Analysis. In: Computational Mathematics. Springer, Berlin (1992)
259. Song, Z., Chen, Y., Sastry, C., Tas, N.: Optimal Observation for Cyber-physical Systems: A Fisher-Information-Matrix-Based Approach. Springer, Heidelberg (2009)
260. Spall, J.C.: Introduction to Stochastic Search and Optimization. Estimation, Simulation, and Control. John Wiley & Sons, Hoboken (2003)

261. von Stryk, O.: User's Guide for DIRCOL, a Direct Collocation Method for the Numerical Solution of Optimal Control Problems. Version 2.1. Fachgebiet Simulation und Systemoptimierung, Technische Universität Darmstadt (1999), `http://www.sim.informatik.tu-darmstadt.de/index/leftnav.html.en`
262. Sturm, J.F.: Primal-dual interior point approach to semidefinite programming. Ph.D. thesis, Erasmus University, Rotterdam, Tinbergen Institute Series 156 (1997)
263. Sturm, P.J., Almbauer, R.A., Kunz, R.: Air quality study for the city of Graz, Austria. In: Power, H., Moussiopoulos, N., Brebbia, C.A. (eds.) Urban Air Pollution, vol. 1, ch. 2, pp. 43–100. Computational Mechanics Publications, Southampton (1994)
264. Sun, N.Z.: Inverse Problems in Groundwater Modeling. In: Theory and Applications of Transport in Porous Media. Kluwer Academic Publishers, Dordrecht (1994)
265. Sun, N.Z.: Mathematical Modeling of Groundwater Pollution. Springer, New York (1996)
266. Sydow, A., Lux, T., Mieth, P., Schmidt, M., Unger, S.: The DYMOS model system for the analysis and simulation of regional air pollution. In: Grützner, R. (ed.) Modellierung und Simulation im Umweltbereich, pp. 209–219. Vieweg-Verlag, Wiesbaden (1997)
267. Sydow, A., Lux, T., Rosé, H., Rufeger, W., Walter, B.: Conceptual design of the branch-oriented simulation system DYMOS (dynamic models for smog analysis). Transactions of the Society for Computer Simulation International 15(3), 95–100 (1998)
268. Tempo, R., Bai, E.W., Dabbene, F.: Probabilistic robustness analysis: Explicit bounds for the minimum number of sampling points. Systems & Control Letters 30, 237–242 (1997)
269. Titterington, D.M.: Aspects of optimal design in dynamic systems. Technometrics 22(3), 287–299 (1980)
270. Torsney, B.: A moment inequality and monotonicity of an algorithm. In: Kortanek, K.O., Fiacco, A.V. (eds.) Proc. of the Int. Symp. on Semi-infinite Programming and Applications, University Texas at Austin. Lecture Notes in Economics and Mathematical Systems, vol. 215, pp. 249–260. Springer, New York (1983)
271. Torsney, B.: Computing optimising distributions with applications in design, estimation and image processing. In: Dodge, Y., Fedorov, V.V., Wynn, H.P. (eds.) Optimal Design and Analysis of Experiments, pp. 316–370. Elsevier, Amsterdam (1988)
272. Torsney, B., Mandal, S.: Multiplicative algorithms for constructing optimizing distributions: Further developments. In: Di Bucchianico, A., Läuter, H., Wynn, H.P. (eds.) Proc. 7th Int. Workshop on Model-Oriented Data Analysis, mODa 7, Heeze, The Netherlands, pp. 163–171. Physica-Verlag, Heidelberg (2004)
273. Tricaud, C., Patan, M., Uciński, D., Chen, Y.Q.: D-optimal trajectory design of heterogenous mobile sensors for parameter estimation of distributed systems. In: Proc. American Control Conference, Seattle, USA, pp. 663–668 (2008)

274. Tseng, P.: Design sensitivity analysis: Overview and review. Journal of Optimization Theory and Applications 109(3), 475–494 (2001)
275. Tuenter, H.J.H.: The minimum L_2-distance projection onto the canonical simplex: A simple algorithm. Algo Research Quarterly, 53–56 (2001)
276. Uciński, D.: Measurement Optimization for Parameter Estimation in Distributed Systems. Technical University Press, Zielona Góra (1999), `http://www.issi.uz.zgora.pl/~ucinski/`
277. Uciński, D.: Optimum design of sensor locations in parameter estimation of distributed systems. Studia z Automatyki i Informatyki 24, 151–167 (1999) (in Polish)
278. Uciński, D.: A technique of robust sensor allocation for parameter estimation in distributed systems. In: Frank, P.M. (ed.) Proc. 5th European Control Conf., EUCA, Karlsruhe, Germany, August 31-September 3 (1999), Published on CD-ROM
279. Uciński, D.: Optimal selection of measurement locations for parameter estimation in distributed processes. International Journal of Applied Mathematics and Computer Science 10(2), 357–379 (2000)
280. Uciński, D.: Optimal sensor location for parameter estimation of distributed processes. International Journal of Control 73(13), 1235–1248 (2000)
281. Uciński, D.: Optimal Measurement Methods for Distributed-Parameter System Identification. CRC Press, Boca Raton (2005)
282. Uciński, D.: D-optimum sensor activity scheduling for distributed parameter systems. In: Proc. 15th IFAC Symposium on System Identification, SYSID 2009, Saint-Malo, France, July 6-8 (2009), Published on CD-ROM
283. Uciński, D., Atkinson, A.C.: Experimental design for time-dependent models with correlated observations. Studies in Nonlinear Dynamics & Econometrics 8(2), article No. 13 (2004), `www.bepress.com/snde/vol8/iss2/art13`
284. Uciński, D., Chen, Y.: Time-optimal path planning of moving sensors for parameter estimation of distributed systems. In: Proc. 44th IEEE Conference on Decision and Control, and the European Control Conference, Seville, Spain (2005), Published on CD-ROM
285. Uciński, D., Demetriou, M.A.: An approach to the optimal scanning measurement problem using optimum experimental design. In: Proc. American Control Conference, Boston, MA (2004), Published on CD-ROM
286. Uciński, D., Demetriou, M.A.: Resource-constrained sensor routing for optimal observation of distributed parameter systems. In: Proc. 18th International Symposium on Mathematical Theory of Networks and Systems, Blacksburg, VA, USA, July 28-August 1 (2008), Published on CD-ROM
287. Uciński, D., Korbicz, J.: Parameter identification of two-dimensional distributed systems. International Journal of Systems Science 21(2), 2441–2456 (1990)
288. Uciński, D., Korbicz, J.: Optimal sensor allocation for parameter estimation in distributed systems. Journal of Inverse and Ill-Posed Problems 9(3), 301–317 (2001)
289. Uciński, D., Patan, M.: Optimal location of discrete scanning sensors for parameter estimation of distributed systems. In: Proc. 15th IFAC World Congress, Barcelona, Spain, July 22-26 (2002), Published on CD-ROM

290. Uciński, D., Patan, M.: Configuring a sensor network for fault detection in distributed parameter system – Part I: Solution of a relaxed problem. In: Proc. 8th Conf. on Diagnostics of Processes and Systems, Subice, Poland, September 10-12 (2007)
291. Uciński, D., Patan, M.: D-optimal design of a monitoring network for parameter estimation of distributed systems. Journal of Global Optimization 39, 291–322 (2007)
292. Uciński, D., Patan, M.: Mobile sensor routing for detection of moving contamination sources - part 1: optimal control formulation. In: Diagnosis of Processes and Systems, pp. 195–202. Pomeranian Science and Technology Publishers PWNT (2009)
293. Uciński, D., Patan, M.: Sensor network design for estimation of spatially distributed processes. In: Proc. 7th Workshop on Advanced Control and Diagnosis - ACD 2009, Zielona Góra, Poland, November 19-20, Institute of Control and Computation Engineering University of Zielona Góra. University of Zielona Góra Press (2009), Published on CD-ROM
294. Uciński, D., Patan, M.: Sensor network design for the estimation on spatially distributed processes. International Journal of Applied Mathematics and Computer Science 20(3), 459–481 (2010)
295. Uciński, D., Patan, M., Kuczewski, B.: Sensor network design for identification of distributed parameter systems. In: Korbicz, J. (ed.) Measurements, Models, Systems and Design, pp. 121–155. Wydawnictwa Komunikacji i Łączności, Warsaw (2007)
296. Uspenskii, A.B., Fedorov, V.V.: Computational Aspects of the Least-Squares Method in the Analysis and Design of Regression Experiments. Moscow University Press, Moscow (1975) (in Russian)
297. Vandenberghe, L., Boyd, S.: Semidefinite programming. SIAM Review 38(1), 49–95 (1996)
298. Vandenberghe, L., Boyd, S., Wu, S.P.: Determinant maximization with linear matrix inequality constraints. SIAM Journal on Matrix Analysis and Applications 19(2), 499–533 (1998)
299. Venkataraman, P.: Applied Optimization with MATLAB Programming. John Wiley & Sons, New York (2002)
300. Ventura, J.A., Hearn, D.W.: Restricted simplicial decomposition for convex constrained problems. Mathematical Programming 59, 71–85 (1993)
301. Vidyasagar, M.: A Theory of Learning and Generalization with Applications to Neural Networks and Control Systems. Springer, London (1997)
302. Vidyasagar, M.: Statistical learning theory and randomized algorithms for control. IEEE Control Systems 18(6), 69–85 (1998)
303. Vidyasagar, M.: Randomized algorithms for robust controller synthesis using statistical learning theory. Automatica 37, 1515–1528 (2001)
304. Vijaykrishnan, N., Irwin, M., Kandemir, M., Li, L., Chen, G., Kang, B.: Designing energy-aware sensor systems. In: Iyengar, S.S., Brooks, R.R. (eds.) Distributed sensor networks, pp. 453–482. Chapman & Hall/CRC, Boca Raton (2005)
305. van de Wal, M., de Jager, B.: A review of methods for input/output selection. Automatica 37, 487–510 (2001)
306. Walter, É., Pronzato, L.: Qualitative and quantitative experiment design for phenomenological models – A survey. Automatica 26(2), 195–213 (1990)

307. Walter, É., Pronzato, L.: Identification of Parametric Models from Experimental Data. In: Communications and Control Engineering. Springer, Berlin (1997)
308. Williams, R.A., Beck, M.S.: Process Tomography: Principles, Techniques, and Applications. Butterworth-Heinemann, Oxford (1995)
309. Wu, J. (ed.): Handbook on Theoretical and Algorithmic Aspects of Sensor, Ad Hoc Wireless, and Peer-to-Peer Networks. Auerbach Publications/Taylor & Francis Group, Boca Raton (2006)
310. Wynn, H.P.: The sequential generation of D-optimum experimental designs. The Annals of Mathematical Statistics 1, 1655–1664 (1970)
311. Xiao, L., Boyd, S.: Fast linear iterations for distributed averaging. Systems and Control Letters 53, 65–78 (2004)
312. Yin, G., Fitzpatrick, B.G.: On invariance principles for distributed parameter identification algorithms. Informatica 3(1), 98–118 (1992)
313. York, T.: Electrical tomography for industrial applications. In: Holder, D.S. (ed.) Electrical Impedance Tomography: Methods, History and Applications, pp. 295–347. Taylor & Francis, Philadelphia (2004)
314. Youdim, K.A., Atkinson, A.C., Patan, M., Bogacka, B., Johnson, P.J.: Potential application of D-optimal designs in the efficient investigation of cytochrome p450 inhibition kinetic models. Drug Metabolism and Disposition 38, 1019–1023 (2010)
315. Zhao, F., Guibas, L.J.: Wireless Sensor Networks: An Information Processing Approach. Morgan Kaufmann Publishers, Amsterdam (2004)
316. Zhao, T., Nehorai, A.: Detecting and estimating biochemical dispersion of a moving source in a semi-infinite medium. IEEE Transactions on Signal Processing 54(6), 2213–2225 (2006)
317. Zlochiver, S.: Induced Current Electrical Impedance Tomography. VDM Verlag (2009)
318. Zwart, H., Bontsema, J.: An application-driven guide through infinite-dimensional systems theory. In: Bastin, G., Gevers, M. (eds.) European Control Conference 1997: Plenaries and Mini-Courses, pp. 289–328. CIACO, Ottignies/Louvain-la-Neuve (1997)

Index

Lecture Notes in Control and Information Sciences

Edited by M. Thoma, F. Allgöwer, M. Morari

Further volumes of this series can be found on our homepage:
springer.com

Vol. 425: Patan, M.:
Optimal Sensor Networks Scheduling in Identification of Distributed Parameter Systems
290 p. 2012 [978-3-642-28229-4]

Vol. 424: Corradini, M.L.; Cristofaro, A.; Giannoni, F.; Orlando, G.:
Control Systems with Saturating Inputs
134 p. 2012 [978-1-4471-2505-1]

Vol. 423: Sipahi, R.; Vyhlídal, T.; Niculescu, S.-I.; Pepe, P. (Eds.)
Time Delay Systems: Methods, Applications and New Trends
442 p. 2012 [978-3-642-25220-4]

Vol. 422: Kozłowski, K.R. (Ed.)
Robot Motion and Control 2011
418 p. 2012 [978-1-4471-2342-2]

Vol. 420: Trinh, H.; Fernando, T.:
Functional Observers for Dynamical Systems
218 p. 2012 [978-3-642-24063-8]

Vol. 419: Samy, I.; Gu, D.-W.:
Fault Detection and Flight Data Measurement
170 p. 2012 [978-3-642-24051-5]

Vol. 418: Alberer, D.; Hjalmarsson, H.; del Re, L.:
Identification for Automotive Systems
348 p. 2012 [978-1-4471-2220-3]

Vol. 417: Johansson, R.; Rantzer A.:
Distributed Decision Making and Control
412 p. 2012 [978-1-4471-2264-7]

Vol. 416: Varga, A.; Hansson, A.; Puyou, G.:
Optimization Based Clearance of Flight Control Laws
451 p. 2012 [978-3-642-22626-7]

Vol. 415: Chesi, G.:
Domain of Attraction
283 p. 2011 [978-0-85729-958-1]

Vol. 414: Imine, H.; Fridman, L.; Shraim, H.; Djemai, M.:
Sliding Mode Based Analysis and Identification of Vehicle Dynamics
127 p. 2011 [978-3-642-22223-8]

Vol. 413: Eleftheriou, E.; Reza Moheimani, S.O.:
Control Technologies for Emerging Micro and Nanoscale Systems
289 p. 2011 [978-3-642-22172-9]

Vol. 412: Fridman, L.; Moreno, J.; Iriarte, R.:
Sliding Modes after the first Decade of the 21st Century
XXX p. 2011 [978-3-642-22163-7]

Vol. 411: Kaczorek, T.;
Selected Problems of Fractional Systems Theory
344 p. 2011 [978-3-642-20501-9]

Vol. 410: Bourlès, H., Marinescu, B.:
Linear Time-Varying Systems
637 p. 2011 [978-3-642-19726-0]

Vol. 409: Xia, Y., Fu, M., Liu, G.-P.:
Analysis and Synthesis of Networked Control Systems
198 p. 2011 [978-3-642-17924-2]

Vol. 408: Richter, J.H.;
Reconfigurable Control of Nonlinear Dynamical Systems
291 p. 2011 [978-3-642-17627-2]

Vol. 407: Lévine, J., Müllhaupt, P.:
Advances in the Theory of Control, Signals and Systems with Physical Modeling
380 p. 2010 [978-3-642-16134-6]

Vol. 406: Bemporad, A., Heemels, M., Johansson, M.:
Networked Control Systems
appro. 371 p. 2010 [978-0-85729-032-8]

Vol. 405: Stefanovic, M., Safonov, M.G.:
Safe Adaptive Control
appro. 153 p. 2010 [978-1-84996-452-4]

Vol. 404: Giri, F.; Bai, E.-W. (Eds.):
Block-oriented Nonlinear System Identification
425 p. 2010 [978-1-84996-512-5]

Vol. 403: Tóth, R.;
Modeling and Identification of Linear Parameter-Varying Systems
319 p. 2010 [978-3-642-13811-9]

Vol. 402: del Re, L.; Allgöwer, F.; Glielmo, L.; Guardiola, C.; Kolmanovsky, I. (Eds.):
Automotive Model Predictive Control
284 p. 2010 [978-1-84996-070-0]

Vol. 401: Chesi, G.; Hashimoto, K. (Eds.):
Visual Servoing via Advanced
Numerical Methods
393 p. 2010 [978-1-84996-088-5]

Vol. 400: Tomás-Rodríguez, M.;
Banks, S.P.:
Linear, Time-varying Approximations
to Nonlinear Dynamical Systems
298 p. 2010 [978-1-84996-100-4]

Vol. 399: Edwards, C.; Lombaerts, T.;
Smaili, H. (Eds.):
Fault Tolerant Flight Control
appro. 350 p. 2010 [978-3-642-11689-6]

Vol. 398: Hara, S.; Ohta, Y.;
Willems, J.C.; Hisaya, F. (Eds.):
Perspectives in Mathematical System
Theory, Control, and Signal Processing
appro. 370 p. 2010 [978-3-540-93917-7]

Vol. 397: Yang, H.; Jiang, B.;
Cocquempot, V.:
Fault Tolerant Control Design for
Hybrid Systems
191 p. 2010 [978-3-642-10680-4]

Vol. 396: Kozlowski, K. (Ed.):
Robot Motion and Control 2009
475 p. 2009 [978-1-84882-984-8]

Vol. 395: Talebi, H.A.; Abdollahi, F.;
Patel, R.V.; Khorasani, K.:
Neural Network-Based State
Estimation of Nonlinear Systems
appro. 175 p. 2010 [978-1-4419-1437-8]

Vol. 394: Pipeleers, G.; Demeulenaere, B.;
Swevers, J.:
Optimal Linear Controller Design for
Periodic Inputs
177 p. 2009 [978-1-84882-974-9]

Vol. 393: Ghosh, B.K.; Martin, C.F.; Zhou, Y.:
Emergent Problems in Nonlinear
Systems and Control
285 p. 2009 [978-3-642-03626-2]

Vol. 392: Bandyopadhyay, B.;
Deepak, F.; Kim, K.-S.:
Sliding Mode Control Using Novel Sliding
Surfaces
137 p. 2009 [978-3-642-03447-3]

Vol. 391: Khaki-Sedigh, A.; Moaveni, B.:
Control Configuration Selection for
Multivariable Plants
232 p. 2009 [978-3-642-03192-2]

Vol. 390: Chesi, G.; Garulli, A.;
Tesi, A.; Vicino, A.:
Homogeneous Polynomial Forms for
Robustness Analysis of Uncertain Systems
197 p. 2009 [978-1-84882-780-6]

Vol. 389: Bru, R.; Romero-Vivó, S. (Eds.):
Positive Systems
398 p. 2009 [978-3-642-02893-9]

Vol. 388: Jacques Loiseau, J.; Michiels, W.;
Niculescu, S-I.; Sipahi, R. (Eds.):
Topics in Time Delay Systems
418 p. 2009 [978-3-642-02896-0]

Vol. 387: Xia, Y.; Fu, M.; Shi, P.:
Analysis and Synthesis of
Dynamical Systems with Time-Delays
283 p. 2009 [978-3-642-02695-9]

Vol. 386: Huang, D.; Nguang, S.K.:
Robust Control for Uncertain
Networked Control Systems with
Random Delays
159 p. 2009 [978-1-84882-677-9]

Vol. 385: Jungers, R.:
The Joint Spectral Radius
144 p. 2009 [978-3-540-95979-3]

Vol. 384: Magni, L.; Raimondo, D.M.;
Allgöwer, F. (Eds.):
Nonlinear Model Predictive Control
572 p. 2009 [978-3-642-01093-4]

Vol. 383: Sobhani-Tehrani E.;
Khorasani K.;
Fault Diagnosis of Nonlinear Systems
Using a Hybrid Approach
360 p. 2009 [978-0-387-92906-4]

Vol. 382: Bartoszewicz A.;
Nowacka-Leverton A.;
Time-Varying Sliding Modes for Second
and Third Order Systems
192 p. 2009 [978-3-540-92216-2]

Vol. 381: Hirsch M.J.; Commander C.W.;
Pardalos P.M.; Murphey R. (Eds.)
Optimization and Cooperative Control
Strategies: Proceedings of the 8th International
Conference on Cooperative Control and
Optimization
459 p. 2009 [978-3-540-88062-2]

Vol. 380: Basin M.
New Trends in Optimal Filtering and Control
for Polynomial and Time-Delay Systems
206 p. 2008 [978-3-540-70802-5]

Vol. 379: Mellodge P.; Kachroo P.;
Model Abstraction in Dynamical Systems:
Application to Mobile Robot Control
116 p. 2008 [978-3-540-70792-9]

Vol. 378: Femat R.; Solis-Perales G.;
Robust Synchronization of Chaotic Systems
Via Feedback
199 p. 2008 [978-3-540-69306-2]